Remote Sensing and Digital Image Processing

Remote Sensing and Digital Image Processing

VOLUME 16

For further volumes:
http://www.springer.com/series/6477

Rosa Lasaponara • Nicola Masini
Editors

Satellite Remote Sensing

A New Tool for Archaeology

Springer

Editors
Rosa Lasaponara
Institute of Methodologies
for Environmental Analysis
CNR-IMAA, C. da S. Loya
85100 Tito Scalo, PZ
Italy
lasaponara@imaa.cnr.it

Nicola Masini
Institute of Archaeological
and Architectural Heritage
CNR-IBAM, C. da S. Loya
85050 Tito Scalo, PZ
Italy
n.masini@ibam.cnr.it

ISSN 1567-3200
ISBN 978-90-481-8800-0 e-ISBN 978-90-481-8801-7
DOI 10.1007/978-90-481-8801-7
Springer Dordrecht Heidelberg London New York

Library of Congress Control Number: 2011946224

Cover illustration: "with courtesy of Rio Nasca (Peru): discovery of a buried settlement with pyramid discovered by the processing of QuickBird-2 imagery"

Printed on acid-free paper

Springer is part of Springer Science+Business Media (www.springer.com)

Foreword

In 2008 I was asked by Rosa Lasaponara and Nicola Masini whether the European Space Agency could support a workshop on the use of remote sensing for archaeology. We decided to do that and this gave a platform to sum up the state of the art in this area. Meanwhile the area has developed and it was time to produce a book which introduces the new methods to a broader community.

Nowadays the quantity and quality of information about the surface of the Earth obtained remotely is increasing at an ever faster rate. In the beginning of the twentieth century, photos were taken from hot air balloons. In the 1970s, multispectral sensors were put into orbit, followed by spaceborne imaging radar (SAR) and airborne LIDAR. In the last decades we have seen remote sensing technology pushing the boundaries of all fronts. Instruments are acquiring data at higher spatial resolution, broader spectral ranges, more radar frequency bands and wave polarisations. Processing techniques are constantly evolving with increasing computing power.

Technological developments in satellite remote sensing have rarely been driven by archaeologists. The instruments they use on the ground for their field surveys, such as magnetometers and ground penetrating radars, were initially developed for geological and civil engineering purposes. In a similar way, the science of remote sensing has largely been shaped by other disciplines. However, the benefits of satellite remote sensing to archaeologists have been quickly recognised. Remote sensing is now an essential tool for a range of different archaeological domains.

The creativity of archaeologists worldwide in using remote sensing for subsurface archaeological feature detection and analysis of the palaeo-environment has been demonstrated in many cases. Crop and soil marks have been identified on air photos, and shadow effects on oblique photos highlight topographical effects potentially caused by buried structures. LIDAR provides evidence of subsurface structures hidden under the forest canopy, by mapping subtle changes in relief at very high resolution. The sensitivity of near infrared radiation to vegetation has greatly enhanced the ability to identify crop marks. SAR images have led to the discovery of ancient roads leading to lost cities through the analysis of linear features. The same type of imagery, at long wavelengths, have given archaeologists clues to the location of formerly inhabited sites with the discovery of ancient river beds beneath the sand of the Sahara. The combination of multispectral image analysis to identify the spectral signature of tropical vegetation growing over buried structures, together with DEM analysis of the canopy topography has lead to the identification of many sites in rainforest regions. Buried archaeological features have been identified on thermal infrared satellite imagery, most notably in Egypt with a remarkable recent discovery of many sites.

The main benefit of remote sensing to managers of cultural heritage sites is due to the fact that instruments are regularly acquiring data onboard satellites continuously orbiting the Earth. Cultural heritage sites require regular monitoring of any potential degradation, looting and changes to the surrounding environment that may threaten their safety. For the many sites which are in inaccessible places satellite remote sensing is the only feasible tool for such monitoring. Remote sensing will be increasingly used with the launch of more satellites, easier access to data, and a greater archive of data becoming available for the analysis of changes over time, many of which are very gradual.

The application of remote sensing to archaeology has also included the use of classical data processing methods, such as image enhancement, edge detection, vegetation indices and principle components analysis. GIS is becoming a means to manage the ever larger volumes of data required for archaeological analysis and cultural heritage documentation.

It is important to acknowledge that limitations do exist, be it in the spatial resolution, the spectral characteristics, or in the availability of data for the requested season or time of day. As the momentum of progress continues, these limitations will be increasingly overcome. In the meantime, archaeologists do the best they can with what is available to them, and what they are able to achieve is impressive indeed.

This book gives an overview of different remote sensing methods of optical satellite data. I hope the systematic overview found in the book will foster the wider utilisation of remote sensing methods for archaeology and for the documentation and monitoring of our cultural heritage. Living myself in a country where archaeological sites are found everywhere the latter seems especially important seeing the fast rate of change in our environment.

Volker Liebig

Foreword

In 2008 the editors of this volume, also authors of the first methodological section and three chapters on case studies, organized the 1st EARSeL International Workshop on Advances in Remote Sensing for Archaeology and Cultural Management convened by the Italian National Research Council (CNR) and UNESCO in Rome and sponsored by the European Space Agency (ESA), Italian Space Agency (ASI), Belgian Science Policy (BELSPO), and Geocart srl.

NASA had an interest as well, as 2007 marked the release of the first NASA opportunity in Space Archaeology. Building on the success of the workshop, several Archaeology sessions were sponsored at the following annual European Association of Remote Sensing Laboratories (EARSeL) Symposium with particular reference to the event held at UNESCO Headquarters (Paris) in 2010.

Throughout the workshop and conference sessions, it was remarkable to see how Geographic Information Systems (GIS) have allowed a fusion of remote sensing and other geographic data with traditional field-based techniques. Even so, it was clear that the application of remote sensing techniques to archaeology was at a very early stage, similar to its early development in geology.

One aspect of these early stages of development is an apparent dichotomy between detailed site-based studies, generally using geophysical techniques to supplement traditional field-based excavations vs. context or prospection applications, which use broader coverage to provide a context to the site-based work and to seek new sites through surveys. Site-based studies focus on low-altitude airborne (or kite and balloon) photography, often oblique, to map subtle surficial indicators while field geophysical techniques such as ground penetrating radar (GPR) and other electromagnetic techniques are used to probe the subsurface in a very limited area in conjunction with traditional surface mapping of the site. Usually these techniques are employed at known sites.

Context or prospection applications represent some of the newer applications of remote sensing techniques, which tend to be lower resolution and hence unfit for detailed site characterization. Studies presented here show how satellite remote sensing can be used to characterize landscapes in a region, allowing prediction of occupation sites or transportation pathways. Airborne multi-sensor remote sensing (as distinguished from air photography) produces better resolution than satellites, but is more costly so is not being used extensively.

One promising remote sensing technique that could bridge the gap between prospection and detailed site characterization is airborne LiDAR. This technique provides highly detailed topographic data (less than 1 m spatial resolution) over large areas. Another 'cross-over' technique discussed in this volume is very high resolution (VHR) satellite images. These may be used to replace air photography.

Both airborne photography and remote sensing techniques suffer from a plethora of possible data-collection possibilities. For both, season and time of day create variations in shadowing, soil moisture, vegetation stress, snow cover, etc. that affect the visibility of the subtle marks betraying buried archaeological sites. In addition, remote sensing techniques range the entire electromagnetic spectrum from ultraviolet to microwaves along with many possibilities for processing to bring out the subtle remote sensing signatures of archaeological remains. This wide variety of possibilities is one reason more traditional archaeologists haven't embraced remote sensing techniques. However, some studies show the possibility of convergence toward more 'popular' spectral bands, seasons, and processing techniques. It is possible that the community may be able to simplify these choices to a set of easily implemented techniques which could be included in image-processing packages. Simplifying the choices of remote sensing acquisitions will also reduce the cost of acquiring many scenes of the same site at different times.

As the papers in this volume and those presented at the past workshop and conferences show, this is an exciting time in the application of remote sensing techniques to archaeology. Much progress has been made, but many dimensions remain to be explored.

Pasadena, CA, USA Tom G. Farr

Acknowledgments

We would like to thank the European Association of Remote Sensing Laboratories (EARSeL) for supporting all the activities carried out in the framework of the EARSeL Special Interest Group (SIG) on Remote Sensing for Archaeology, Natural and Cultural Heritage (Re.Se.Ar.C.H.) including this book.

Special thanks to:

Rudi Goossens former EARSeL Chairman who strongly encouraged the proposal and all the activities of the Re.Se.Ar.C.H SIG mainly focused on fostering interaction among archaeologists, remote sensing researchers and managers interested in the use of remote sensing data (from ground, aerial and satellite) and Information Technologies to improve traditional approach for archaeological investigation, protection and management of natural and cultural heritage;

Rainer Reuter EARSeL Chairman who encouraged and supported this book inside the framework of Springer Series on Remote Sensing and Digital Image Processing and for promoting a strong and profitable interaction among archaeologists and remote sensing community.

Enzo Lapenna, Director of CNR-IMAA, who strongly promoted scientific applications of Earth Observation technologies in the field of Cultural heritage.

We thank also CNR, European Space Agency (ESA), Italian Space Agency (ASI), Belgian Science Policy (BELSPO), and Geocart srl which funded our scientific activities in the field of Satellite Remote Sensing in Archaeology.

Sincere thanks to the reviewers of manuscripts and to all colleagues and institutions (ESA, NASA, CNR) for their helpful contribution and support.

Tito Scalo, PZ, Italy

Rosa Lasaponara
Nicola Masini

Introduction

The importance of applying satellite remote sensing as a non destructive tool to uncover remains of ancient human occupation and past archaeological landscapes has been highlighted by national and international space agencies such as Nasa and ESA, promoted by UNESCO and now recognized by archaeologists and remote sensing scientific community. In this exciting cultural framework and in collaboration with UNESCO, in June 2007 EARSeL launched the Special Interest Group (SIG) on Remote Sensing for Archaeology, Natural and Cultural Heritage (Re.Se.Ar.C.H.) chaired by Rosa Lasaponara, Nicola Masini from CNR, and Mario Hernadez from UNESCO. Over the years the Re.Se.Ar.C.H SIG organized international conferences and workshops on the topic of "archaeology from space" to explore the emerging capability and strategic challenges in this field.

During these years two books approached this topic, one by Wiseman and Farouk (2007), mainly focused on the use of satellite radar data, and the other is a handbook on "a survey of the history and development of the field" (Parcak 2009).

"Satellite Remote Sensing: A New Tool for Archaeology" book is the first comprehensive overview of "optical" satellite remote sensing for archaeology focused on both methodological approaches and practical applications, obtained as result from a long and intense cooperation between archaeologists and remote sensing community. This cooperation provided vital lymph to compare diverse needs, perspectives and thinking very useful to go on the emerging capability of remote sensing archaeology and face new strategic challenges to preserve the past for the future.

This feeling is in this volume which is harmonically organized to deal with advanced technological/methodological developments and practical results obtained in archaeological projects and palaeo-environmental studies.

The book is divided into three sections, grouping chapters dealing with similar topics.

The first section (by Lasaponara and Masini) divided in four chapters focuses on technical developments and provides a resumel of the elaboration and computational strategies which may be suitable exploited for practical uses of satellite remote sensing in archaeology spanning from discovery of unknown sites, to documentation, management and site preservation, facing natural and anthropogenic risks such as agricultural mechanization, looting, pollution and conflict/war devastations.

In particular the first chapter provides a brief overview on qualitative and quantitative data analysis from visual interpretation to digital manipulation. In the following chapter image enhancement and features extraction methodologies along with geospatial analysis have been discussed in an archaeological perspective. Chapter 3 deals with the methods and the potential of classification and pattern recognition using satellite data for archaeological purposes. The first section ends with an overview of pan-sharpening techniques and quantitative evaluation of their capability in preserving spectral fidelity and spatial and textural content associated to archaeological information.

The second is focused on the use of Satellite Remote Sensing to support the survey, documentation and management of cultural resources.

The third part explores the potentiality of VHR satellite remote sensing data for recording ancient features as new invaluable tools in archaeological projects and palaeo-environmental studies.

Magnificent and highly significant case studies are presented and discussed, selected from within Europe, Asia, South America, and the Middle East including Hierapolis in Phrigia, Nasca (Peru), Angkor (Cambodia), Jabali silver mines (Yemen), Sri Ksetra (Central Burma), Dobrogea (Romania), Arge-Bam (Iran).

In particular Bitelli G. reviewed some of the issues related to Satellite Remote Sensing and Geomatics in the archaeological context. Moreover, a risk management application in the Arge-Bam Citadel, south-eastern Iran, based on object-oriented approach has been presented.

The reconstruction of the urban fabric of Hierapolis and the study of the ancient topography of its territory by means of the integration of VHR satellite data, declassified satellite photos, field works and other ancillary information is described by Scardozzi G.

Marco Giardino provides an overview of NASA experience in the field of aerial and satellite remote sensing in Archaeology. The chapter include the some projects such as the identification of prehistoric roads in Chaco canyon and prehistoric settlement patterns in southeast Louisiana, the Coast 2050 Cultural Resources Survey, the Lewis and Clark Expedition project.

In Chap. 8, a method based on geostatistics and satellite remote sensing for the assessment of archaeological looting is proposed by Lasaponara et al. The rate of success of such method has been computed test site is in Cahuachi near Nasca (Peru), thus providing satisfactory results.

Evans and Traviglia within the Greater Angkor Project provide important information on the rise and fall of medieval urbanism at Angkor, in Cambodia, by extensive use of remotely sensed imagery which allowed to find, map and analyse elements of urban form. In particular, the authors focused on the role of Angkor's water management system and on human modifications to the natural hydrology and topography in the demise of the urban complex.

Deroin et al shows the results of a multidisciplinary research, based on the use of satellite remote sensing applied to mining archaeology. The ALOS-AVNIR-2 data along with QuickBird images have been fruitfully used for the teleanalytical geological mapping of the ancient Jabali silver mines in northern Yemen.

The chapter by Stargardt et al. documents the sub-surface and surface movement of moisture after the end of the monsoon season, across the urban site of Sri Ksetra, on the southern edge of the Dry Zone of Central Burma. Multi-spectral and multi-temporal satellite imagery reveal how surface and sub-surface water still flows along the ancient irrigation channels whose study is crucial for the analysis of urban development of Sri Ksetra founded by the Pyu people, ca. fourth century AD.

In Chap. 12 by Lasaponara and Masini, Landsat and ASTER data have been analyzed for some areas near Nasca river within the drainage basin of the Rio Grande (Peru), densely settled over the centuries and millennia even if the physical environment presented serious obstacles to human occupation. The results allowed to identify areas for further investigations aimed at detecting ancient underground aqueducts called puquios which were constructed for water control and retrieval during the Nasca age.

Oltean and Abell perform a comparative analysis of a rich aerial and satellite dataset, including declassified images, in order to discuss the potential of satellite imagery to reveal buried archaeological features in the area of southern Dobrogea (Romania) and to assist in the reconstruction of past landscapes. It outlines some of the methodological approaches employed, highlights areas for further research and presents some preliminary outcomes.

Finally, Masini et al showed the archaeogeophysical approach of ITACA Mission based on the use of satellite and ground remote sensing and the results obtained in the Ceremonial Center of Cahuachi, near Nasca (Peru). In particular a large buried settlement characterized by a pyramid has been detected in the Nasca riverbed as well as tombs and ritual offerings have been identified and unearthed by archaeologists in Piramide Naranjada.

Tito Scalo, PZ, Italy

Rosa Lasaponara
Nicola Masini

Contents

Contributors

Lauren L. Abell Remote Sensing Center, Naval Postgraduate School, Monterey, CA 93943, USA

Gabriel Amable Unit for Landscape Modelling, Department of Geography, University of Cambridge, Cambridge CB2 1QB, UK

Gabriele Bitelli Department of Civil, Environmental and Materials Engineering, University of Bologna, Viale Risorgimento, 2, Bologna, Italy

Maria Danese Institute of Archaeological and Architectural Heritage, CNR-IBAM, C. da S. Loya, 85050 Tito Scalo, PZ, Italy

Jean-Paul Deroin GEGENAA EA 3795, Université de Reims Champagne-Ardenne, 2 esplanade Roland Garros, 51100 Reims, France

Bernard Devereux Unit for Landscape Modelling, Department of Geography, University of Cambridge, Cambridge, CB2 1QB, UK

Damian Evans Department of Archaeology, School of Philosophical and Historical Inquiry, Faculty of Arts, University of Sydney, A18 Brennan-MacCallum Building, Sydney, NSW 2006, Australia

Marco J. Giardino National Aeronautics and Space Administration, Stennis Space Center, Mississippi, USA

Jürgen Heckes Deutsches Bergbau Museum, Am Bergbaumuseum 28, 44791 Bochum, Germany

Rosa Lasaponara Institute of Methodologies for Environmental Analysis, CNR-IMAA, C. da S. Loya, 85100 Tito Scalo, PZ, Italy

Nicola Masini Institute of Archaeological and Architectural Heritage, CNR-IBAM, C. da S. Loya, 85050 Tito Scalo, PZ, Italy

Ioana A. Oltean Department of Archaeology, University of Exeter, Exeter EX4 4QE, UK

Giuseppe Orefici Centro de Estudios Arqueológicos Precolombinos, Avenida de la Cultura, 600 (Bisambra), Nasca, Peru

Enzo Rizzo Institute of Methodologies for Environmental Analysis, CNR-IMAA, C. da S. Loya, 85100 Tito Scalo, PZ, Italy

Giuseppe Scardozzi Institute of Archaeological and Monumental Heritage, CNR-IBAM, Prov.le Lecce-Monteroni, 73100 Lecce, Italy

Janice Stargardt Department of Geography and Sidney Sussex College, University of Cambridge, Cambridge, CB2 3HU, UK

Florian Téreygeol UMR 5060, Institut de Recherche sur les Archéomatériaux and CEA, Saclay, France

Arianna Traviglia Department of Archaeology, School of Philosophical and Historical Inquiry, Faculty of Arts, University of Sydney, A18 Brennan-MacCallum Building, Sydney, NSW 2006, Australia

Part I
Optical Satellite Remote Sensing in Archaeology: An Overview

Chapter 1
Remote Sensing in Archaeology: From Visual Data Interpretation to Digital Data Manipulation

Rosa Lasaponara and Nicola Masini

Abstract Satellite remote sensing technologies have triggered improvements in archaeological research and developments of new tools in archaeological prospection from discovery to monitoring, from documentation to preservation of cultural resources. Nevertheless, this increasing interest in remote sensing has not been accompanied by new perspectives in data processing, analysis and interpretation. Specific methodologies must be developed ad hoc for archaeology in order to optimize the extraction and understanding of the information content from the numerous active and passive satellite data sets. This chapter provides a brief overview on qualitative and quantitative data analysis from visual interpretation to digital manipulation.

Keywords Visual interpretation • Digital processing • Archaeological-marks • History of satellite remote sensing

1.1 Introduction

The importance of applying space technology to archaeological research has been paid great attention worldwide, due to the following aspects:

(i) the improvement in spectral and spatial resolution reveals increasing detailed information for archaeological purposes;

R. Lasaponara (✉)
Institute of Methodologies for Environmental Analysis, CNR-IMAA,
C. da S. Loya, 85100 Tito Scalo, PZ, Italy
e-mail: lasaponara@imaa.cnr.it

N. Masini
Institute of Archaeological and Architectural Heritage, CNR-IBAM,
C. da S. Loya, 85050 Tito Scalo, PZ, Italy
e-mail: n.masini@ibam.cnr.it

R. Lasaponara and N. Masini (eds.), *Satellite Remote Sensing: A New Tool for Archaeology*, Remote Sensing and Digital Image Processing 16,
DOI 10.1007/978-90-481-8801-7_1, © Springer Science+Business Media B.V. 2012

(ii) the synoptic view offered by satellite data helps us to understand the complexity of archaeological investigations at a variety of different scales;
(iii) satellite-based digital elevation models (DEMs) are widely used in archaeology for several purposes to considerably improve data analysis and interpretation;
(iv) the availability of long satellite time series allows the monitoring of hazard and risk in archaeological sites;
(v) remotely sensed data enable us to carry out both inter and intra site prospection and data analysis.

The availability of high resolution satellite data has been so rapidly growing that new problems have arisen mainly linked with methodological aspects of data analyses. In this context, the main concern is the lack of correspondence between the great amount of remote sensing image and effective data processing methods not only for archaeology but also for different field of applications.

This is particularly pressing in archaeology and palaeo-environmental studies, for which the data processing currently applied need to be adjusted and adapted for archaeological purposes.

The starting point is that ordinary data processing routines may provide outputs without any relevant information for archaeology. The traces of ancient human transformations of landscape create very subtle spatial features, namely surface anomalies, that are only visible from a bird view. The characteristics of these archaeological features strongly depend on vegetation cover and phenology, pedology, soil types and topography. These features are generally named soil, shadow and crop marks. Soil-marks can appear on bare soil as changes of color or texture. Shadow marks can be seen in presence of variations of micro-topographic relief visible by shadowing. Crop marks can be evident for vegetated areas, covered by crops or weeds. They can appear as differences in height or color in crops which are under stress due to lack of water or deficiencies in nutrients. Crop marks can be formed as *negative* marks above wall foundations as well as *positive* marks above pits and ditches (Crawford 1929; Adamesteanu 1973; Dassie 1978; Wilson 1982; Masini 1998; Piccarreta and Ceraudo 2000; see Fig. 1.1).

So, the possibility of locating unknown small sites as well as large scale cultural features from an aerial or space view is highly dependent on the image spatial resolution, extension of buried sites, ground characteristics, illumination conditions, view geometry etc.

The practical potential of satellite in archaeology needs to be tested in relation to the local specific different surface conditions, sites (extension, building materials), expected features (soil, shadow, crop marks), and environmental setting (Grøn et al. 2011). Early investigations, based on the use of satellite VHR data for archaeological studies in different environmental conditions (see state of art in Wiseman and El-Baz 2008; Parcak 2009; Lasaponara and Masini 2011) clearly pointed out that the subtle signals of archaeological features pose a serious challenge for digital image processing, so that satellite-based identification of archaeological features is recognized as one of the most complex tasks faced by computer vision and photogrammetry communities.

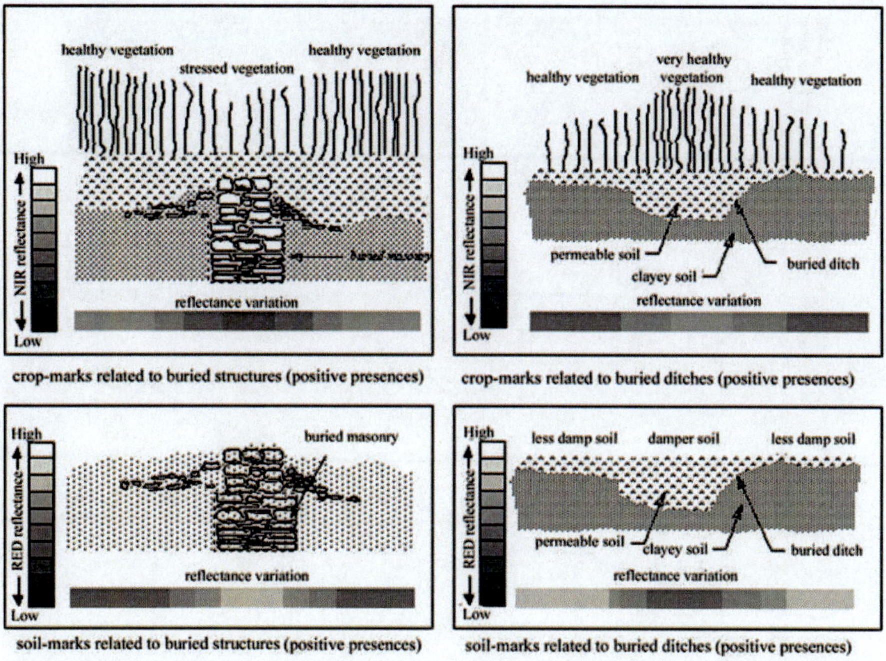

Fig. 1.1 Reflectance variation in presence of crop marks, caused by buried structures (*upper left*) and ditches (*upper right*) and soil marks (*lower left* and *right*, referable to shallow masonry structures and ditches, respectively)

Moreover, the integration of satellite data with traditional data sources, such as historical documentation and data collected on field at various scales, inevitably involves new challenges to be addressed and ad hoc processing strategies (see, for example, Fig. 1.2), such as:

(i) data fusion;
(ii) image enhancement, edge detection and pattern recognition;
(iii) spatial and spectral analysis that combines the advantages of two approaches based on spectral information and contextual information from a given pixel and its neighboring (geospatial data analysis);
(iv) image segmentation and classification, correlation between spectral data derived from pixel information and libraries as the basis for a new classification routine.
(v) satellite time series data analysis (temporal and spatial autocorrelation).

The listed methods are only a first step towards the optimization and integration of various components of data analysis (for additional details on points (i) to (v), the reader is referred to Chaps. 2, 3 and 4 of this book). Effective approaches must be devised, implemented and quantitatively evaluated to address key challenges of what can be called ‘digital light archaeology’ (namely non invasive archaeological research by digital *manipulation* of remotely sensed data).

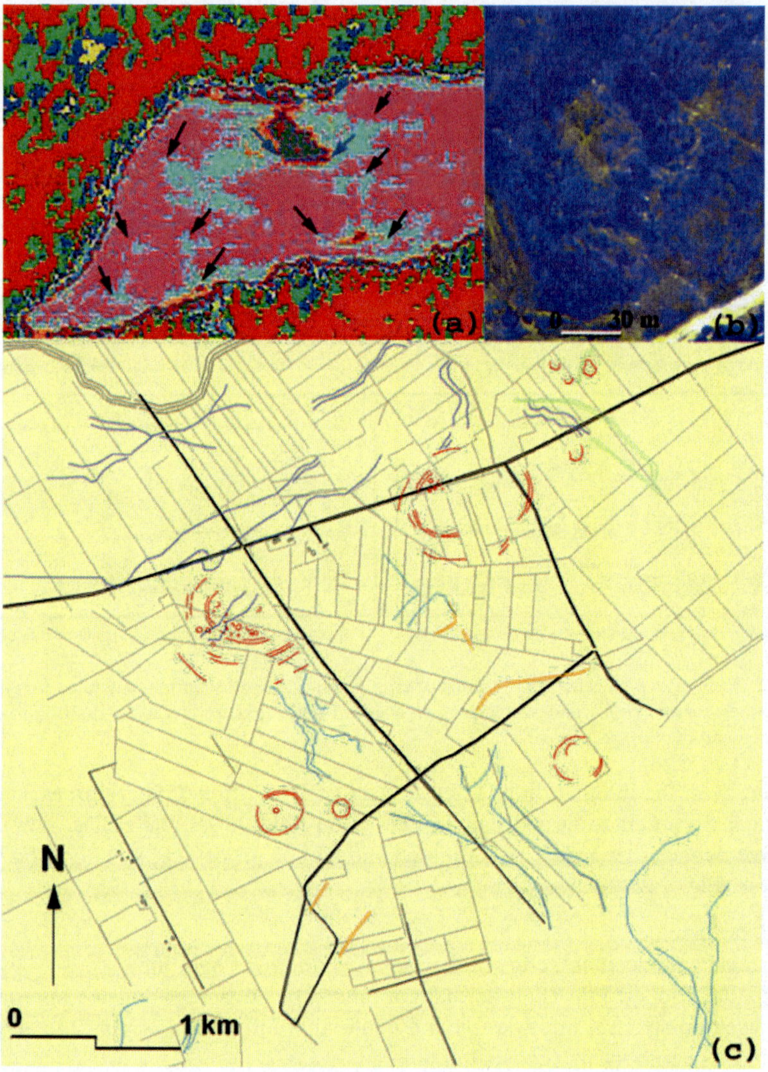

Fig. 1.2 Detection of archaeological sites by processing QuickBird imagery. (**a**) Etruscan site in San Giovenale (Viterbo, Italy): unsupervised classification to support the extraction of archaeological features; (**b**) Hierapolis (Turkey): discovery of a farm dating back between the early Imperial Roman and the proto-Byzantine age; RGB composition of red, green and NIR channel; (**c**) map of a Neolithic settlement near Foggia (Apulia region, Italy) derived from the processing of QuickBird imagery (see Backe Forsberg et al. 2008; Lasaponara et al. 2008; Ciminale et al. 2009, respectively)

The reconstruction of ancient landscape represents an important issue not only in the field of archaeology, but, also for botany, forestry, soil science and hydrology. Information on the impacts of human actions upon the environment can be widely used to address issues in human settlement, to better understand environmental interactions, climate change and the Earth's system.

1.2 Visual Data Interpretation Versus Digital Processing

Image interpretation is defined as the qualitative and quantitative examination of images recorded by various sensor systems (panchromatic, multispectral and hyperspectral cameras), in order to identify objects and evaluate their significance. The recognition of various targets can be carried out on the basis of the comparison of the visual elements which characterize them such as, tone, shape, size, pattern, texture, shadow, and spatial association, which are generally strongly dependent on the scale of observation:

(i) **Tone** denotes the relative brightness or colour of objects in an image, whose variations enable photo interpreters to well discriminate shape, texture, and patterns.
(ii) **Texture** denotes the arrangement and frequency of tonal variation and it is strongly dependent on the observation scale. For example, rough textures are related to irregular surfaces (such as a forest canopy), whereas smooth textures are linked to more uniform surfaces (such as fields, asphalt, ancient land divisions).
(iii) **Shape** denotes the general form of individual objects (for example Neolithic curvilinear ditches, square pyramids).
(iv) **Size** is a function of scale; both absolute and relative size can help in the interpretation of object/target (from buried wall structures to urban fabric).
(v) **Pattern** denotes the spatial arrangement of objects. Distinctive patterns are generally a repetition of similar tones and textures (for example, Roman centuriations).
(vi) **Shadow** provides an idea of the profile and relative height of targets which may make their identification easier (i.e. microrelief in medieval hilly settlements).
(vii) **Association** takes into account the relationship between objects or features in proximity to the target of interest (palaeoriverbeds close to ancient settlements which allow to study the relationship between environmental changes with human frequentation).

The following factors influence image quality, and therefore, object recognition:

- sensor characteristics (film types, digital systems);
- acquisition time (season of the year and time of day);
- atmospheric effects;
- resolution of the imaging system and scale;

- image motion;
- stereoscopic parallax.

In order to extract meaningful information on the basis of a visual inspection, the image interpreter has to identify and allocate objects into known categories. This implies the knowledge of local environments and a deeper understanding of the processes and phenomena under investigation. The result from recognition process is the identification of objects and features in the area as well as the delineation of areas having homogeneous patterns and characteristics.

Visual interpretation is qualitative, inexpensive and simple. Its main advantage, compared to digital data processing, is the fact that it can be carried out also when the object features are not easily distinguishable. Obviously, the main disadvantage is that visual interpretation can only be carried out for small investigation areas.

In the last decades, the advancement of technology has provided a revolution which was at the first the availability of remote sensing data in digital format and later digital processing. This requires a performing computer system, with dedicated hardware and software for image processing. The latter can be categorized into the following two groups: (1) pre-processing and (2) processing (Lillesand and Kiefer 2000).

(1) Pre-processing generally indicates radiometric or geometric corrections along with all the operations required prior to carry out the data analysis and information extraction. Radiometric corrections aim at: (i) correcting the data for sensor irregularities and reducing noise due to sensor and/or atmospheric effects, and (ii) converting the data in calibrated measurements of the reflected or emitted radiation. Geometric corrections aim at: (i) reducing geometric distortions due to view geometry, and (ii) converting data to coordinates (e.g. latitude and longitude) of a given reference system.
(2) Processing procedures assist visual interpretation and analysis, according to investigation purposes, for example, image enhancement, edge detection and extraction, pattern recognition, classification, data fusion may be applied to emphasize subtle details, etc. (see Fig. 1.3). Data processing can be automatic or semi-automatic depending on the absence or presence of human intervention, respectively. The semi-automatic data processing methods are generally more common and effective than fully-automated procedures.

Obviously, the computer cannot replace the knowledge, experience and understanding of the image interpreter (archaeologist) but can allow quantitative analysis of huge data sets and make the information extraction and interpretation easier also for large areas under investigation.

Compared to visual data inspection, digital processing offers several advantages such as, the possibility to: (i) perform repetitive and cost effective data analyses for large areas of cultural interest, (ii) obtain consistent results based on "objective" instead of subjective evaluations, (iii) facilitate the integration of imagery with other data source (archaeological record, documentary sources, etc.), (iv) explore alternative data processing methods and, (v) if required, also to apply complex algorithms to make

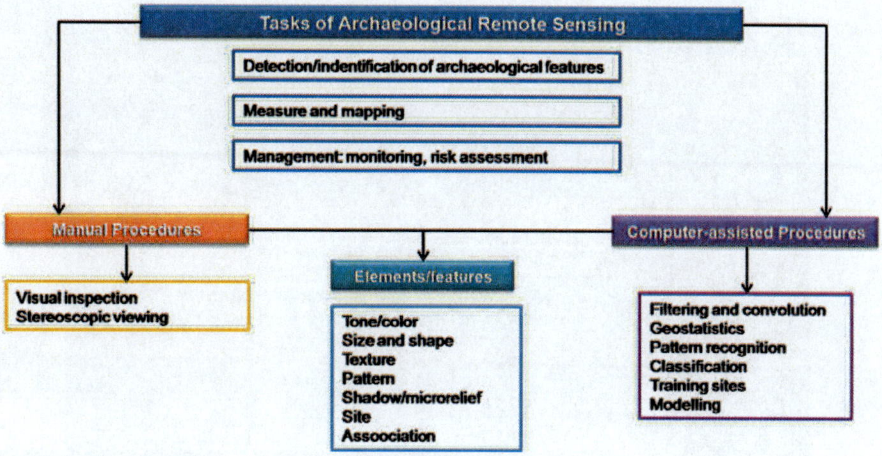

Fig. 1.3 Rational basis of archaeological remote sensing

archaeological information extraction and interpretation easier. Therefore, digital data processing and manipulation in archaeology can allow the development of effective tools for facilitating the detection of sites/features linked to past landscapes which are generally quite subtle and obscured by marks of the modern landscape. The use of digital filtering can easily enable us to leave out details of modern landscapes and enhance the subtle ones related to past human activities.

Obviously some disadvantages are present. Digital processing: may be: (i) expensive for one-time interpretation or small areas under investigation, (ii) have high start-up costs, (iii) and require elaborated and dedicated equipment systems.

1.3 Satellite Remote Sensing in Archaeology: From Early Applications to Recent Developments

Early applications of satellite for studies on past human activities were attempted starting from the 1980s using Landsat Thematic Mapper (TM), which was the highest (30 m) spatial resolution sensor available at that time for civilian applications. TM data were quite successfully used for landscape archaeological investigations, as in the case of the identification ancient field division and agricultural systems (Clark et al. 1998; Sever 1998), and also in palaeo-environmental studies (Parry 1992; Drake 1997; White and El Asmar 1999), carried out mainly based on photointerpretation.

The French Spot data (10 m) were generally not used in archaeology even if they offered higher resolution, but they were much more expensive than TM. Moreover, they still offered a spatial detail not enough to enlarge the field of application to smaller features of cultural interest.

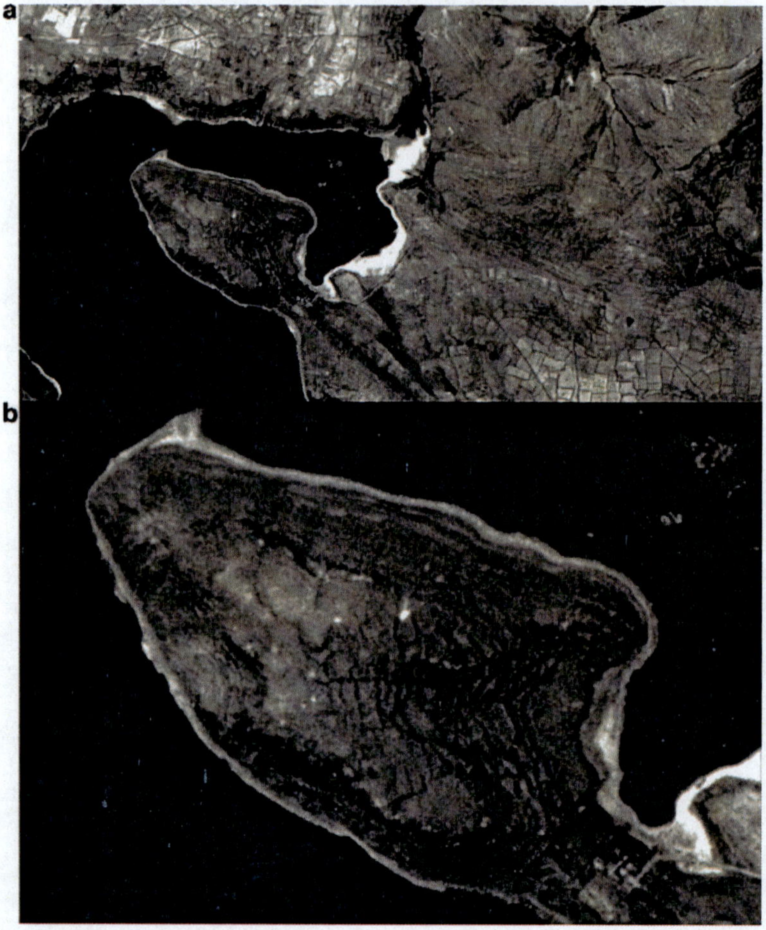

Fig. 1.4 Corona satellite image taken in 11 April 1966. (**a**) Shores of Lake Umayo near Puno in Peru; (**b**) detail of Sillustani, a burial site dating back to pre-Incas age

After the end of the Cold War, in the 1990s, Russian and American intelligence satellite photographs were made commercially available for civilian purposes. This strongly pushed archaeologists to use this extensive archive of photographs acquired in the 1960s and 1970s. For some regions of the Middle East and Southern America (see Fig. 1.4) there is a huge dataset available at a high spatial resolution (up to 2 m) which represents the unique data source useful to detect archaeological features later destroyed by mechanized agriculture and other anthropogenic activities.

Fowler (1996) exploited Russian declassified KVR-1000 imagery to identify crop and soil marks in the surrounding of Stonehenge. KVR-1000 pictures were also employed by Comfort (1997) for archaeological investigations in the Greek and Roman city of Zeugma on the Euphrates in Turkey. The study of ancient irrigated and cultivated areas was carried out in Yemen by Marcolongo and Morandi Bonacossi (1997) using Russian Soyuz Kate-200 images.

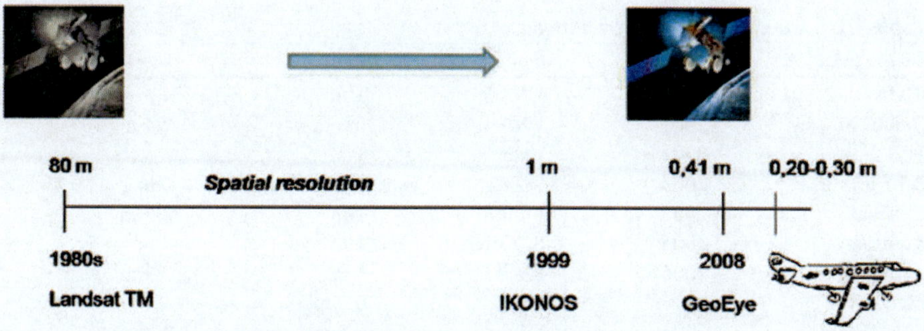

Fig. 1.5 The Great Run of satellite technology in reaching the spatial resolution of aerial images

The Russian declassified data were available for only 4 years. Therefore, the use of American declassified KH-4B Corona increased over time.

Kennedy (1998a, b) and Fowler (1997) conducted the first applications of Corona images for investigating the Euphrates valley (Turkey) and an Iron Age hill fort in Hampshire, respectively.

Even today Corona is the unique data source for archaeological prospection in countries where aerial photography is restricted for military reasons, as in the case of the Upper Khabur basin in North-eastern Syria, where Ur (2003) discovered an ancient road system dating to the Early Bronze Age.

More recently, CORONA has been integrated with ASTER multispectral satellite imagery by Altaweel (2005) to identify hollow ways, canals and sites in North Iraq. Other researchers exploited the information content of Corona photographs along with other multispectral satellite data such as IKONOS, used by Beck et al. (2007), for studying tell settlements and field systems in Western Syria.

Goossens et al. (2006) highlighted the problems linked to the geometric rectification of Corona images due to the fact that the latter were collected by a non-metric panoramic camera on board a satellite with a decaying orbit. To overcome these drawbacks, they developed a method to reduce geometric distortions, which was applied to the Altai Mountains.

The availability of the first commercial VHR satellite data (IKONOS in 1999), opened new research perspectives for archaeological applications. Later the spatial resolution has been strongly increased (see Fig. 1.5), up today with GeoEye1 which acquire 41 cm panchromatic and 1.65 m multispectral imagery (see Table 1.1).

The access to VHR satellite images is different, depending on the satellites owners, in the case of private companies such as IKONOS, QuickBird and OrbView the images are well distributed. A good distribution network also exists for SPOT, the Indian Satellites and EROS.

The advantages of VHR satellite imagery are not only linked to the availability of multispectral, georeferenced and very high resolution images but also to the possibility of using effective data processing for extracting valuable information from site level up to historical landscapes.

Table 1.1 List of optical VHR satellite data

Satellite data	Launch	Country	Pan	Ms
IKONOS 2	1999	USA	1 m	4 m
QuickBird	2001	USA	0.6 m	2.4 m
TES	2001	India	1 m	
OrbView 3	2003	USA	1 m	4 m
Cartosat 1	2005	India	1 m	2.5 m
Kompsat 2	2006	S. Korea	1 m	4 m
Resurs DK2	2006	Russia	1 m	2:3 m
EROS B	2006	Israel	0.7 m	
WorldView-2	2007	USA	0.5 m	2 m
Cartosat 2	2007	India	0.8 m	
RapidEye	2008	Germany	5 m	5 m
GeoEye-1[a]	2008	USA	0.41/0.5 m	1.65/2 m

[a] GeoEye1 collects images with a ground sample distance of 0.41-m or 16 inches in the panchromatic and multispectral imagery at 1.65-m resolution. Unfortunately, GeoEye's operating license from the U.S. Government requires re-sampling the imagery to 0.5-m

Nevertheless, such potential capability has not been adequately exploited due to the lack of data processing procedures developed ad hoc to address the archaeological needs. Early investigations clearly pointed out that the use of image enhancement such as pan-sharpening, edge detection, pattern recognition and classification can improve and facilitate the identification of archaeological features (e.g., Lasaponara and Masini 2007; Grøn et al. 2008).

Moreover, great effort must be devoted to set up automatic and/or semi automatic approaches for archaeology.

Presently, no effective automatic procedures are available for archaeological purposes for both VHR and medium/high resolution satellite images. Semiautomatic approaches work quite well, but their performance are "site specific". For example, pre-Hispanic pathways, in Aztec cities within and outside the Valley of Mexico have been identified by Principal Components Analysis (PCA), texture segmentation, linear pattern detection and spatial filtering applied to Landsat 7 data (Argote-Espino and Chavez 2005). Other examples include the discrimination of surface archaeological remains in Hisar (southwest Turkey) using supervised classifications (De Laet et al. 2007), the extraction of land patterns, useful for palaeo-geographic and palaeo-environmental investigations in Metaponto (Ionian coast of Southern Italy) using edge enhancement techniques (Masini and Lasaponara 2006), and the detection of change over time in Southern Peru by Lasaponara and Masini (2010) using spatial autocorrelation statistics for looting monitoring (see Chap. 8). Another approach, mainly based on e-cognition and mathematical morphology, has been applied to evaluate changes in archaeological excavated sites of Nisa (Turkmenistan) and Babylon (Iraq) (Jahjah and Ulivieri 2010). The latter was investigated with the aim to evaluate changes before and after the second gulf war.

VHR satellite images also offer a stereo view, and, therefore, the possibility to extract high resolution Digital Elevation Models (DEMs) useful for studying

Fig. 1.6 Territory of Hierapolis (Turkey): 3d pan-sharpened QuickBird multispectral image (courtesy of Giuseppe Scardozzi)

ancient landscapes (see Fig. 1.6). DEM products can be also obtained from high to medium resolution satellite images (SPOT, ASTER). DEM obtained from optical satellite images has several advantages, including relatively low costs (compared to field GPS survey or photogrammetric campaigns), high spatial resolution, good correlation over vegetated areas, whereas, being passive sensors, the main disadvantages include mainly the potential masking by clouds.

Currently the research in the field of DEM generation from optical images is quite active in the different application fields, such as city modelling and landslide monitoring, and maybe, in the near future, results from these studies will be also useful for archaeology. To obtain improved and inexpensive DEM from ASTER images, multi-temporal and multi- sensor satellite data can be used. By contrast it should be noted that the cost for acquisition of VHR stereo images is more than double and therefore, also the availability of stereo mode acquisition from the archive is much lower than the single mode acquisition.

The potential of satellite VHR imagery is better exploited if they are used in combination with other data sources, such as historical documentation and records, along with geophysical prospection aimed to approach different fields of site investigations, from geosciences (geoarchaeology, geomorphology) to archaeology (field survey, excavations, etc.).

An integrated approach, based on GIS and remote sensing for geomorphology and DEM analysis, provides added value and precious contribution, from site discovery to historical landscape analyses. This was adopted by Alexakis et al. (2011), for locating settlements, modelling habitation and reconstructing landscape of the Neolithic age in Thessaly.

Finally, a multi-scale and multi-sensor data integration has been adopted by Ciminale et al. (2009) for a Neolithic settlement in Apulia region (Southern Italy). In particular, satellite data allowed the reconstruction of palaeo-environmental pattern, whereas aerial images and geomagnetic maps were used to identify the circular ditched enclosures of the Neolithic village, and other smaller features related to circular and semi-circular compounds.

1.4 Conclusion and Outlook

The application of aerial photographs had been long appreciated by archaeologists. In fact, over the last century, aerial reconnaissance has been one of the most important ways in which new archaeological sites have been discovered through the world. The advantages of aerial photographs are manifold: they can be taken vertically or obliquely, easily interpreted, used for photogrammetric application and also to provide a three-dimensional view.

Presently, the great amount of multispectral VHR satellite images, even available free of charge in Google earth, opened new strategic challenges in the field of remote sensing in archaeology. These challenges substantial deal with the exploitation of such data as much as possible, and, in turn, with the setting up of effective and reliable automatic and/or semiautomatic data processing strategies and the integration of the traditional ground truthing activity with numerical scientific testing (i.e. in-situ spectro-radiometric measurements).

Nowadays, the use of EO for archaeology is still an open issue and additional strategic challenges deals with the integration of remote sensing with other traditional archaeological data sources, such as field surveys, trials, excavations and historical documentation.

The integration of diverse data source can strongly improve our capacity to uncover unique and invaluable information, from site discovery to investigations focused on dynamics of human frequentation in relation to environmental changes.

This strategic integration requires a strong interaction among archaeologists, scientists and cultural heritage managers to improve traditional approach for archaeological investigation, protection and conservation of archaeological heritage.

Data coming from diverse non invasive remote sensing data sources can support a scalable and modular approach in the improvement of knowledge as a continuous process oriented to collect and puzzle pieces of information on past human activities, thus should enable us to better understand the past, manage the present and support modelling for future forecasting.

References

Adamesteanu D (1973) Le suddivisioni di terra nel metapontino. In: Finley MI (ed) Problèmes de la terre en Grèce Ancienne. Mouton, Paris, pp 49–61

Alexakis D, Sarris A, The A, Albanakis K (2011) Integrated GIS, remote sensing and geomorphologic approaches for the reconstruction of the landscape habitation of Thessaly during the Neolithic period. J Archaeol Sci 38:89–100

Altaweel M (2005) The use of ASTER satellite imagery in archaeological contexts. Archaeol Prospect 12:151–166

Argote-Espino D, Chavez RE (2005) Detection of possible archaeological pathways in Central Mexico through digital processing of remote sensing images. Archaeol Prospect 12:105–114

Backe Forsberg Y, Holmgren H, Lanorte A, Lasaponara N, Masini N (2008) Airborne and satellite multispectral imagery at the Etruscan site of San Giovenale, Blera (Lazio) – Preliminary results. In: Lasaponara N, Masini N (eds) Advances in remote sensing for archaeology and cultural heritage management. Aracne, Roma

Beck A, Philip G, Abdulkarim M, Donoghue D (2007) Evaluation of Corona and Ikonos high resolution satellite imagery for archaeological prospection in western Syria. Antiquity 81:161–175

Ciminale M, Gallo D, Lasaponara R, Masini N (2009) A multiscale approach for reconstructing archaeological landscapes: applications in Northern Apulia (Italy). Archaeol Prospect 16:143–153

Clark CD, Garrod SM, Parker Pearson M (1998) Landscape archaeology and remote sensing in southern Madagascar. Int J Remote Sens 19(8):1461–1477

Comfort A (1997) Satellite remote sensing and archaeological survey on the Euphrates. Archaeol Comput Newslett 48:1–8

Crawford OGS (1929) Air photography for archaeologists. Ordnance survey professional papers, new series, 12, HMSO, Southampton

Dassie J (1978) Manuel d'archéologie aérienne. Technip, Paris

De Laet V, Paulissen E, Waelkens M (2007) Methods for the extraction of archaeological features from very high-resolution Ikonos-2 remote sensing imagery, Hisar (southwest Turkey). J Archaeol Sci 34:830–841

Drake NA (1997) Recent aeolian origin of superficial gypsum crusts in Southern Tunisia: geomorphological, archaeological and remote sensing evidence. Earth Surf Proc Land 22:641–656

Fowler MJF (1996) High resolution satellite imagery in archaeological application: a Russian satellite photograph of the Stonehenge region. Antiquity 70:667–671

Fowler MJF (1997) A Cold War spy satellite image of Bury Hill near Andover, Hampshire. Hamps Field Club Archaeol Soc Newslett 27:5–7

Goossens R, De Wulf A, Bourgeois J, Gheyle W, Willems T (2006) Satellite imagery and archaeology: the example of CORONA in the Altai Mountains. J Archaeol Sci 33(6):745–755

Grøn O, Stylegar F-A, Palmer S, Aase S, Orlando P, Esbensen K, Kucheryavski S (2008) Practical use of multispectral satellite images in general Norwegian Cultural Heritage Management and focused Viking Age research. Experiences from South-Western Norway. In: Lasaponara R, Masini N (eds) Remote sensing for archaeology and cultural heritage management. Aracne Editrice, Rome, pp 285–288

Grøn O, Palmer S, Stylegar F-A, Aase S, Esbensen K, Kucheryavski S, Sigurd A (2011) Interpretation of archaeological small-scale features in spectral images. J Archaeol Sci 38:2024–2030

Jahjah M, Ulivieri C (2010) Automatic archaeological feature extraction from satellite VHR images. Acta Astronaut 66:1302–1310

Kennedy D (1998a) Declassified satellite photographs and archaeology in the Middle East: case studies from Turkey. Antiquity 72:553–561

Kennedy D (1998b) The twin towns of Zeugma on the Euphrates: rescue work and historical studies. J Roman Archaeol Suppl Ser 27:1–247

Lasaponara R, Masini N (2007) Detection of archaeological crop marks by using satellite QuickBird multispectral imagery. J Archaeol Sci 34:214–221

Lasaponara R, Masini N (2010) Facing the archaeological looting in Peru by local spatial autocorrelation statistics of very high resolution satellite imagery. In: Gervasi O, Murgante B, Pardede E, Apduhan BO (eds) Proceedings of ICSSA, The 2010 international conference on computational science and its application, Fukuoka-Japan, 23–26 Mar 2010, Taniar. Springer, Berlin, pp 261–269

Lasaponara R, Masini N (2011) Satellite remote sensing in archaeology: past, present and future. J Archaeol Sci 38:1995–2002

Lasaponara R, Masini N, Scardozzi G (2008) New perspectives for satellite-based archaeological research in the ancient territory of Hierapolis (Turkey). Adv Geosci 19:87–96

Lillesand TM, Kiefer RW (2000) Remote sensing and image interpretation. Wiley, New York

Marcolongo M, Morandi Bonacossi D (1997) Abandonment of the Qatabanian irrigation system in the Wadi Bayhan valley (Yemen): a geoarchaeological analysis. C R Acad Sci 325:79–86

Masini N (1998) La fotointerpretazione aerea finalizzata allo studio morfologico dei siti urbani e fortificati medioevali della Basilicata. In: Fonseca CD (ed) "Castra ipsa possunt et debent reparari." Indagini conoscitive e metodologie di restauro delle strutture castellane normanno-sveve. Edizioni De Luca, Roma, pp 205–250

Masini N, Lasaponara R (2006) Satellite-based recognition of landscape archaeological features related to ancient human transformation. J Geophys Eng 3:230–235. doi:10.1088/1742-2132/3/3/004

Parcak S (2009) Satellite remote sensing for archaeology. Routledge, Abingdon/New York

Parry JT (1992) The investigative role of Landsat TM in the examination of pre-proto-historic water management sites in northeast Thailand. Geocarto Int 4:5–24

Piccarreta F, Ceraudo G (2000) Manuale di aerofotografia archeologica. Metodologia, tecniche e applicazioni. Edipuglia, Bari

Sever TL (1998) Validating prehistoric and current social phenomena upon the landscape of Peten, Guatemala. In: Liverman D, Moran EF, Rinfuss RR, Stern PC (eds) People and pixels: linking remote sensing and social science. National Academy Press, Washington, DC

Ur J (2003) CORONA satellite photography and ancient road networks: a northern Mesopotamian case study. Antiquity 77:102–115

White K, El Asmar HM (1999) Monitoring changing position of coastlines using Thematic Mapper imagery, an example from the Nile Delta. Geomorphology 29:93–105

Wilson DR (1982) Air photo interpretation for archaeologists. St. Martin's Press, London

Wiseman J, El-Baz F (2008) Remote sensing in archaeology. Springer, New York

Chapter 2
Image Enhancement, Feature Extraction and Geospatial Analysis in an Archaeological Perspective

Rosa Lasaponara and Nicola Masini

Abstract The goal of image processing for archaeological applications is to enhance spatial patterns and/or local anomalies linked to ancient human activities and traces of palaeo-environments still fossilized in the modern landscape. In order to make the satellite data more meaningful for archaeologists and more exploitable for investigations, reliable data processing may be carried out. Over the years a great variety of digital image enhancement techniques have been devised for specific application fields according to data availability. Nevertheless, only recently these methods have captured great attention also in the field of archaeology for an easier extraction of quantitative information using effective and reliable semiautomatic data processing. The setting up of fully-automatic methodologies is a big challenge to be strategically addressed by research communities in the next years.

Keywords Radiometric and spectral enhancement • Spectral indices • PCA • TCT • Spatial enhancement • Fourier • Wavelet • Filtering • Geospatial analysis

2.1 Introduction

The application of aerial photographs had been long appreciated in archaeological investigations (see, for example, Reeves 1936; Miller 1957; Stone 1964; Estes et al. 1977; Spennemann 1987; Weber and Yool 1999; Bewley et al. 1999; Wilson 2000).

R. Lasaponara (✉)
Institute of Methodologies for Environmental Analysis, CNR-IMAA,
C. da S. Loya, 85100 Tito Scalo, PZ, Italy
e-mail: lasaponara@imaa.cnr.it

N. Masini
Institute of Archaeological and Architectural Heritage, CNR-IBAM,
C. da S. Loya, 85050 Tito Scalo, PZ, Italy
e-mail: n.masini@ibam.cnr.it

R. Lasaponara and N. Masini (eds.), *Satellite Remote Sensing: A New Tool for Archaeology*, Remote Sensing and Digital Image Processing 16,
DOI 10.1007/978-90-481-8801-7_2,

In fact, over the last century, aerial reconnaissance has been one of the most important ways in which new archaeological sites have been discovered throughout the world. The advantages of aerial photographs are manifold: they can be taken vertically or obliquely, easily interpreted, used for photogrammetric applications. They also provide a three-dimensional view.

Presently, the great amount of multispectral VHR satellite images, even available free of charge in Google earth (see, for example, Beck 2006; Brown et al. 2011; Kennedy and Bishop 2011) opened new strategic challenges in the field of remote sensing in archaeology (Campana and Forte 2006; Parcak 2009; Lasaponara and Masini 2011; Giardino 2011). These challenges substantially deal with: (i) the strategic exploitation of both aerial and satellite data as much as possible, (ii) the setting up of effective and reliable automatic and/or semiautomatic data processing strategies and (iii) the integration with other data sources from documentary resources to the traditional ground truthing activity including geophysical prospection.

VHR satellites provide high resolution data which can improve knowledge on past human activities providing precious qualitative and quantitative information developed to such an extent that currently they share many of the physical characteristics of aerial imagery. This makes them ideal for investigations ranging from a local to a regional scale (see. for example, Urwin and Ireland 1992; Sussman et al. 1994; Stein and Cullen 1994; Clark et al. 1998; Fowler 1996; Fritz 1996; Drake 1997; Strahler and Strahler 1997; Sever 1998; Sheets and Sever 1988; Sarris and Jones 2000; Toutin 2001, 2002; Kouchoukos 2001; Davis and Wang 2003; Lasaponara and Masini 2006a, b, 2007a, 2011; Masini and Lasaponara 2006, 2007). Moreover, satellite data are still the only data source for research performed in areas where aerial photography is restricted because of military or political reasons. Among the main advantages of using satellite remote sensing compared to traditional field archaeology herein we briefly summarized the following:

(a) **Multiscale/Multispectral capability**. Satellite remote sensing provides panchromatic and multispectral images acquired simultaneously at different scales and resolutions for the study area. This allows us to perform inter-site and intra-site analysis, ranging from detailed archaeological investigations to regional environmental and landscape studies.
(b) **High temporal resolution**. Satellite remote sensing assures the acquisition of imagery at high repeated re-visit cycle (1–3 days for VHR), thus allowing the possibility of performing multitemporal observations useful for systematic risk monitoring and for revealing changes over time.
(c) **Non-invasivity**. Remote sensing enables us to detect unknown sites and study the structure and distribution of ancient remains without exposing them, saving money, time and maintaining archaeological relics "as they are" thus preserving them for additional future investigations.
(d) **Data manipulation and information extraction**. Data manipulation enables the enhancement and extraction of features, patterns and their recognition, thus providing improved information on cultural heritage.

Among all the data manipulation techniques, image enhancement can make subtle features more visible to the human eye and, therefore, easier to interpret than raw data and fruitfully manageable for further data processing.

The choice of particular digital image enhancement approaches, among the great number of today's available techniques, depends on the application, data availability, experience and preferences of the image analyst (Richards and Jia 2006). It is really important to highlight that the image enhancement algorithms are data set and application-dependent, therefore, their performance can strongly vary from one application to another or from one data set to another.

We can categorize the image enhancement approaches into three main groups:

- Radiometric enhancement based on the values of individual pixels;
- Spectral enhancement based on the transformation of the values of each pixel on a multi-band basis;
- Spatial enhancement techniques based on the values of individual given pixels and their neighboring areas.

Moreover the enhancement may be carried out on: (i) temporary mode, when an image is only displayed in a viewer and all edits are not saved; or (ii) permanent mode, when all operations are saved as another image file.

2.2 Radiometric Enhancement

The radiometric enhancement techniques are based on the values of individual pixels within each band. To emphasize targets/objects/features of interest, these techniques are based on the use of histograms, namely the graph of the number of pixels with a given brightness versus brightness values. The histogram provides a discrete probability distribution being that the relative height of a given bar indicates the chance to find a pixel in the image with that particular brightness value, generally called the Digital Number (DN). The histogram based analysis is very useful since the tonal or radiometric quality of an image can be assessed from its histogram (see in Fig. 2.1 the histogram of a study area in Cauchy in Peru).

Pixels related to different materials may have similar brightness values. This is because different materials or objects may reflect or emit similar amounts of radiant flux. In this condition, it is highly probable to have an image with a poor contrast, which can be enhanced by stretching the image data to re-arrange the brightness values and obtain a new histogram over the full available range. Contrast enhancement is only intended to improve the visual quality of an image by using the whole available radiometric range (usually 0–255). It does not change the pixel values, unless it is saved as a new image. Contrast enhancement is very helpful for

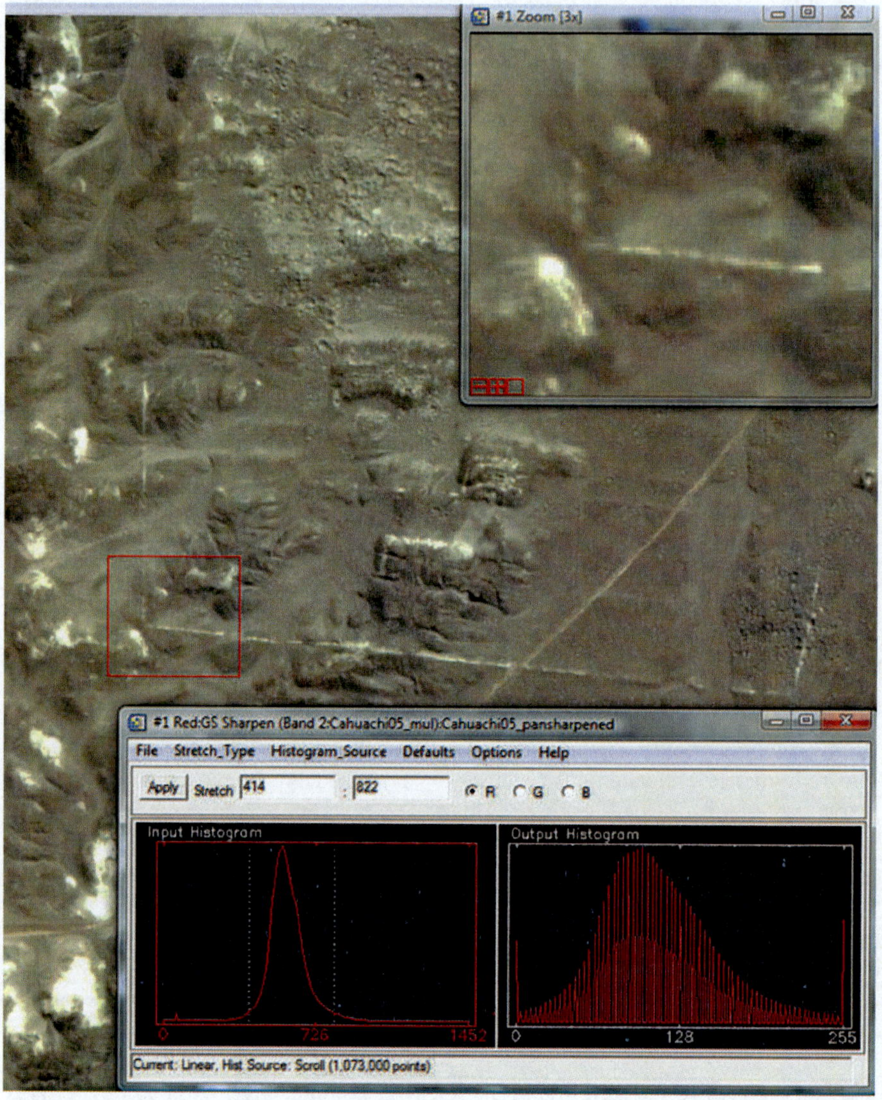

Fig. 2.1 Histogram of QuickBird (2005 March) *red* channel of a subset related to the *square block* in the upper right which includes a portion of the enclosure of a sacred area in the Ceremonial Centre of Cahuachi (Peru)

sharpening subtle details, but it is not a good practice to use the saved image for classification and/or change detection, etc.

Several contrast enhancement techniques are frequently used in archaeology (see for example Beck et al. 2007; Campana 2004; Campana and Francovich 2003) and implemented in common image processing routines and in open-source softwares. A comprehensive analysis is in Richards and Jia (2006), herein we only outline some of them below:

- Linear techniques:
  - Minimum-maximum contrast stretch (including percentage or standard deviation);
  - Piecewise linear contrast stretch.
- Non-linear techniques:
  - Histogram equalization;
  - Logarithmic and Exponential Contrast Enhancement.

2.2.1 Linear Enhancement Techniques

The minimum-maximum contrast is expressed by the following mathematical formulation.

$$BV_{out} = \frac{(BV_{in} - min_k)}{(max_k - min_k)} quant \tag{2.1}$$

where:

- BV_{out} is the output brightness value,
- BV_{in} is the original input brightness value,
- min_k is the minimum value in the image,
- max_k is the maximum value in the image, and
- $quant_k$ is the range of the brightness values that can be displayed (generally 255)

The linear contrast stretching with percentage or standard deviation defines a given percentage at the top or low values (generally 2% or 5% and similarly in $\pm 1\sigma$, $\pm 2\sigma$, $\pm 3\sigma$), of the image that will be set to 0 or 255; whereas the rest of the values are linearly stretched. It is also possible to apply a piecewise linear contrast stretch with specified break points, as shown in Fig. 2.2, to flexibly fit the non-Gaussian histograms (i.e. bimodal, trimodal). The user can also generally specify the number of break points.

2.2.2 Non Linear Enhancement Techniques

One of the most common non linear enhancement techniques is the Histogram equalization. Dark and light features can be easily enhanced by using logarithmic and exponential contrast stretch, respectively.

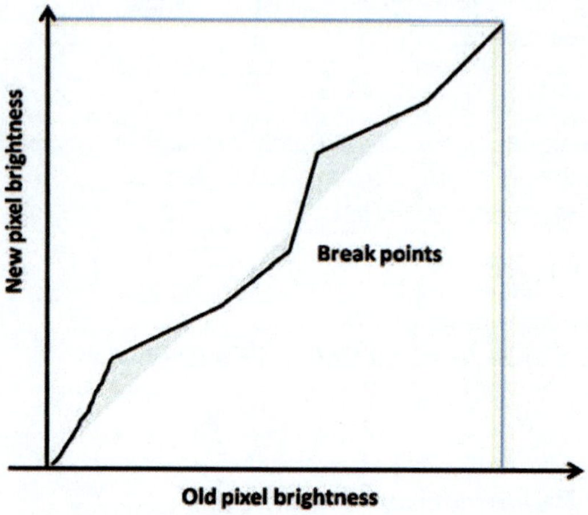

Fig. 2.2 Linear contrast stretch with specified break points

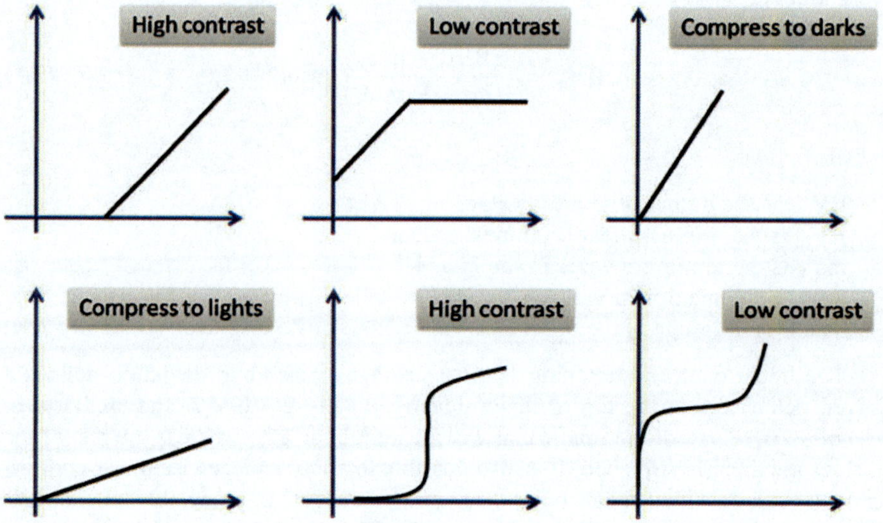

Fig. 2.3 Examples of histogram equalization functions

Figure 2.3 shows some examples of functions. Figure 2.4 depicts the results from linear and equalization enhancement for an archaeological site.

According to the mathematical formulation, the output values are generally scaled to lie within the range of the device used to display the image and that the output values are rounded to allow discrete values.

Please note that contrast modification simply locates the new histogram at different locations without altering size and number of bars of the old histogram.

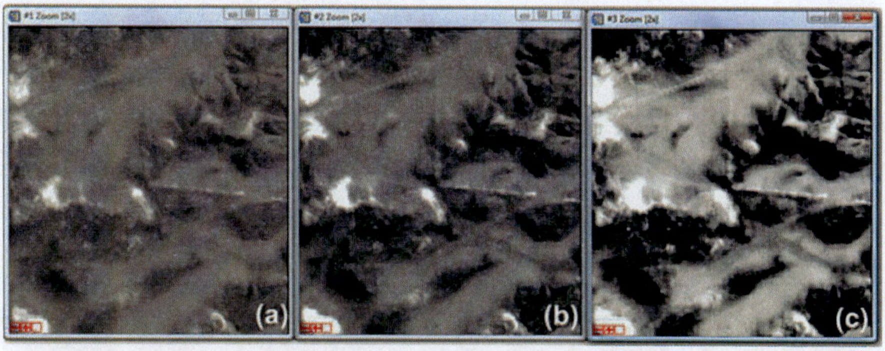

Fig. 2.4 **(a)** Same subset as in Fig. 2.2. **(b–c)** Results from linear and equalization enhancement, respectively. The latter emphasize the contrast between the linear features and its surrounding

2.2.3 Red-Green-Blue Colour Composites

The use of colour composites can facilitate the interpretation of multi-channel images by exploiting the fact that the visual overlapping of three images enhances the different types of surface/features/objects characterized by different grey levels in the three bands. The possibilities of varying the number of colour compositions are linked directly to the number of available spectral channels being the number of the mathematical combinations of them. In particular, the RGB (Red-Green-Blue Colour Composites) provides color visualization very close to the human colour perception and this can make the identification of different types of information easier. Figure 2.5 shows the result from RGB Composites. The Colour Composites visualization can be employed in conjunction with a Lookup- Table (LUT) used to transform the input data into a more interpretable output format. For example, a greyscale picture of the sea and land will be transformed into a colour image to emphasize the differences in the surfaces.

The RGB can be used not only for visualization but it may also be saved as a new image in a jpg file format. Moreover, slice may provide a more efficient distribution of the available 0–255 grey values to the range of the data. The RGB composition as well as false colour composition is widely used in archaeology (Aminzadeh and Samani 2006; Lasaponara et al. 2008; Alexakis et al. 2009). See for example, Fig. 2.6a, b which show two RGB compositions of ASTER images depicting the drainage basin of the Rio Nasca. They are the composition result of bands 3-2-1 (NIR/Red/Green channels) and bands 6-3-2 (SWIR/NIR/Red), respectively. In the false colour of Fig. 2.6b it is worth to note the red pixels which put in evidence a high reflectance of SWIR, related to a higher moisture content of the river and the tributaries respect to the surrounding arid area (for additional information on this case study the reader is referred to Chap. 12).

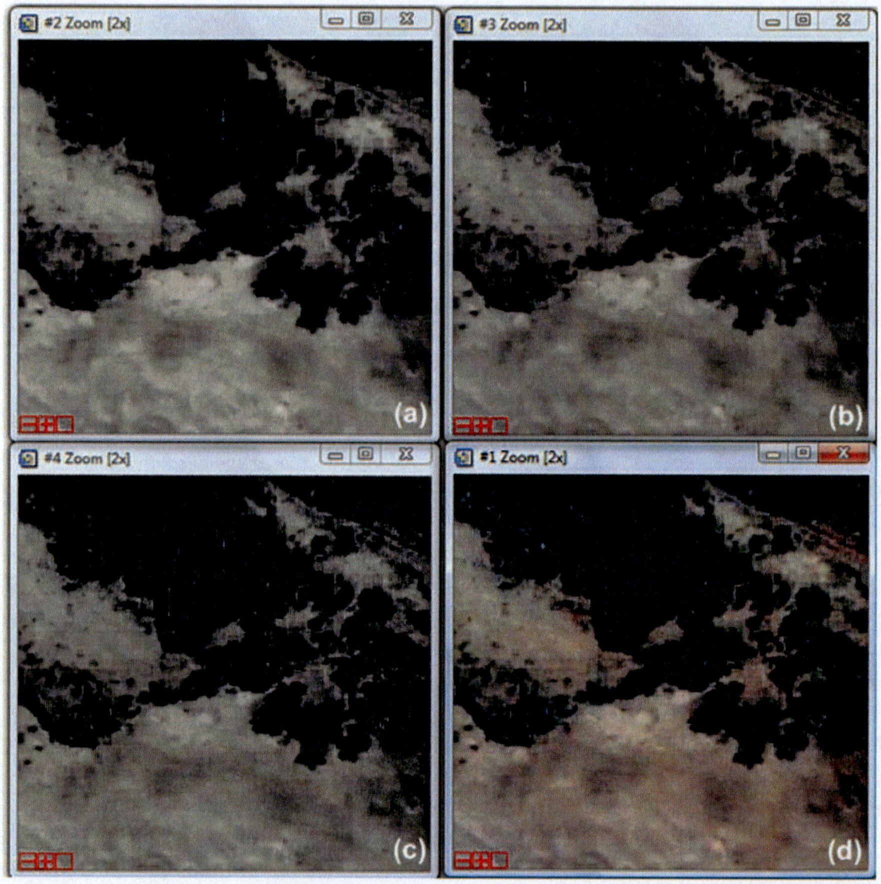

Fig. 2.5 Subset of a QuickBird image (2005 March) of Cahuachi. (**a**) Red band; (**b**) Green band; (**c**) Blue band; (**d**) RGB composition of the three bands Red-Green-Blue

2.3 Spectral Enhancement

2.3.1 Spectral Behaviour of Archaeological Marks

In order to exploit the multispectral capability of satellite imagery for the identification of archaeological marks all the available spectral bands (from visible to short wave infrared) may be considered.

Additionally, on the basis of multispectral properties of satellite remotely sensed data, archaeological-marks (crop or weed-marks) can be successfully identified by exploiting spectral indices that are mathematical combinations of different bands. Such indices are quantitative measures, based on the vegetation/soil/snow spectral properties that attempt to enhance the target under investigation.

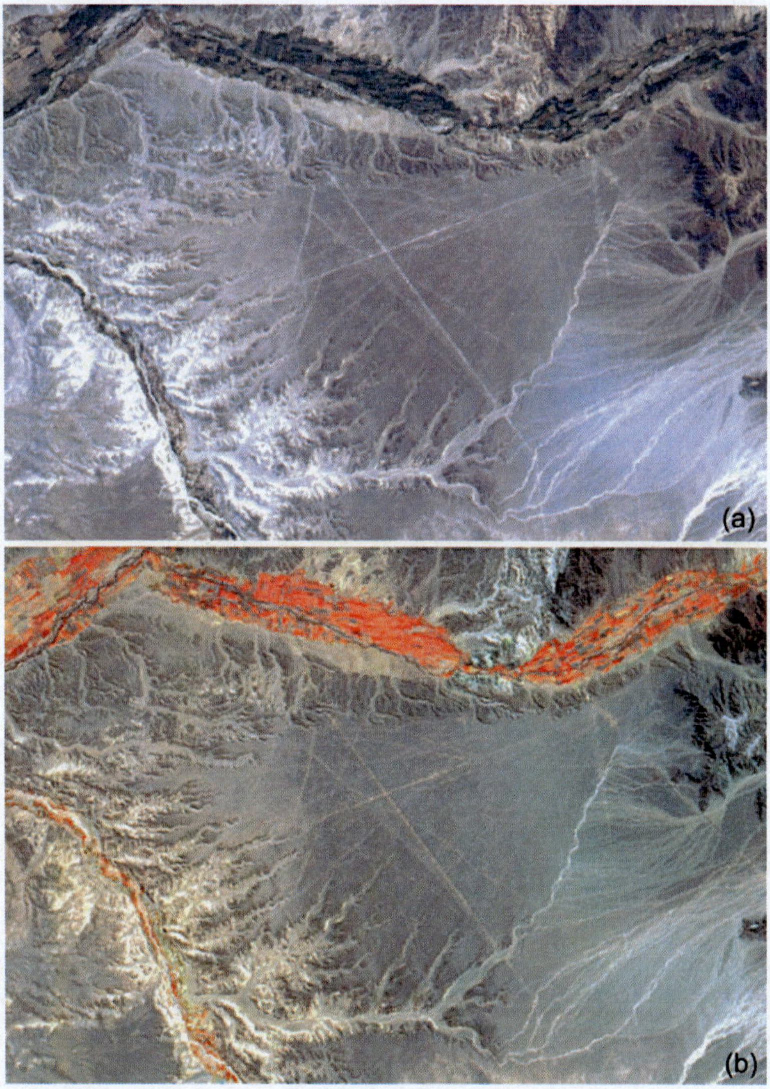

Fig. 2.6 ASTER images of the drainage Basin of the Rio Nasca. (**a**) RGB composition of bands 3-2-1 (NIR/Red/Green channels) and (**b**) bands 6-3-2 (SWIR/NIR/Red)

It is expected that, as in the case of aerial photos, satellite images at high spatial resolution should be able to identify the typical archaeological marks due the presence of buried structures. Recently, some evaluations concerning the capability of VHR satellite (QuickBird) spectral channels for archaeological applications

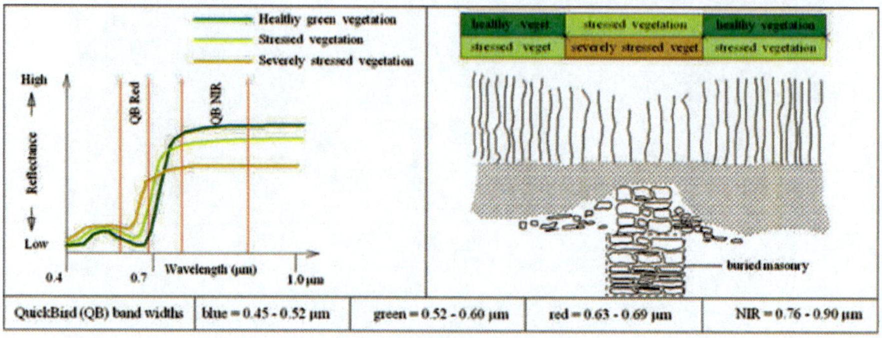

Fig. 2.7 Phenological and spectral characteristics linked to the presence of crop-marks. In the lower part of figure the QuickBird spectral range are reported

were performed (Lasaponara and Masini 2007a, b; Masini and Lasaponara 2007; Grøn et al. 2011). These investigations pointed out that the use of the

(i) red channel enhances the shadow and soil marks
(ii) NIR channel makes the detection of crop-marks easier

This is due to the fact that the presence of archaeological deposits under bare surface, lacking in vegetation, can influence the soil constituents and, above all, the moisture content. This causes a different spectral response of the Red band that can be used for the identification and spatial characterization of buried structures. Moreover, archaeological remains under vegetated surfaces strongly influence the state of plants, enhancing or reducing their growth. These subtle effects are only evident from an aerial view. The small differences in the plant growth and/or colour can be characterized by different spectral responses that, on the basis of VHR satellite spectral images, are enhanced in the NIR channel. Green and healthy plants tend to exhibit high NIR reflectance values; whereas vegetation under stress, due to lack of water or nutrients, is characterized by low NIR reflectance values.

As an example, yellow and brown curves, drawn in the left part of Fig. 2.7, are related to stressed and severely stressed plants, respectively. The Red and NIR bands are indicated by vertical red lines; thus, showing that the NIR offers a larger spectral separability than the Red band. This should make the NIR band ideal for identifying archaeological crop-marks linked to variations in vegetation growth and/or colour in comparison to the surrounding area (see, right part of Fig. 2.7).

For HR satellite data, SWIR bands are also available. SWIR spectral range is strongly sensitive to moisture content (soil moisture and leaf water content), thus allowing us to analyze moisture variations in space and time which are very useful for palaeo-geographic studies (see Chaps. 9, 11 and 12 in this book).

2.3.2 *Spectral Indices*

On the basis of remotely-sensed data, crop marks may be suitably identified by exploiting vegetation indices that are spectral combinations of different bands.

Such indices are quantitative measures, based on vegetation spectral properties that attempt to quantify biomass or vegetative vigor. The rationale for spectral vegetation indices is to exploit the spectral signatures of green and healthy vegetation as compared to those of stressed/unhealthy vegetation and other earth materials. Earth materials, such as, bare soil, sand, exposed rock, concrete, asphalt generally exhibit a steady rise in reflectance, without strong variations in the visible to the near-infrared. Whereas, green vegetation exhibits an increasing spectral reflectance from the visible to the near-infrared. Vegetation reflectance is very low in the blue and red regions, tends to be slightly higher in the green band and, finally, shows the greatest spectral response in the near-infrared (NIR).

Vegetation indices are spectral combinations of two or more bands devised to enhance the spectral signal of vegetation, mainly derived Red and near-infrared bands. For this reason they should allow a more reliable evaluation of photosynthetic activity and structural variations of the vegetation cover.

Vegetation indices operate by contrasting intense chlorophyll pigment absorption in the red with the high NIR reflectance of leaf mesophyll. The simplest form of vegetation index is the ratio between two digital values from the red and near-infrared spectral bands. The most widely used index is the well-known Normalized Difference Vegetation Index (NDVI) obtained by using the following formula:

$$\mathrm{NDVI} = \frac{(\mathrm{NIR} - \mathrm{RED})}{(\mathrm{NIR} + \mathrm{RED})} \tag{2.2}$$

The arithmetic combination of the NDVI enables us to exploit the different spectral response of vegetation cover in the visible spectral (red) and near infrared bands. It provides a dimensionless numerical value. The formula is designed as a ratio, in order to normalize its variability field between -1 and +1, and in particular takes on NDVI values less than 0 for water, slightly higher than 0 for soils and between 0.4 and 0.7 for vegetation, dense vegetation can exceed 0.8 and be close to saturation (1) for a rainforest.

The normalization of the NDVI reduces the effects of variations caused by atmospheric contaminations. NDVI is indicative of plant photosynthetic activity and has been found to be related to the green leaf area index and the fraction of photosynthetically active radiation absorbed by vegetation. High (low) values of the vegetation index identify pixels covered by substantial proportions of healthy (disease or stressed) vegetation. It is expected that crop-marks created by vegetation patterns should be emphasized by using NDVI.

The NDVI (see Eq. 2.2) is able to account for vegetation vigor, and therefore, to detect the deterioration in pigments and leaf structure caused by the presence of subsurface archaeological materials. Figure 2.8 show the red, NIR and NDVI map for some archaeological areas.

NDVI has been firstly applied by Lasaponara and Masini (2006b) for enhancing crop marks in a medieval settlement, thus allowing to reconstruct its urban fabric.

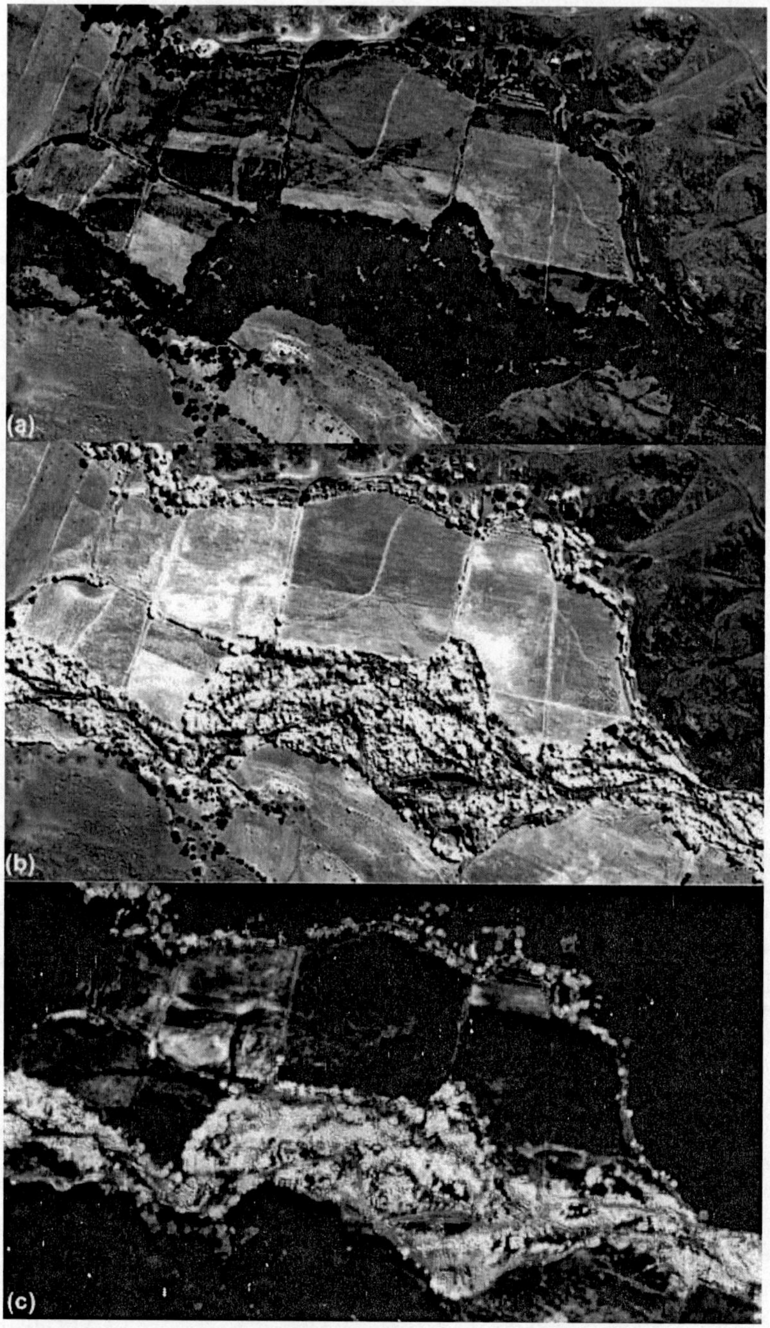

Fig. 2.8 QuickBird of Rio Nasca (2005 March). (**a**) Red band; (**b**) NIR band; (**c**) NDVI map

Other spectral bands and/or spectral combinations may be suitable for the enhancement of archaeological marks, single channels and spectral indices useful/or specifically designed for surface characterization (vegetation, soil, snow) can be considered for the analysis.

Spectral combinations of different bands are widely used since they generally tend to emphasize the spectral features caused by moisture content, the presence/type/status of vegetation cover, as well as by soil types and characteristics. Although the NDVI has been one of the most widely used vegetation indices, further spectral combinations were proposed to increase the sensitivity over a wide range of vegetation conditions and to reduce soil background influence and residual atmospheric contamination.

Among all the spectral indices devised over the years, we cite (i) albedo (Saunders 1990), Simple Ratio (SR), GEMI (Pinty and Verstraete 1992), SAVI (Huete 1988), Enhanced Vegetation Index (EVI), ARVI, Green NDVI, moisture index and snow index. Quite recently, the potentiality of using satellite SWIR spectral bands for moisture content estimation has been investigated.

Albedo is an indicator of the surface characteristics including brightness and darkness. The empirical approximation of surface albedo generally used is shown in formula 2.2 (Saunders 1990).

$$\text{Albedo} = (\text{NIR} + \text{RED})/2 \tag{2.3}$$

The simple ratio SR (see formula 2.4) is calculated using the ratio between NIR and red (or between NIR and Visible if no red band is available). Therefore, if these bands have the same or similar reflectance, the SR is 1 or close to 1. SR values for bare soils generally are near 1. When, for a given pixel, the amount of green vegetation increases the SR increases. Very high SR values are generally around 30.

$$\text{SR} = \text{NIR}/\text{RED} \tag{2.4}$$

Gitelson et al. (1996) proposed the use of a green channel instead of the red one for the remote sensing of vegetation. They devised the "Green"NDVI (see Eq. 2.5) that was applied to MODIS data and was found to be more sensitive to concentration in chlorophyll.

$$\text{Green NDVI} = \frac{(\text{NIR} - \text{GREEN})}{(\text{NIR} + \text{GREEN})} \tag{2.5}$$

The "Green"NDVI exhibits a wider range values than the original NDVI and is, on average, at least five times more sensitive to chlorophyll concentration.

In order to reduce the soil background influence, the soil-adjusted vegetation index (SAVI) was proposed by Huete (1988). This index is computed by using the red and near-infrared reflectance with some added terms (see Eq. 2.6). SAVI should be able to optimize the NDVI on the basis of Beer's law.

$$\mathrm{SAVI} = \frac{(1+\mathrm{L}) \cdot (\mathrm{NIR} - \mathrm{RED})}{(\mathrm{NIR} + \mathrm{RED} + \mathrm{L})} \tag{2.6}$$

Where the term L can vary from 0 to 1 depending on the amount of visible soil; L = 1 is generally used when the amount of soil is unknown.

The SAVI is not sensitive to soil noise within a certain range of (Leaf Area Index) LAI values. The perfect range depends on the choice of the "L" value in the SAVI formulation. The authors suggested that, over the full range of canopy covers, L = 0.5 can be considered to be the optimal adjustment factor in reducing soil noise. Finally, the multiplication factor (1 + L) is used to maintain the dynamic range of the index.

In order to minimize the influence of atmospheric effects Pinty and Verstraete (1992) proposed formula 2.7:

$$\mathrm{GEMI} = \gamma \times (1 - 0.25\gamma) - \frac{(\mathrm{RED} - 0.25)}{(1 - \mathrm{RED})} \tag{2.7}$$

Where

$$\gamma = \frac{(2 \times (\mathrm{NIR} - \mathrm{RED}) + 1.5 \times \mathrm{NIR} + 0.5 \times \mathrm{RED})}{(\mathrm{NIR} + \mathrm{RED} + 0.5)}$$

The GEMI should be transparent to the atmosphere. It represents plant biological information similar to the NDVI. The main problem that is encountered with this index is mainly due to its complex formulation.

In order to reduce aerosol effects, the atmospheric resistant vegetation index (ARVI) was devised by Kaufman and Tanrer (1992).

This index is computed using formula 2.8. It is based on a function that should stabilize the index when there are temporal/spatial variations in the concentration of sand, gaseous, and particulate pollutants.

$$\mathrm{ARVI} = \frac{(\mathrm{NIR} - \mathrm{RB})}{(\mathrm{NIR} + \mathrm{RB})} \tag{2.8}$$

where RB is a combination of the reflectances in the BLUE and in the RED channels: $RB = RED - \gamma \times (BLUE - RED)$ and γ depends on the aerosol type.

In order to optimize the vegetation signal from deserts to rainforests, the enhanced vegetation index (EVI) was developed.

It is obtained using formula 2.9.

$$\mathrm{EVI} = \frac{(1+\mathrm{L}) \times (\mathrm{NIR} - \mathrm{RED})}{(\mathrm{NIR} + \mathrm{C}_1 \times \mathrm{RED} - \mathrm{C}_2 \times \mathrm{BLUE} + \mathrm{L})} \tag{2.9}$$

Where C_1, C_2, and L are constants empirically determined. The values generally used are as $C_1 = 6.0$; $C_2 = 7.5$; and $L = 1$.

The coefficient L denotes the canopy background and snow correction caused by differential NIR and red radiant transfer (transmittance) through a canopy; C1 and C2 denote the coefficient of the aerosol "resistance" term, which uses the blue band to correct for aerosol effects in the red band (Kaufman and Tanrer 1992). Therefore, the use of EVI should considerably reduce aerosol variations via the self-correcting combination of the red and blue channels.

Increased sensitivity should be achieved by placing more weight to the NIR reflectance component of the EVI formulation (see Eq. 2.9). Thus, allowing the EVI improved vegetation studies for densely vegetated areas as well as sparsely vegetated, arid and semi arid regions.

Recently, a number of satellite sensors at medium resolution (15–90 m) offer SWIR band acquisition which is quite sensitive to moisture content and moisture variation in both soil and vegetation cover. This capability has been supported by both modelling (Ceccato et al. 2001; Fourty and Baret 1998; Tucker 1980; Ustin et al. 1998; Zarco-Tejada et al. 2003) and experimental studies based on the available multispectral satellite datasets (Ceccato et al. 2002a, b; Chen et al. 2003; Jackson et al. 2004).

Actually, for more than three decades the NDVI (see formula 2.2) has been used as a surrogate to estimate vegetation water content even with strong limitations mainly encountered when vegetation coverage is dense and the index is close to the saturation level.

NDVI provides information closer to the amount and greenness of vegetation rather than moisture content and it is generally limited by soil reflection.

Using multispectral satellite data the estimation of moisture content in soil and vegetation may be improved using vegetation indices based upon NIR and SWIR and in general on the longer wave-lengths (1,240–3,000 nm), for example, the short-wave infrared (SWIR) reflectance (1,300–2,500 nm).

Several spectral indices, such as Normalized Vegetation Moisture Index (NVMI) or Normalized Difference Water Index (NDWI), mainly based on SWIR bands, can be computed to estimate moisture content for both soil and vegetation (see for example Fig. 2.9). The mathematical formulation of these indices (see formula 2.2 and 2.3) is very similar to the NDVI, but based on specific bands related to water absorption.

$$\mathrm{NVMI} = \frac{(\mathrm{NIR} - \mathrm{SWIR})}{(\mathrm{NIR} + \mathrm{SWIR})} \tag{2.10}$$

$$\mathrm{NDWI} = \frac{(\mathrm{RED} - \mathrm{SWIR})}{(\mathrm{RED} + \mathrm{SWIR})} \tag{2.11}$$

Both of these indices, NVMI and NDWI, are sensitive to water content in vegetation and soil, respectively being that the absorption of water content of

Fig. 2.9 ASTER images (2003 June) of the Rio Nasca drainage basin. (**a**) Red channel (band 2), (**b**) SWIR channel (band 6); (**c**) NDWI map

vegetation is negligible close to the NIR band (or close to the red band for soil), whereas a small absorption is present in the SWIR range.

Moreover, in comparison with NDVI, both NVMI and NDWI are less sensitive to the effects of the atmosphere, but effects of soil reflection are still present.

To date, there are two main satellite sensors which offer information in the SWIR spectral bands from medium to low spatial resolution, they are (i) Landsat MSS (available at 80 m), Landsat TM/ETM (available at 30 m) and (iii) ASTER images (with a spatial resolution ranging from 15 to 90 m).

Furthermore, in order to estimate the spectral separability (see Sect. 2.2) of all the considered spectral indicators, a sensitive analysis can be performed using an objective image quality index. This index can automatically predict perceived image quality.

As a whole, we can argue that the question is to assess which of the available vegetation indices, namely spectral indicators, is more suitable and adequate for the identification of the given archaeological marks (crop, soil, snow marks).

2.3.3 *Spectral Separability of Vegetation Indices*

In order to quantitatively assess the performance obtained from different vegetation indices we can use a universal quality index. This index, computed using formula 2.11, combines the correlation coefficient with luminance and contrast distortion (Wang and Bovik 2002). It was devised for image fusion to assess the quality of output image, as well as for image processing systems and algorithms. The quality index can be applied to qualitatively evaluate the capability of each satellite based index to enhance archaeological marks.

$$Q = \frac{(2 \times \mu_x \times \mu_y) \times (2 \times \sigma_{xy} + C_2)}{\left(\mu_x^2 + \mu_y^2 + C_1\right) \times (\sigma_x^2 + \sigma_y^2 + C_2)} \tag{2.12}$$

where $C_1 = (k_1 \times L)^2$ and $C_2 = (k_2 \times L)^2$.

σ_{xy} is the covariance between the two images x and y, σ_x and σ_y are the standard deviations, μ_x and μ_y are the means. L is the dynamic range for the image pixel values, $k_1 \ll 1$ and $k_2 \ll 1$ are two constants chosen equal to 0.01 and 0.03, respectively. Although the values selected for k_1 and k_2 are arbitrary, It was experienced that the quality index is insensitive to variations of k_1 and k_2. Note that C_1 and C_{12} are solely introduced to stabilize the measure. In other word, just to avoid the denominator approaches zero values for flat regions.

The quality index compares two images by using three parameters: luminance, contrast and structure. Note that Q is scaled between -1 and 1. The value 1 is achieved when the two images x and y are the same. To account for local variation, a moving window sized 8 × 8 pixels was suggested by the Wang and Bovik (2002).

To measure the overall image quality the mean quality index can be rewritten as a three factor product, that can be regarded are relatively independent.

$$Q(x,y) = f(l(x,y), c(x,y), s(x,y))$$
$$= \frac{\left(\sigma_{xy} + \frac{C_2}{2}\right) \times \left(2 \times \mu_x \mu_y + C_1\right) \times 2 \times \left(\sigma_x \sigma_y + C_2\right)}{\left(\sigma_x \sigma_y \frac{C_2}{2}\right) \times \left(\mu_x^2 + \mu_{y+}^2 C_1\right) \times 2 \times \left(\sigma_x^2 + \sigma_y^2 + C_2\right)} \tag{2.13}$$

In particular, among the three factor of Eq. 2.13, the first (varying between −1 and 1) represents the correlation coefficient between the two image x and y; the second (varying between 0 and 1) measures the similarity between the mean luminance values of x and y and finally, the third (varying between 0 and 1) measures the contrast similarity.

Since traditionally the archaeological crop-mark identification has been performed by using black and white aerial photos, the performance evaluation may be performed comparing satellite panchromatic image with each considered multispectral index.

Of course other spectral separability indices may be considered, for any additional information reader is referred to Chap. 3 and references therein quoted.

2.3.4 Spectral Combinations Based on Linear Transformation

2.3.4.1 Tasseled Cap Transformation

Tasseled Cap Transformation (TCT) could be adopted for extracting archaeological features linked to ancient human transformations of landscape. Several authors have successfully applied TCT In archaeological studies, see, for example, Sarris and Jones (2000), Beck et al. (2007), Lasaponara and Masini (2007b). The TCT, also known as Kauth-Thomas technique, was devised for enhancing spectral information content of satellite data. The reader is referred to Crist and Kauth (1986) for an introduction.

The TCT is a linear affine transformation substantially based on the conversion of given input channel data set (or cartesian reference system) in a new data set (or cartesian reference system) of composite values; i.e., obtained via weighted sums of the input channels. The TCT is performed on a pixel basis (by using formula 2.14) to better show the underlying structure of the image.

$$TC = W_{TC} DN + B \tag{2.14}$$

where TC is the W_{Tc} specific transformation (coefficients), DN Digital Number and B Bias.

The transformation W_{Tc} depends on the considered sensor, because different sensors have different numbers of bands which, in turn have different spectral responses.

The original TCT was derived (Kauth and Thomas 1976) for the four bands of the Landsat MSS sensor. Later, the TCT was extended to the Landsat TM (Crist and Cicone 1984), ETM (as available in a routine of PCI Geomatics) and IKONOS sensor (Horne 2003).

All the existing TCTs decorrelate the multispectral data by using weighted sums of the input channels to extract a greater amount of information that can allow an easier identification of distinct surface types.

There are just three composite variables of TCT bands (axes) which are traditionally used:

- TCT-band 1 (brightness, measure of soil)
- TCT-band 2 (greenness, measure of vegetation)
- TCT-band 3 (wetness, interrelationship of soil and canopy moisture)

The weighted sums adapted from the ETM input channels:

$$\mathrm{TCT}_{\mathrm{ETM-band1}} = 0.1544\,\mathrm{BLU} + 0.2552\,\mathrm{GREEN} + 0.3592\,\mathrm{RED} + 0.5494\,\mathrm{NIR} \tag{2.15}$$

$$TCT_{ETM-band2} = -0.1099\,BLU - 0.1255\,GREEN - 0.2866\,RED + 0.8226\,NIR \tag{2.16}$$

$$TCT_{ETM-band3} = 0.3191\,BLU + 0.5061\,GREEN + 0.5534\,RED + 0.0301\,NIR \tag{2.17}$$

The weighted sums developed by Horne (2003) for IKONOS channels:

$$TCT_{IKONOS-bans1} = 0.326\,BLU + 0.509\,GREEN + 0.560\,RED + 0.567\,NIR \tag{2.18}$$

$$TCT_{IKONOS-band2} = -0.311\,BLU - 0.356\,GREEN - 0.325\,RED + 0.819\,NIR \tag{2.19}$$

$$TCT_{IKONOS-band3} = -0.612\,BLU - 0.312\,GREEN + 0.722\,RED - 0.081\,NIR \tag{2.20}$$

$$TCT_{IKONOS-band4} = -0.650\,BLU + 0.719\,GREEN - 0.243\,RED - 0.031\,NIR \tag{2.21}$$

TCT has been adopted in archaeology by several authors, see, for example Campana 2003; Lasaponara and Masini 2007b; Traviglia 2008.

Figure 2.10a shows a composition (RGB) of the first three TCT components computed from Landsat ETM 7 (2000) related to the drainage basin of the river Rio Grande (Southern Peru). Figure 2.10b depicts a subset of Fig. 2.10a focusing the famous Nasca lines. The colours enable us to discriminate the desert area in red (brightness), the vegetation cover mainly in the fluvial oases in green and, finally, the area with higher moisture content in blue and/or light blue (wetness).

2.3.4.2 Principal Component Analysis (PCA)

In order to identify the small spectral signals linked to archaeological features, some authors (see for example Zhang et al. 2010; Deroin et al. 2011; Lasaponara et al. 2011; Traviglia and Cottica 2011) adopted Principal Component Analysis (PCA).

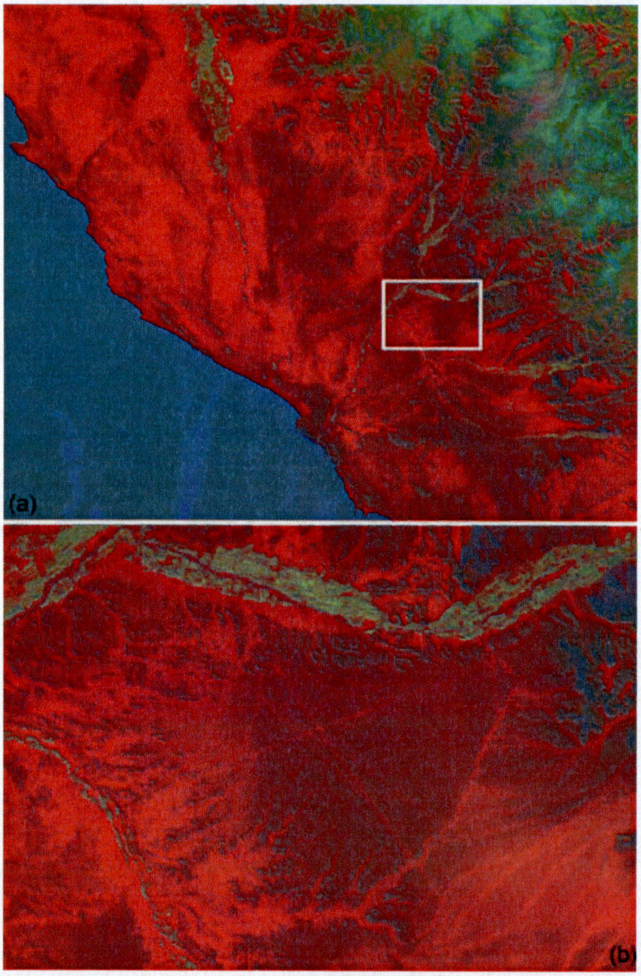

Fig. 2.10 RGB composition of the first three TCT components (brightness, greenness and wetness) computed from Landsat ETM 7 (2000). (**a**) Drainage basin of the river Rio Grande (Southern Peru); (**b**) subset including the Nasca lines. The colours enable us to discriminate the desert area in *red* (brightness), the vegetation cover mainly in the fluvial oases in *green* and the area with higher moisture content in *blue* and/or *light blue* (wetness)

It is a linear transformation which decorrelates multivariate data by translating and/ or rotating the axes of the original feature space, so that the data can be represented without correlation in a new component space (see Fig. 2.11). In order to do this, It is first computed: (i) the covariance matrix (S) among all input spectra bands (each element of S is calculated by using formula 2.22), then (ii) eigenvalues and eigenvectors of S in order to obtain the new feature components.

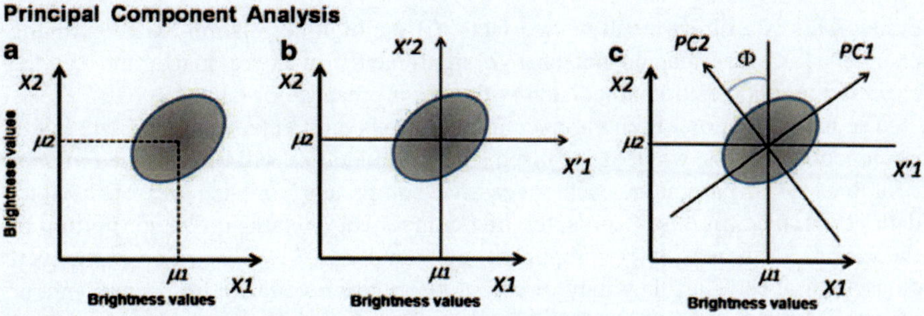

Fig. 2.11 PCA decorrelates multivariate data by translating and/ or rotating the axes of the original feature space. The data can be represented in a new component space without correlation

$$\mathrm{covk_1k_2} = \frac{1}{\mathrm{n \times m}} \sum_{\mathrm{i=1}}^{\mathrm{n}} \sum_{\mathrm{j=1}}^{\mathrm{m}} \left(\mathrm{SB_{i,j,k_1}} - \mu_{\mathrm{k_2}}\right) \times \left(\mathrm{SB_{i,j,k_2}} - \mu_{\mathrm{k_2}}\right) \tag{2.22}$$

where k_1, k_2 are two input spectral channels, $SB_{i,j}$ is the spectral value of the given channel in row *i* and column *j*, *n* number of row, *m* number of columns, μ is the mean of all pixel $SB_{i,j}$ values in the subscripted input channels.

The percent of total dataset variance explained by each component is obtained by formula 2.23:

$$\%\, i = 100 \times \vartheta_{\mathrm{i}} \Big/ \sum\nolimits_{\mathrm{i=1}}^{\mathrm{k}} \lambda_{\mathrm{i}} \tag{2.23}$$

where λ_i are eigenvalues of S

Finally, a series of new image layers (called eigenchannels or components) are computed (using formula 2.24) by multiplying, for each pixel, the eigenvector of S for the original value of a given pixel in the input bands

$$\mathrm{P_i} = \sum_{\mathrm{i=1}}^{\mathrm{n}} \mathrm{P_k} \times \mathrm{u_{k,i}} \tag{2.24}$$

where P_i indicates a spectral channel in component i, $u_{k,i}$ eigenvector element for component i in input band k, P_k spectral value for channel *k*, number of input band.

A loading, or correlation *R*, of each component *i* with each input band *k* can be calculated by using formula 2.25.

$$\mathrm{R_{k,i}} = \mathrm{u_{k,i}} \times (\lambda_{\mathrm{i}})^{1/2} \times (\mathrm{var_k})^{1/2} \tag{2.25}$$

where var_k is the variance of input date *k* (obtained by reading the *k*th diagonal of the covariance matrix)

The PCA transforms the input multispectral bands in new components that should be able to make the identification of distinct features and surface types easier. This is a direct result of two facts: (i) the high correlation existing among channels for areas that do not change significantly over the space; and (ii) the expected low correlation associated with higher presence of noise.

The major portion of the variance in a multi-spectral data set is associated with homogeneous areas, whereas localised surface anomalies will be enhanced in later components. In particular, each successive component contains less of the total dataset variance. In other words, the first component contains the major portion of the variance, whereas, later components contain a very low proportion of the total dataset variance. Thus they may represent information variance for a small area or essentially noise and, in this case, it must be disregarded. Some problems can arise from the fact that eigenvectors cannot have general and universal meaning since they are extracted from the series (Fig. 2.12).

The main difference between TCT and PCA is that the latter computes the weight coefficients from the data and this is the reason why the meaning of the PCA components is dataset dependent. TCT has for each component a specific physical meaning because the transformation is based on weight coefficients which are empirically obtained for each sensor and surface types and characteristics.

2.3.4.3 IHS Transform

The RGB color space represents the amount of red, green and blue components in a color, and therefore it is not easy to distinguish one colour from another only using the RGB colour coordinates. To overcome this limitation, some authors suggest the use of IHS transform which converts three images from the RGB space into the Intensity, Hue and Saturation (IHS) color space. The IHS space provides parameters and information closer to the human colour perception, in particular:

(i) Intensity is the overall brightness and informs us about how close the image is to black or white.
(ii) Hue provides information about the actual perceived color (i.e. the dominant or average wavelength of light)
(iii) Saturation informs us about the purity of the color

Similarly to TCT and PCA, IHS performs a rotation from the RGB axis to a new orthogonal IHS. This axis rotation can be obtained by using different formulas which enable us to separate spatial (I) and spectral (H, S) information from a standard RGB image. Among the number of diverse combinations to obtain the new space, herein we report the Pellemans' equations (Pellemans et al. 1993) in formula 2.26.

$$\begin{bmatrix} x \\ y \\ z \end{bmatrix} = \begin{bmatrix} \frac{1}{\sqrt{3}} & \frac{2}{\sqrt{3}} & 0 \\ -\frac{2}{\sqrt{3}} & \frac{1}{\sqrt{3}} & 0 \\ 0 & 0 & 1 \end{bmatrix} \begin{bmatrix} \frac{1}{2} & 0 & \frac{1}{\sqrt{2}} \\ 0 & 1 & 0 \\ -\frac{1}{\sqrt{2}} & 0 & \frac{1}{\sqrt{2}} \end{bmatrix} \begin{bmatrix} B \\ G \\ R \end{bmatrix} \tag{2.26}$$

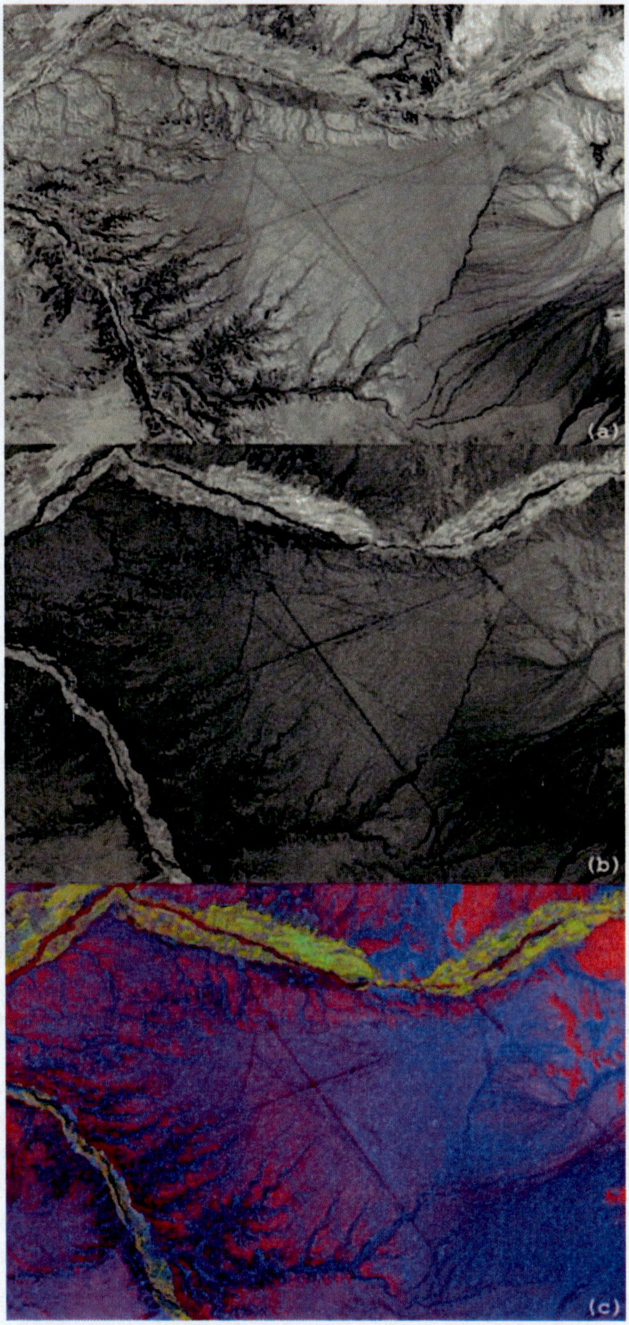

Fig. 2.12 Results from PCA obtained for an area at South of Rio Ingenio characterized by the presence of Nasca Lines. PCA is computed using Landsat ETM 7 (2000). (**a**) PC2; (**b**) PC3; (**c**) RGB composition (R = PC2, G = PC3, B = PC4). PC2 and RGB emphasize the geoglyphs, PC3 puts in evidence the vegetation in the fluvial oasis

The value of H, S and I can then be computed as:

$$H = \tan^{-1}\left(-\frac{\sqrt{z}}{\sqrt{x}}\right) \tag{2.27}$$

$$S = \cos^{-1}\left(\frac{\frac{\sqrt{y}}{\sqrt{x+y+z}}}{\Phi_m(H)}\right) \tag{2.28}$$

$$I = \frac{(x+y+z)}{I_M(H,S)} \tag{2.29}$$

Where $\Phi_m(H)$ is the maximum colatitude for a given hue, $I_M(H,S)$ is the maximum intensity for a given hue and colatitude.

The transformation from the RGB color space to a given IHS space is characterized by the following properties: (i) nonlinearity (ii) reversibility and (iii) independent of each component from the others i.e. it may be varied without affecting the others.

Thanks to these properties, IHS is considered an ideal tool for image processing ranging from contrast enhancement to image fusion aimed at (i) enhancing the spatial content and (ii) preserving the spectral properties of the data set to be processed.

This transformation has been used in archaeology, see for example Campana (2003) and Aminzadeh and Samani (2006).

2.4 Spatial Enhancement

2.4.1 Image Enhancement in the Spatial Domain

The spatial domain refers to pixel aggregation which composes an image that may be manipulated according to the "investigation needs". Within this domain, pixels are the "*spatial variables*" being that they are a function (Treitz and Howarth 2000) of spatial position. Therefore, pixel values and position inform us about the spatial detail namely providing the information content of the scene. Image enhancement in the spatial domain operates directly (i) on the given pixels, which are the basic building block of a remotely sensed image and also on (ii) the neighbouring areas, which provide the spatial arrangement of information. As an example, features linked to buried remains may refer to a roman villa if they are displaced with a rectangular shape, or to Neolithic settlement in the case of circular arrangement, etc.

To extract information content from a digital image, we can use spatial analyses to capture features, patterns and trends using distance and spatial relationships, in particular: (i) measuring the interdependence of brightness values, (ii) quantifying

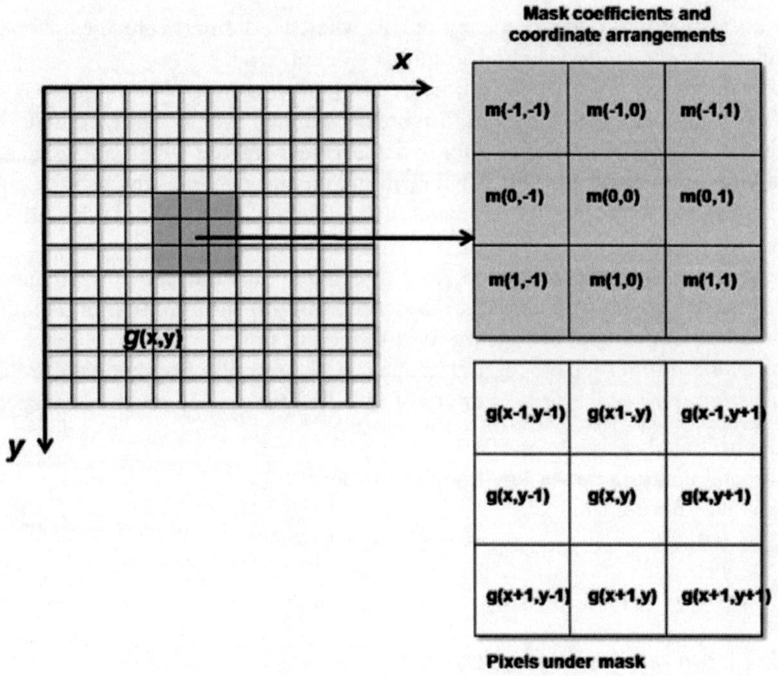

Fig. 2.13 *Grey* indicates a (moving) mask, also known as kernel, used to assign given weights to the pixels under mask in order to carry out local mathematical computation to obtain a new output image

spatial distribution and (iii) describing spatial clustering or random distribution of features.

Spatial enhancement highlights geometric detail in a digital scene linked to changes in brightness values which are generally influenced by the surrounding pixels. The use of filtering determines variations in the geometric detail at both local (contextual) and global (whole image) level in the perception of the information content. Figure 2.13 shows the spatial analysis which is generally conducted using a (moving) mask also known as kernel to assign given weights to the pixels under mask.

The spatial filtering process may enhance or smooth these image characteristics to extract or leave out pattern features from the background. The spatial filtering process is based on a 'local' mathematical operation generally applied using a two dimensional moving window ('kernel') to obtain a new output image from the results of the given mathematical operation (kernel, averaging process or whatever).

The spatial information content is also evident in the frequency domain and, therefore, it may be also captured in this domain.

For a given scene (or a subset), spatial frequency is defined as the number of changes in brightness value per unit distance. Therefore, few changes in brightness value corresponds to low-frequency detail, whereas, dramatic changes over short distances are referred to as a high-frequency detail.

Low frequency image characteristics refer to areas quite homogeneous with relatively little changes in pixel brightness values over a large region. High frequency image characteristics refer to heterogeneous areas with dramatic changes over a very short distance. This is when abrupt changes are present, as for example for scan line errors or fragmented landscapes, defined as 'roughness' by Lillesand and Kiefer (2000).

Generally within an image we may find high, medium and low frequencies which can be emphasized or suppressed according to the information of interest and/or the object, shape, pattern we need to hide or unveil.

The spatial properties of a digital image can be promptly improved according to the *investigation needs* using a variety of different processing techniques, such as:

(i) Fourier analysis and wavelet;
(ii) Fractal analysis power law transformation;
(iii) spatial filtering;
(iv) geospatial analysis.

2.4.2 Fourier Transformation

Detailed analysis on the frequency content of an image can be attained by using the Fourier transform which involves the transformation of the original coordinate space into a new two dimension space spectrum. Image enhancement or smoothing may be arranged by suppressing low or high spatial frequencies. Low, medium and high frequency components are first identified through the direct Fourier transform (formula 2.30), then removed, according to the data processing aims. Finally, a new image is generated using the inverse Fourier transform, which is expressed in formula 2.31:

$$F(x,y) = \sum_{u=0}^{N-1} \sum_{v=0}^{N-1} f(u,v) e^{i2\pi(xu+yv)/N} \tag{2.30}$$

In formula 2.30 f(u,v) is the image in the spatial domain, N is raw and line pixel number, the exponential expresses the basis functions of Fourier transformation as sine and cosine waves with increasing frequencies, finally F(x,y) is the f(u,v) in the Fourier transformation.

As an example, F(0,0) represents the component of the image which corresponds to the average brightness and F(N-1,N-1) is the highest frequency. The Fourier image can be re-transformed to the spatial domain, by using Eq. 2.31:

$$\mathrm{f(u,v)} = \frac{1}{\mathrm{N}^2} \sum_{\mathrm{x=0}}^{\mathrm{N-1}} \sum_{\mathrm{y=0}}^{\mathrm{N-1}} \mathrm{F(x,y)e}^{-\mathrm{i}2\pi\left(\frac{\mathrm{xu}}{\mathrm{N}} + \frac{\mathrm{yv}}{\mathrm{N}}\right)} \tag{2.31}$$

Fig. 2.14 Fourier transform (magnitude or real part) for ASTER image of Nasca Lines

where the $\frac{1}{N^2}$ is a normalization term in the inverse transformation. It can be applied to the forward transform instead of the inverse transform, but it should not be used for both of them.

The results of the Fourier Transform are complex numbers which are displayed using two images for the *real* and *imaginary* part, respectively or, equivalently, with *magnitude* and *phase*.

The Fourier Transform enables us to access the geometric characteristics of a spatial domain image, being easy to identify and process given frequencies of the image, so modifying the geometric structure in the spatial domain. In image processing, the Fourier image is often shifted in order to display F(0,0) (i.e. the image mean) in the center of the image and the higher frequency components further away from the center. Moreover, being that the dynamic range of the Fourier coefficients is too large, a logarithmic transformation is generally applied to display it on the screen. Figure 2.14 shows an example of Fourier transform.

This type of analysis and processing may be very useful for noise removal and image restoration, low, high frequency filtering, edge detection, radiometric correction, and image to image registration. In particular, noise is easier identified in frequency domain than in spatial domain. This is the main reason for the increasing use of Fourier in the pre-processing data analysis. As an example, ASTER SWIR imagery present a radiometric error caused by deterioration of the scanning system (known as 'six-line banding'). This data set cannot be employed without a de-striping method which can be easily performed using Fourier. This type of noise is clearly visible in the spectrum as features at 90° to their direction in the spatial domain of the original scene.

One more example can be the presence in the spatial domain of stationary periodic noise, which produces, in the frequency domain, a single-frequency sinusoidal function appearing as bright points. A line connecting the points is always perpendicular to the orientation of the noise lines in the original image.

Nevertheless, significant limitations in the use of Fourier analysis are: (i) the fact that it works globally and (ii) it is not possible to identify direct relationship between a specific components and frequencies.

2.4.3 Wavelet Transform

Recently, wavelet-based data analysis has become one of the fastest growing research areas, being used for a number of different fields and applications. Wavelets are mathematical functions which were developed independently in diverse application fields ranging from mathematics, electrical engineering, quantum physics, and seismic geology. In the last years, wavelet has rapidly covered new application fields, spanning from communication systems to imaging including satellite data analysis.

Signal wavelet decomposition using Discrete Wavelet Transform (DWT) provides an alternative to the Discrete Fourier Transform (DFT). Wavelet transform offers a number of advantages compared to traditional Fourier methods especially in searching for discontinuities and sharp spikes in analyzing signal. The main ability of DWT, which is also the benefit of DWT over DFT, is its multiresolution time-scale analysis. Moreover, wavelet transform can be used to analyze signals which contain non-stationary power at many different frequencies (Daubechies 1990).

Given a discrete a discrete sequence x_m, the continuous wavelet transform is defined as the convolution of x_m with a scaled and translated version of the function $F_{0n}(\varepsilon)$, which is called mather wavelet, as in formula 2.32:

$$W_n(s) = \sum_{m=0}^{N-1} x_m F^* \left[\frac{(m-n)\delta t}{s} \right] \qquad (2.32)$$

where F* indicates the complex conjugate.

It is really important to note that variations in the transformed signal is function of two variables, n and s, the translation and scale parameters, respectively. Through the variations of wavelet scale (s and n), it is possible to obtain information about the amplitude of any features versus the scale and how this amplitude varies with time/space. The function F is the wavelet Function. To be "admissible" as a wavelet, F must have zero mean and be localized in both time (space, in the case of image processing) and frequency domain (Farge 1992).

One criticism of wavelet analysis is the arbitrary choice of the mother wavelet function, which is obviously linked with the data analysis purposes.

Figure 2.15 shows an example of the wavelet transform applied to the same image as in Fig. 2.14.

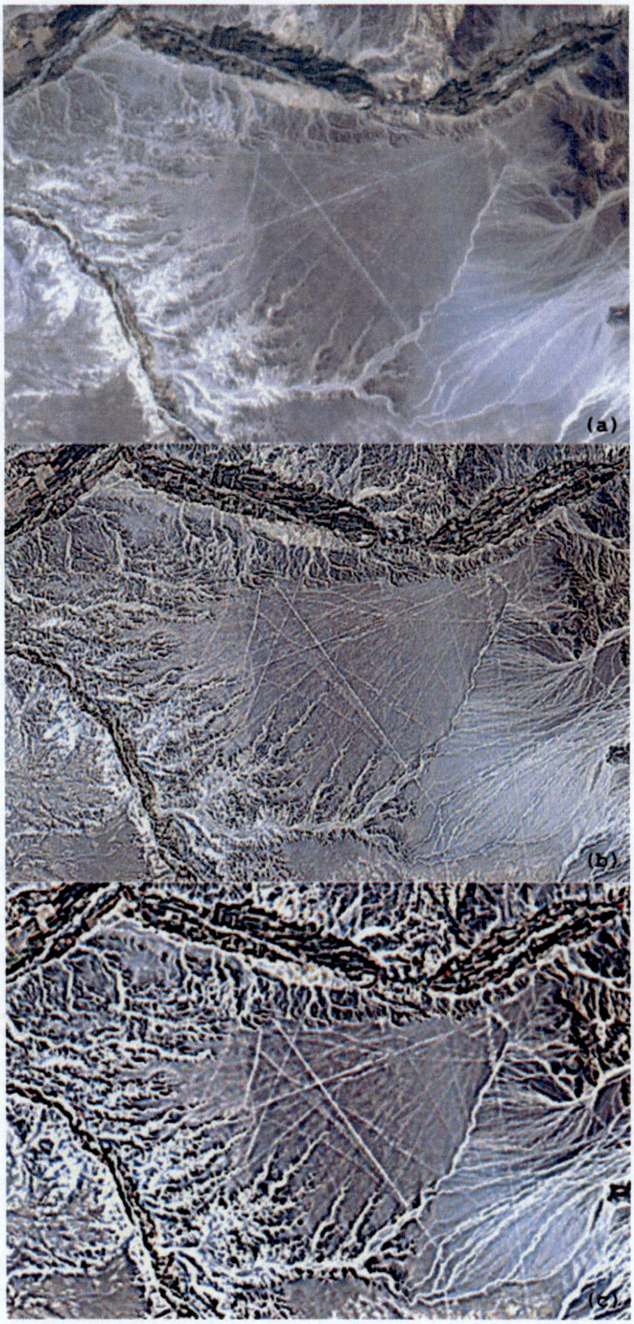

Fig. 2.15 Wavelet transform performed by using open source software IRIS for the same image as in Fig. 2.14

The shape of the wavelet function should reflect the type of features present in signal/image under processing. As an example, for abrupt changes, such as sharp jumps or steps, the Harr function is the most recommended, whereas in the case of smooth variations a damped cosine function is more adequate.

Wavelet function may be complex or real. In the first case it provides information relating to both amplitude and phase, whereas a real wavelet only provides a single component which is very useful to identify and to isolate peaks or discontinuities.

Wavelets split the data into different frequency components, and then study each component with a resolution matched to its scale. In the field of image processing, DWT acts in this way: an image is decomposed, with each level corresponding to a coarser resolution band.

There are two basic types of wavelet transforms: (i) one, widely used for image compression and cleaning (noise and blur reduction), designed to be easily reversible (invertible); (ii) the second type of wavelet transform is designed for signal analysis; for example, to detect noise, to study signals, to determine how the frequency content of a signal evolves.

Some techniques only focus on the image enhancement without any reduction of noise, such as the method proposed by Fu (Wang et al. 2010) based on the histogram equalization in wavelet domain; whereas others mainly focus on the noise reduction. Many algorithms consider both detail enhancement and noise reduction, as in the case of Pengxin Zeng (Zeng et al. 2004), who proposed an algorithm which offers a good compromise between noise reduction and subtle detail enhancement.

Wavelet transformations are also used in field of cultural heritage mainly in the processing of georadar and geomagnetic data (Bloisi et al. 2010). Another application is the data fusion between panchromatic image and multispectral data or between satellite radar and multispectral images (see Chap. 4 and references therein quoted).

2.4.4 Fractal Analysis

Fractal concept is quite close to self-similarity which means that a small part of a given object/image/pattern exhibits the same statistical properties as the whole, and, therefore, it appears as "scale invariant". In other words, it is the same (in terms of mathematical properties) at all scales of observation. The self-similar elements are also related in scale by a non-integral power-law. Self-similar objects with parameters N and S are described by a power law such as

$$N = s^d \tag{2.33}$$

where $\mathrm{d} = \dfrac{\mathrm{In\ N}}{\mathrm{In\ s}}$ is the "dimension" of the scaling law, known as the Hausdorff dimension.

Fractals are a particularly interesting class of self-similar objects. It is important to highlight that a pattern is self-similar if it is made up of smaller-scale copies of itself, but this is not enough to be fractal, which requires "fractional dimension". In other words the fractal dimension *d* must be a fraction, not an integer.

Numerous studies have been performed to analyze the fractal characteristics of modern human settlement mainly focusing on boundaries for both large and small urban areas (see for example Telesca et al. 2009 and references therein quoted).

Within archaeological sites, Brown and Witschey (2003) investigated the fractal characteristics of Maya settlement considered fractal at both intra-site and inter site levels. At the intrasite level Brown's analysis pointed out that "buildings form a pattern of repeated, complex, nested clusters of clusters". At the intersite organization they found that ": (1) the size–frequency distribution of settlements is fractal; (2) the rank–size relation among sites is fractal; and (3) the geographical clustering of sites is fractal." The main point suggested by Brown was the consideration that the fractal model may explain the cycle of rising, flourishing and collapse of early South American states. They were self-organized critical systems therefore, quite far from equilibrium and this may explain why some of them were highly susceptible to major changes in the face of minor forces.

Brown and Witschey (2003) also provided examples of other cultural settlements where this model and distribution is untenable, because not supported by empirical data. For example, the orthogonal grid pattern of a Roman city tends to be Euclidean rather than fractal, although its fractality depends on the details of the grid squares. Brown and Witschey (2003) mainly focused on the empirical verification of the fractal behavior and also in the establishing relationships between the Pareto exponents and social, economic, geographic processes and geometric distribution of archaeological settlement. In order to assess if the ancient settlement distribution exhibits a fractal geometry, the spatial organization of the buildings must be logically and geometrically self-similar and mathematically of fractional dimension, namely it must also be related in scale by a non-integral power-law.

Over the years, numerous techniques for fractal dimension computation have been developed and also applied to digital image domain. Fractal dimension can be applied to image data analysis for estimating various parameters, such as, roughness, texture, segmentation, surface roughness, shape irregularity, etc.

Any kind of set, such as points, lines, surfaces, multi-dimensional data, or time series may be fractal. For a regular 2-dimensional curve the fractal dimension is 1. Spatial fractal curves are characterized by fractal dimension larger than 2. The fractal dimension can be computed using a variety of different methodologies, herein we focus on the box-counting method which is one of the most widely used approaches.

The box-counting method computes the degree of irregularity of borders; therefore higher the fractal dimension, more irregular the border. The image of the settlements under investigation must be divided in non-overlapping cells of size L. Then the number of cells occupied by at least one point belonging to the border is computed. The procedure is iterated for different sizes L and the number of cells as a function of the size L behaves as a power-law for fractal borders:

$$N(L) \approx L^{-df_{BOX}} \tag{2.34}$$

where df_{BOX} is the box-counting fractal dimension:

$$df_{box} = \lim_{L \to 0} \frac{\ln(N(L))}{\ln(1/L)} \tag{2.35}$$

$$df_{BOX} = \lim_{L \to 0} \frac{\ln\,(N(L))}{\ln\,(\frac{1}{L})} \tag{2.36}$$

The estimate of the fractal dimension is performed calculating the slope of the line fitting the Eq. 2.1 in its linear part.

2.4.5 *Spatial Filtering*

All the previous techniques, including Fourier, Wavelet, and Fractal analysis may be time consuming and may also require an intense user-interaction. There are a number of other techniques for feature enhancement and extraction which are more user- friendly and also implemented in a number of commercial software (such ENVI, ERDAS, PCI) and also in open source programs, such as GRASS, IRIS, etc.

Spatial filtering can be used to highlight or subdue edges, features and spatial relationships among neighboring pixels. High pass filters enhance differences, whereas low pass filters smooth them.

Spatial domain process can be described as

$$g(x, y) = T[f(x, y)] \tag{2.37}$$

where g(x,y) is the output image, T is an operator, $f(x, y)$ is the input image.

The spatial filtering generally involves some types of convolution techniques that evaluate pixels on the basis of the spectral values of its neighboring. As we are only concerned with digital images, we will restrict the discussion to the discrete convolution into two dimensions (see Eq. 2.38). where the sum is taken over all values of x and h for which non zero results exists

$$y(m, n) = \sum_{i=-k}^{k} \sum_{j=-k}^{k} x(m+i,\ n+j)(h(i, j)) \tag{2.38}$$

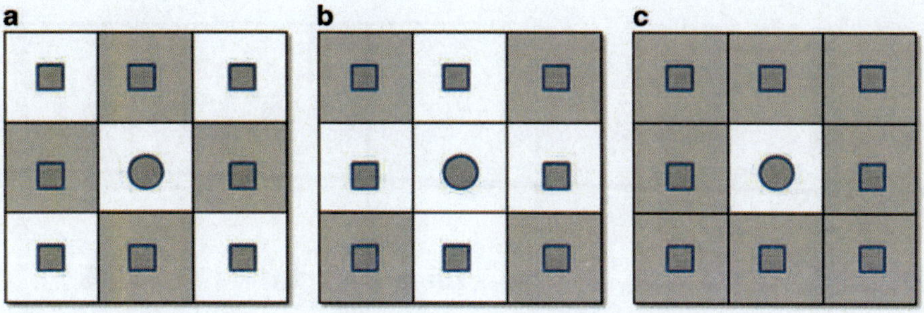

Fig. 2.16 (**a**) Tower contiguity, (**b**) bishop contiguity, (**c**) queen contiguity

It is really important to highlight that the convolution corresponds to the application of a template (or kernel matrix), which can be directly found in the image processing software, such as ENVI, PCI, ERDAS (1999) or can also be promptly customized according to specific computation needs. In the following sub-sections the most common filtering techniques are briefly presented and discussed along with the most common corresponding templates which can be also varied to improve the data processing.

All the image processing software programs offer the possibility to use pre-defined kernels but it is also possible to customize shape, size and weight values of the kernel. The pre-defined kernel matrix has specific dimensions generally there are odd-number sized kernels starting from 3 × 3 upwards to larger window size. Different shapes may be used, for example directional or contiguity kernel matrix. Figure 2.16 shows some possible kernel different form the existing methods which take their name from the game of chess. They are called tower contiguity (Fig. 2.16a), bishop contiguity (Fig. 2.16b) and queen contiguity (Fig. 2.16c).

2.4.5.1 Low Pass Filters

Low pass filters are also known as smoothing filters because they act to remove high spatial frequency details and preserves the low frequency components. In its simples formulation the pass filter contains the same weights in each kernel element, replacing the center pixel value with an average of the surrounding values. Figure 2.17 describes the nature of the mathematical operation of this moving kernel matrix.

Low pass filters are employed to emphasize the broader changes present within the image scene, but is very useful for noise removal or reduction and as an intermediate step of data processing oriented to image enhancement.

Fig. 2.17 The low pass filter kernel

1/9	1/9	1/9
1/9	1/9	1/9
1/9	1/9	1/9

Fig. 2.18 The high pass filter kernel

-1	-1	-1
-1	8	-1
-1	-1	-1

2.4.5.2 High Pass Filters

High pass filters act to remove low spatial frequency details and preserve the high frequency components linked to local variations. This operation can be obtained by subtracting the results of the low pass filter from the original image so obtaining a new image with enhanced local contrast (Richards and Jia 2006). This method has been used as the simplest way to obtain high spatial frequency image. Figure 2.18 describes the nature of the mathematical operation of this moving kernel matrix characterized by a high central value, typically surrounded by negative weights.

High pass filters can only have odd kernel dimensions. They are used for image sharpening or edge enhance.

The high pass filter enhances spatial frequencies that are less than the size of the kernel matrix used in the operation. Therefore the selection of kernel size produces a strong control on the level of detail generated from the process. For instance, very small anomalies (archaeological remains, or linear pattern anomalies) require small kernels size; whereas very large features, such as building, geological faults, etc. require much larger kernels in order to effectively enhance their properties within the imagery.

2.4.5.3 Median Filters

Median filter replaces each center pixel with the median value (not the average) within the neighborhood specified by the Kernel size. The median filter search for the "middle" value (i.e. median pixel value) in a given set of data. Therefore, it acts as a smoothing in an image while preserving edges larger than the kernel dimensions. Weighted Median Filter (more general Median filter) may be defined ad hoc to remove or retain a predefine set of features

$$CT_{filtered} = \frac{\sum\{[\mathrm{ABS}(\mathrm{CT} - \mathrm{a}) + \cdots + \mathrm{ABS}(\mathrm{CT} - \mathrm{a8})]\}}{8} \tag{2.39}$$

Where:

a1,. . .,a8 are grey pixel values into the template
CT is the grey value for pixel at the center of the template

2.4.5.4 Gaussian Filters

A Gaussian filter is a smoothing operator quite similar to the mean filter, but based on a different kernel (with Gaussian shape), which smooths preserving edges more than a mean filter with a similar size.

The Gaussian convolution function must be applied to an odd dimension kernel. In the image processing, the Gaussian distribution is a 2-D function formalized as in formula:

$$\mathrm{G(x, y)} = \frac{1}{2\pi\sigma^2}\mathrm{e}^{-\frac{x^2+y^2}{2\sigma^2}} \tag{2.40}$$

It is assumed that the distribution has a zero mean.

Actually, it is possible to use the Gaussian operator for both smoothing and enhancement purposes even using a corresponding template. The standard deviation of the Gaussian determines the degree of smoothing and, for example Fig. 2.19 shows the equivalent convolution kernel that approximates a Gaussian low pass filter.

Fig. 2.19 Discrete approximation to Gaussian function

0.0007	0.0256	0.0007
0.0256	0.8948	0.0256
0.0007	0.0256	0.0007

2.4.5.5 Derivative Edge Detection Methods

For a given image, the pixel value, such as brightness, or temperature, or vegetation indices, can be expressed as a discrete function ϕ of coordinates x and y. For this function we can compute the first, the second derivative and the vector gradient according to formulas 2.41, 2.42, 2.43, 2.44

$$\nabla\Phi(x,y) = \frac{\partial}{\partial x}\phi(x,y)i + \frac{\partial}{\partial y}\phi(x,y)i \tag{2.41}$$

$$\nabla^2\Phi(x,y) = \left[\frac{\partial^2}{\partial x}\Phi(x,y)\hat{\imath}\right] + \left[\frac{\partial^2}{\partial y}\Phi(x,y)\hat{\jmath}\right] \tag{2.42}$$

$$\alpha(x,y) = \tan^{-1}\frac{\nabla_2}{\nabla_1} \tag{2.43}$$

$$|\nabla| = \sqrt{\nabla_1^2 + \nabla_2^2} = \sqrt{\left[\frac{\partial}{\partial x}\phi(x,y)\right]^2 + \left[\frac{\partial}{\partial y}\phi(x,y)\right]^2} \tag{2.44}$$

where i,j are a pair of unit vectors.

A digital image defines a discrete space therefore its derivative and gradient can be easily expressed as differences.

A number of filters are obtained using the first derivative or the second derivative and considering the spatial variations in all the possible directions or in specific directions, as in the so called directional filters which selectively enhances image features having specific direction components (gradients).

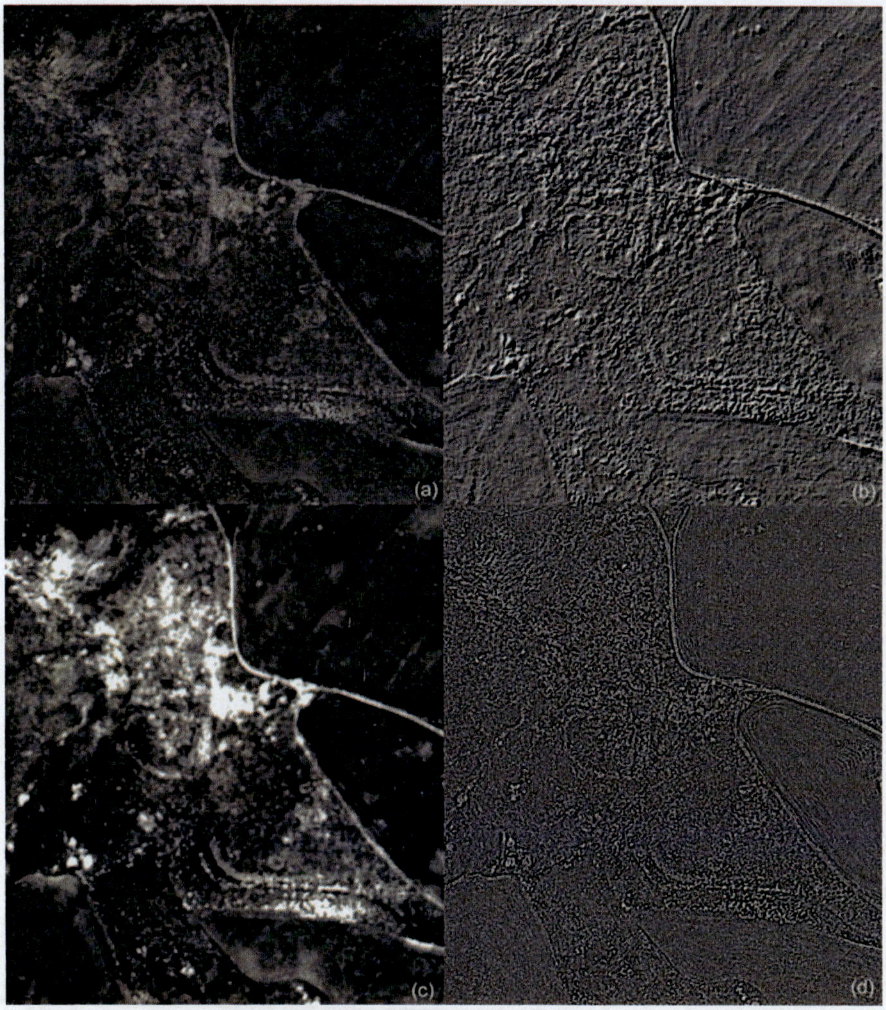

Fig. 2.20 Results of spatial filters on a panchromatic image (**a**) related to the medieval village of Irsi (Southern Italy): (**b**) directional filter, (**c**) Gaussian low pass, (**d**) High pass filter

As an example, one of the most used filter is the Laplacian filter based on a second derivative edge enhancement which operates without regard to edge direction.

Finally, directional filters have been specifically devised to detect edges running in given direction such as: (i) vertically and horizontally relative to the pixel grid, or (ii) at a given angle relative to the pixel grid.

Edge detection has became one of the fastest growing research areas, being used for a number of different fields and applications, among the others we only cite as examples signal imaging, biomedical data analysis, face recognition and satellite image processing including Archaeology, as shown in Fig. 2.20.

2.4.5.6 Mathematical Morphological Filtering

Mathematical morphology is a non-linear image processing method, which is based on the use of two-dimensional convolution operation and includes a big number of different operators, spanning from the simplest erosion, dilation, opening operation and closing, which are the basis operations of mathematical, to the more complex, such as geodetic transformations or hit-and-miss transformations (see, for example Soille 2003).

Mathematical morphology enables the analysis of the spatial relationships between groups of pixels, thus providing a complementary strategy to the analysis of the spectral signature of single pixels. Starting from the mid-1980s (see for example Haralick et al. 1987) successful applications of MM have been reported in geoscience and remote sensing for noise reduction, edge detection, feature extraction and classification. Nevertheless, recent advances in the theory of mathematical morphology are still largely unexplored.

2.5 Remote Sensing and Geospatial Data Analysis

Geospatial data analysis includes a set of geostatistic tools useful for the characterization of spatial variation, prediction, simulation and autocorrelation. Applications of geostatistics are found in a wide range of fields including biology, environmental science, geography, geology, data mining and remote sensing.

In the field of remote sensing variogram analysis (also known as semivariogram) has been the most widely used geostatistic tool applied to describe the correlation between image pixels in close proximity to each other. Variance is obviously related to the size of objects and spatial resolution. It is considered a powerful tool to describe the spatial dependency of data using measures of local variance namely the frequency of changes within the image scene.

Geospatial data analysis in archaeology has been approached by some researchers (Burke et al. 2008; Meghan and Howey 2011), but up to now only few examples deal with the application of geostatistics to image processing for archaeological investigations (Lasaponara and Masini 2010).

2.5.1 The Spatial Autocorrelation: Basic Concepts

Spatial autocorrelations take into account the spatial attributes of geographical objects under investigation, evaluate and describe their relationship and spatial patterns also including the possibility to infer such patterns at different times for the study area. The spatial patterns are defined by the arrangement of individual entities in space and the spatial relationships among them. Spatial autocorrelations in the field of archaeological investigations measure the extent to which the occurrence of one object/feature/site is influenced by similar objects/features/sites in the adjacent areas.

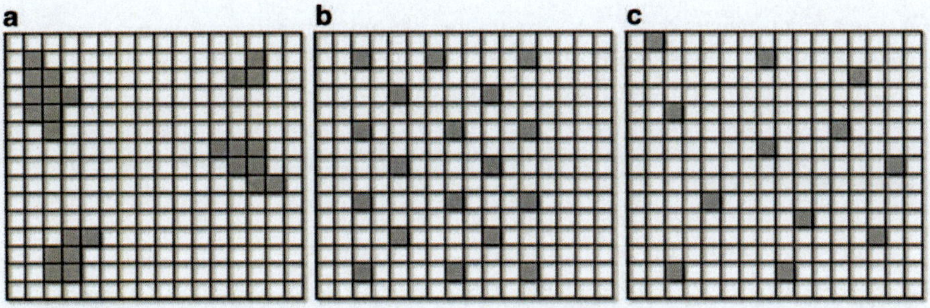

Fig. 2.21 Possible spatial distributions (**a**) positive spatial autocorrelation, (**b**) negative spatial autocorrelation, (**c**) null autocorrelation

As such, statistics of spatial autocorrelation provide (i) indicators of spatial patterns and (ii) key information for understanding the spatial processes underlying the distribution of object/feature/site and/or a given phenomenon under observation.

Geographical observations should be arranged in spatial and temporal order, by latitude and longitude, and historical periods. In this context time series data, such aerial and satellite images can provide useful traces of past human activities and, therefore, can enable us: (i) to some extent predict the amount and types of interaction, (ii) to investigate spatial predictions between objects/features/sites and also to infer potential relations considering different "historical" time windows being that "Everything is related to everything else, but nearest things are more related than distant things" (Tobler 1990).

In absence of spatial autocorrelation the complete spatial randomness hypothesis is valid: the probability to have an event in one point with defined (x, y) coordinates is independent of the probability to have another event belonging to the same variable. The presence of spatial autocorrelation modifies that probability; fixed a neighbourhood for each event, it is possible to understand how much it is modified from the presence of other elements inside that neighbourhood.

A distribution can show three types of spatial autocorrelation: (i) the variable exhibits positive spatial autocorrelation (Fig. 2.21a) when events are near and similar (clustered distribution); (ii) the variable exhibits negative spatial autocorrelation (Fig. 2.21b) when, even if events are near, they are not similar (uniform distribution); the variable exhibits null autocorrelation (Fig. 2.21c) when there are no spatial effects, neither about the position of events, or their properties (random distribution).

2.5.2 *The Spatial Autocorrelation: Basic Formulations*

The presence of autocorrelation in a spatial distribution is caused by two effects, that could be clearly defined (Gatrell et al. 1996), but not separately studied in the practice: (i) first order; and (ii) second order effect.

(i) First order effects depend on the region of study properties and measure how the expected value (mean of the quantitative value associated to each spatial event) varies in the space with the following expression:

$$\hat{\lambda}_\tau(\mathrm{s}) = \lim_{\mathrm{ds}\to 0}\left\{\frac{\mathrm{E(Y(ds))}}{\mathrm{ds}}\right\} \tag{2.45}$$

where *ds* is the neighbourhood around *s*, *E* is the expected mean and *Y(ds)* is the events number in the neighbourhood.

(ii) Second order effects express local interactions between events in a fixed neighbourhood, that tends to the distance between events i and j. These effects are measured with covariance variations expressed by the limit:

$$\gamma(s_i s_j) = \lim_{ds_i ds_j \to 0}\left\{\frac{E(Y(ds_i)Y(ds_j))}{ds_i ds_j}\right\} \tag{2.46}$$

where symbols are similar to those used in Eq. 2.1.

The characterization of spatial autocorrelation requires detailed knowledge on (i) quantitative nature, (ii) geometric nature of the data set under investigation; in detail:

(i) quantitative nature, also called intensity of the spatial process, provides information on space occurrence of a given variable (Murgante et al. 2008), to understand if events are similar or dissimilar.
(ii) geometric nature needs the conceptualization of geometric relationships, usually done using the following matrixes:
   (a) a contiguity matrix to assess if events influence each other;
   (b) a distance matrix to consider at which distance the events influence each other (distance band);
   (c) a matrix of spatial weights to estimate how strong this influence is.

Concerning the distance matrix, a method should be established to calculate distances in module and direction. The Euclidean distance (3), is the most used.

$$d_E(s_i, s_j) = \sqrt{(x_i - x_j)^2 + (y_i - y_j)^2} \tag{2.47}$$

As for any type of dataset also in the case of digital image analysis there are many indicators of spatial autocorrelation, that can be distinguished into the following: Global indicators, Local indicators.

2.5.2.1 Global Indicators of Spatial Autocorrelation

Global statistics summarize the magnitude of spatial autocorrelation for the entire region by a single value. The Global indicators of autocorrelation utilize distance to define the neighbourhood of a region and measure if and how much the dataset is autocorrelated in the entire study region.

One of the principal global indicator of autocorrelation is the Moran's index I (Moran 1948), defined in formula (2.48)

$$I = \frac{N \sum_i \sum_j w_{ij}(X_i - \bar{X})(X_j - \bar{X})}{(\sum_i \sum_j w_{ij}) \sum_i (X_i - \bar{X})^2} \tag{2.48}$$

where, N is the total pixel number, Xi and Xj are intensity in points i and j (with $i \neq j$), $\bar{X}$ is the average value, wij is an element of the weight matrix.

$I \in [-1; 1]$; if $I \in [-1; 0)$ there's negative autocorrelation; if $I \in (0; 1]$ there's positive autocorrelation.

Theoretically, if I converges to 0 there's null autocorrelation, in most of the cases, instead of 0, the value used to affirm the presence of null autocorrelation is given by Eq. 2.48:

$$E(I) = -\frac{1}{N-1} \tag{2.49}$$

where N is the number of events in the whole distribution.

The second global indicator of spatial autocorrelation is the Geary's C (Geary 1954), expressed by formula 2.50:

$$C = \frac{(N-1) \sum_i \sum_j w_{ij}(X_i - X_j)^2}{2w_{ij}(\sum_i (X_i - \bar{X})^2} \tag{2.50}$$

Where symbols have the same meaning than expression 2.48.

$C \in [0; 2]$; if $C \in [0; 1)$ there's positive autocorrelation; if $C \in (0; 2]$ there's negative autocorrelation; if C converges to 1 there's null autocorrelation.

2.5.2.2 Local Indicators of Spatial Autocorrelation

The local version of statistic utilizes distance information to identify local clusters and relies on the distance information captured in Distance matrix. Values indicating the magnitude of spatial association can be derived for each areal unit, namely for each pixel in the case of digital image.

The most common Local indicators of spatial autocorrelation are: Local Moran's I (Anselin 1995), Local Geary's C (Cliff and Ord 1981), and Getis-Ord Local Gi (Getis and Ord 1994; Illian et al. 2008).

Local Moran's I index is defined according to formula 2.51.

$$I_i = \frac{(X_i - \bar{X})}{{S_X}^2} \sum_{j=1}^{N} (w_{ij}(X_j - \bar{X})) \tag{2.51}$$

Local Geary's C Index is defined according to formula 2.52.

$$C = \frac{n-1}{\sum_{i=1}^{n} (X_i - \bar{X})^2} \frac{\sum_{i=1}^{n}\sum_{j=1}^{n} w_{ij}(X_i - X_j)^2}{2\sum_{i=1}^{n}\sum_{j=1}^{n} w_{ij}} \tag{2.52}$$

Getis and Ord's Gi is defined according to formula 2.53.

$$G_i(d) = \frac{\sum_{i=1}^{n} w_i(d)x_i - x_i \sum_{i=1}^{n} w_i(d)}{S(i)\sqrt{\frac{\left[(N-1)\sum_{i=1}^{n} w_i(d) - \left(\sum_{i=1}^{n} w_i(d)\right)^2\right]}{N-2}}} \tag{2.53}$$

These indicators show a different concept of spatial association:

(1) local Moran's I: a high value of the index means positive correlation both for high values and for low values of intensity;
(2) Local Geary's C detects areas of dissimilarity between events;
(3) Getis and Ord's Gi: a high value of the index means positive correlation for high values of intensity, while a low value of the index means positive correlation for low values of intensity.

Such geostatistical analysis tools are available in several commercial softwares, ranging from Geographic Information System (GIS) and image processing ones.

2.5.3 *Spatial Autocorrelation Applied to Satellite Data*

Spatial autocorrelation examines changes in homogeneity and measures the strength of the relationship between values of the same variables. In the analysis of satellite image it is a very useful tool since it not only considers the value of the pixel (reflectance, temperature, spectral index), but also the relationships between a pixel and its surrounding in a given window size.

Global measures of spatial autocorrelation provide a single value that indicates the level of spatial autocorrelation within the variable distribution, namely the homogeneity of a given values within the image under investigation.

Local measures of spatial autocorrelation provide a value for each location within the variable distribution and, therefore, they are able to identify discrete spatial patterns that may not otherwise be apparent. The statistics output is an image for each calculated index, which contains a measure of autocorrelation around that pixel.

Global and Local statistics can be calculated using spectral channels, spectral combinations and/or multi- temporal combination as intensity.

For all these cases, the first step is to find the optimal lag distance using global Moran's I. It provides different values of lag distances: the optimal value is the lag that maximizes I and captures the autocorrelation of the image in the best way.

At this point, the local indicators of spatial association should be calculated using the optimal lag distance and the selected contiguity.

Results from this assessment must be interpreted and/or further elaborated (generally using classifications) before interpretation. As a general rule, we can argue that the Getis-Ord Gi index permits the identification of areas characterized by very high or very low values (hot spots) compared to those of neighboring pixels.

It should be noted that the interpretation of Gi is different from that of Moran's I. In detail the Getis-Ord Gi enables us to distinguish the clustering of high and low values, but does not capture the presence of negative spatial correlation. It permits the identification of areas characterized by very high or very low values (hot spots) compared to those of neighboring pixels.

The Moran's I is able to detect both positive and negative spatial correlations, but clustering of high or low values are not distinguished. In particular, it identifies pixel clustering. It has values that typically range from approximately +1, representing complete positive spatial autocorrelation, to approximately −1, representing complete negative spatial autocorrelation.

The Local Geary's C index allows us to identify edges and areas characterized by a high variability between a pixel value and its neighboring pixels.

Geostatistical analysis tools are available in several commercial software, such as GIS and image processing (see for example ENVI packages).

References

Alexakis D, Sarris A, Astaras Th, Albanakis K (2009) Detection of Neolithic settlements in Thessaly (Greece) through multispectral and hyperspectral satellite imagery. Sensors 9:1167–1187

Aminzadeh B, Samani F (2006) Identifying the boundaries of the historical site of Persepolis using remote sensing. Remote Sens Environ 102:52–62

Anselin L (1995) Local Indicators of Spatial Association LISA. Geogr Anal 27:93–115

Beck A (2006) Google earth and world wind: remote sensing for the masses? Antiquity 80:308

Beck A, Philip A, Abdulkarim M, Donoghue D (2007) Evaluation of Corona and Ikonos high resolution satellite imagery for archaeological prospection in western Syria. Antiquity 81:161–175

Bewley R, Donoghue D, Gaffney V, Van Leusen M, Wise A (1999) Archiving aerial photography and remote sensing data: a guide to good practice. Oxbow, Oxford

Bloisi F, Ebanista C, Falcone L, Vicari L (2010) Infrared image analysis and elaboration for archaeology: the case study of a medieval "capsella" from Cimitile, Italy. Appl Phys B Laser Opt 101(1–2):471–479

Brown Vega M, Craig N, Asencios Lindo G (2011) Ground truthing of remotely identified fortifications on the Central Coast of Perú. J Archaeol Sci 38:1680–1689

Brown CT, Witschey WRT (2003) The fractal geometry of ancient Maya settlement. J Archaeol Sci 30:1619–1632

Burke A, Eberta D, Cardille J, Dauth D (2008) Paleoethology as a tool for the development of archaeological models of land-use: the Crimean Middle Palaeolithic. J Archaeol Sci 35 (4):894–904

Campana S (2003) Ikonos-2 multispectral satellite imagery to the study of archaeological landscapes: an integrated multi-sensor approach in combination with "Traditional" methods. In: Proceedings of the 30th conference CAA, Heraklion, 2–6 April 2002, pp 219–225

Campana S (2004) Le immagini da satellite nell'indagine archeologica: stato dell'arte, casi di studio, prospettive. Archeologia Aerea. Studi di Aerotopogr Archeol 1:279–299

Campana S, Forte M (eds) (2006) From space to place. In: Proceedings of the 2nd international conference on remote sensing in archaeology, Roma, 4–7 Dec 2006, BAR International Series 1568. Archaeopress, Oxford

Campana S, Francovich R (2003) Landscape archaeology in Tuscany: cultural resource management, remotely sensed techniques, GIS based data integration and interpretation. In: The reconstruction of archaeological landscapes through digital technologies, Boston, 2001, BAR International Series 1151. Archaeopress, Oxford, pp 15–28

Ceccato P, Flasse S, Tarantola S, Jacquemoud S, Gregoire JM (2001) Detecting vegetation leaf water content using reflectance in the optical domain. Remote Sens Environ 77(1):22–33

Ceccato P, Flasse S, Gregoire J-M (2002a) Designing a spectral index to estimate vegetation water content from remote sensing data: part 2: validation and applications. Remote Sens Environ 82 (2–3):198–207

Ceccato P, Gobron N, Flasse S, Pinty B, Tarantola S (2002b) Designing a spectral index to estimate vegetation water content from remote sensing data: part 1: theoretical approach. Remote Sens Environ 82(2–3):188–197

Chen CM, Hepner GF, Forster RR (2003) Fusion of hyperspectral and radar data using the IHS transformation to enhance urban surface features. J Photogramm Remote Sens 58:19–30

Clark CD, Garrod SM, Parker Pearson M (1998) Landscape archaeology and remote sensing in southern Madagascar. Int J Remote Sens 19(8):1461–1477

Cliff AD, Ord JK (1981) Spatial processes, models, and applications. Pion, London

Crist EP, Cicone RC (1984) A physically-based transformation of thematic mapper data: the TM tasseled cap. IEEE T Geosci Remote Sens GE22(33):256–263

Crist EP, Kauth RJ (1986) The tasseled cap de mystified. Photogramm Eng Remote Sens 52(1):81–86

Daubechies I (1990) The wavelet transform, time-frequency localization and signal analysis. IEEE Trans Inf Theory 36(5):961–1005

Davis CH, Wang X (2003) Planimetric accuracy of Ikonos 1 m panchromatic orthoimage products and their utility for local government GIS basemap applications. Int J Remote Sens 24 (22):4267–4288

Deroin J-P, Téreygeol F, Heckes J (2011) Evaluation of very high to medium resolution multispectral satellite imagery for geoarchaeology in arid regions – case study from Jabali, Yemen. J Archaeol Sci 38:101–114

Drake NA (1997) Recent aeolian origin of superficial gypsum crusts in Southern Tunisia: geomorphological, archaeological and remote sensing evidence. Earth Surf Proc Land 22:641–656

ENVI (1999) Users guide, research systems. Boulder, Colorado

ERDAS (1999) Field guide, 5th edn. ERDAS, Inc., Atlanta

Estes JE, Jensen JR, Tinney LR (1977) The use of historical photography for mapping archaeological sites. J Field Archaeol 4(4):441–447

Farge M (1992) Wavelet transform and their applications to turbolence. Ann Rev Fluid Mech 24:395–457

Fourty T, Baret F (1998) On spectral estimates of fresh leaf biochemistry. Int J Remote Sens 19 (7):1283–1297

Fowler MJF (1996) High resolution satellite imagery in archaeological application: a Russian satellite photograph of the Stonehenge region. Antiquity 70:667–671

Fritz LW (1996) The era of commercial earth observation satellites. Photogramm Eng Remote Sens 62(1):39–45

Gatrell AC, Bailey TC, Diggle PJ, Rowlingson BS (1996) Spatial point pattern analysis and its application in geographical epidemiology. Trans Inst Br Geogr 21:256–271

Geary RC (1954) The contiguity ratio and statistical mapping. Inc Stat 5:115–145

Getis A, Ord JK (1994) The analysis of spatial association by use of distance statistics. Geogr Anal 24:189–206

Giardino M (2011) A history of NASA remote sensing contributions to archaeology. J Archaeol Sci 38:2003–2009

Gitelson AA, Kaufman Y, Merzlyak MN (1996) Use of green channel in remote sensing of global vegetation from EOS-MODIS. Remote Sens Environ 58:289–298

Grøn O, Palmer S, Stylegar F-A, Aase S, Esbensen K, Kucheryavski S, Sigurd A (2011) Interpretation of archaeological small-scale features in spectral images. J Archaeol Sci 38:2024–2030

Haralick RM, Sternberg SR, Zhuang X (1987) Image analysis using mathematical morphology. IEEE Trans Pattern Anal Mach Intell 4:532–550

Horne JH (2003) A tasselled cap transformation for IKONOS images. In: ASPRS 2003 annual conference proceedings, Anchorage, 5–9 May 2003

Howey MCL (2011) Multiple pathways across past landscapes: circuit theory as a complementary geospatial method to least cost path for modeling past. J Archaeol Sci. doi:10.1016/j.jas.2011.03.024

Huete AR (1988) A soil-adjusted vegetation index (SAVI). Remote Sens Environ 25:295–309

Illian J, Penttinen A, Stoyan H, Stoyan D (2008) Statistical analysis and modelling of spatial point patterns. Wiley, West Sussex, 534 p

Jackson TJ, Chen D, Cosh M et al (2004) Vegetation water content mapping using Landsat data derived normalized difference water index for corn and soybean. Remote Sens Environ 92:475–482

Kaufman YJ, Tanrer D (1992) Atmospherically resistant vegetation index (ARVI) for EOS-MODIS. IEEE Geosci Remote Sens 30:261–270

Kauth RJ, Thomas GS (1976) The Tasseled cap – a graphical description of the spectral-temporal development of agricultural crops as seen by Landsat. In: Proceedings of the symposium on machine processing of remotely sensed data, Purdue University, West Lafayette, 29 June–1 July 1976, pp 4B41–4B51

Kennedy D, Bishop MC (2011) Google earth and the archaeology of Saudi Arabia. A case study from the Jeddah area. J Archaeol Sci 38:1284–1293

Kouchoukos N (2001) Satellite images and Near Eastern landscapes. Near East Archaeol 64 (1–2):80–91

Lasaponara R, Masini N (2006a) On the potential of panchromatic and multispectral Quickbird data for archaeological prospection. Int J Remote Sens 27:3607–3614

Lasaponara R, Masini N (2006b) Identification of archaeological buried remains based on Normalized Difference Vegetation Index (NDVI) from Quickbird satellite data. IEEE Geosci Remote Sens 3(3):325–328

Lasaponara R, Masini N (2007a) Detection of archaeological crop marks by using satellite QuickBird multispectral imagery. J Archaeol Sci 34:214–221

Lasaponara R, Masini N (2007b) Improving satellite Quickbird – based identification of landscape archaeological features trough tasselled cup transformation and PCA. In: 21st CIPA symposium, Atene, 1–6 giugno 2007

Lasaponara R, Masini N (2010) Facing the archaeological looting in Peru by local spatial autocorrelation statistics of very high resolution satellite imagery. In: Taniar D et al (eds) Proceedings of ICSSA, the 2010 international conference on computational science and its application, Fukuoka, 23–26 Mar 2010. Springer, Berlin, pp 261–269

Lasaponara R, Masini N (2011) Satellite remote sensing in archaeology: past, present and future. J Archaeol Sci 38:1995–2002

Lasaponara R, Masini N, Scardozzi G (2008) Satellite based archaeological research in ancient territory of Hierapolis. In: 1st international EARSeL workshop. Advances in remote sensing for archaeology and cultural heritage management, CNR, Rome, 30 Sept–4 Oct 2008. Aracne, Rome, pp 11–16

Lasaponara R, Masini N, Rizzo E, Coluzzi R, Orefici G (2011) New discoveries in the Piramide Naranjada in Cahuachi (Peru) using satellite, Ground Probing Radar and magnetic investigations. J Archaeol Sci 38:2031–2039

Lillesand TM, Kiefer RW (2000) Remote sensing and image interpretation. Wiley, New York

Masini N, Lasaponara R (2006) Satellite-based recognition of landscape archaeological features related to ancient human transformation. J Geophys Eng 3:230–235. doi:10.1088/1742-2132/3/3/004

Masini N, Lasaponara R (2007) Investigating the spectral capability of QuickBird data to detect archaeological remains buried under vegetated and not vegetated areas. J Cult Herit 8(1):53–60

Miller WC (1957) Uses of aerial photographs in archaeological field work. Am Antiq 23(1):46–62

Moran P (1948) The interpretation of statistical maps. J R Stat Soc A 10:243–251

Murgante B, Las Casas G, Danese M (2008) The periurban city: geo-statistical methods for its definition, Urban and regional data management. Taylor & Francis Group, London, pp 473–485

Parcak S (2009) Satellite remote sensing for archaeology. Routledge, Abingdon/New York

Pellemans AH, Jordans RW, Allewijn R (1993) Merging multispectral and panchromatic spot images with respect to the radiometric properties of the sensor. Photogramm Eng Remote Sens 59(1):81–87

Pinty B, Verstraete MM (1992) GEMI: a non-linear index to monitor global vegetation from satellites. Vegetatio 101:15–20

Reeves DM (1936) Aerial photography and archaeology. Am Antiq 2(2):102–107

Richards JA, Jia X (2006) Remote sensing digital image analysis – hardback, 4th edn. Springer, Berlin/Hiedelberg, 476 p

Sarris A, Jones R (2000) Geophysical and related techniques applied to archaeological survey in the Mediterranean: a review. J Mediterr Archaeol 13(1):3–75

Saunders RW (1990) The determination of broad band surface albedo from AVHRR visible and near-infrared radiances. Int J Remote Sens 11:49–67

Sever TL (1998) Validating prehistoric and current social phenomena upon the landscape of Peten, Guatemala. In: Liverman D, Moran EF, Rinfuss RR, Stern PC (eds) People and pixels: linking remote sensing and social science. National Academy Press, Washington, DC

Sheets P, Sever T (1988) High tech wizardry. Archaeology 41(6):28–35

Soille P (2003) Morphological image analysis: principles and applications. Springer, Berlin

Spennemann DHR (1987) Experiences with mapping sites on aerial photographs. J Field Archaeol 14(2):255

Stein C, Cullen B (1994) Satellite imagery and archaeology – a case study from Nikopolis. Am J Archaeol 98(2):326

Stone KH (1964) A guide to the interpretation and analysis of aerial photographs. Ann Assoc Am Geogr 54(3):318–328

Strahler A, Strahler A (1997) Physical geography: science and systems of the human environment. Wiley, New York

Sussman R, Green G, Sussman I (1994) Satellite imagery, human ecology, anthropology, and deforestation in Madagascar. Human Ecol 22(3):333–354

Telesca L, Coluzzi R, Lasaponara R (2009) Urban pattern morphology time variation in Southern Italy by using Landsat imagery. In: Murgante B, Borruso G, Lapucci A (eds) Geocomputation & urban planning, vol SCI 176. Springer, Heidelberg, pp 209–222

Toutin T (2001) DEM generation from new VIR sensors: IKONOS, ASTER and Landsat-7. In: IEEE-IGARSS proceedings, Sydney, 9–13 July 2001

Toutin T (2002) DEM from stereo Landsat 7 ETM + data over high relief areas. Int J Remote Sens 23(10):2133–2139

Traviglia A (2008) The combinatorial explosion: defining procedures to reduce data redundancy and to validate the results of processed hyperspectral images. In: Proceedings of the 1st international EARSeL workshop. Advances in remote sensing for archaeology and cultural heritage management, CNR, Rome, 30 Sept–4 Oct 2008. Aracne, Rome, pp 23–26

Traviglia A, Cottica D (2011) Remote sensing applications and archaeological research in the Northern Lagoon of Venice: the case of the lost settlement of constanciacus. J Archaeol Sci 38:2040–2050

Treitz PM, Howarth PJ (2000) High spatial resolution remote sensing data for forest ecosystem classification: – an examination of spatial scale. Remote Sens Environ 52:268–289

Tucker CJ (1980) Remote sensing of leaf water content in the near infrared. Remote Sens Environ 10:23–32

Urwin N, Ireland T (1992) Satellite imagery and landscape archaeology: an interim report on the environmental component of the Vinhais Landscape Archaeology Project, North Portugal. Mediterr Archaeol 5:121–131

Ustin SL, Roberts DA, Jacquemoud S, Pinzon J, Gardner M, Scheer GJ, Castaneda CM, Palacios A (1998) Estimating canopy water content of chaparral shrubs using optical methods. Remote Sens Environ 65:280–291

Wang Z, Bovik AC (2002) A universal image quality index. IEEE Signal Proc Lett 9(3):81–84

Wang X-Y, Yang H-Y, Fu Z-K (2010) A new wavelet-based image denoising using undecimated discrete wavelet transform and least squares support vector machine. Expert Syst Appl 37 (10):7040–7049

Weber SA, Yool SR (1999) Detection of subsurface archaeological architecture by computer assisted airphoto interpretation. Geoarchaeology 14(6):481–493

Wilson DR (2000) Air photo interpretation for archaeologists. Tempus, Stroud

Zarco-Tejada PJ, Rueda CA, Ustin SL (2003) Water content estimation in vegetation with MODIS reflectance data and model inversion methods. Remote Sens Environ 85:109–124

Zeng P, Dong H, Chi J, Xu X (2004) An approach for wavelet based image enhancement. In: Proceedings of IEEE international conference on robotics and biomimetics, ROBIO, Shenyang, 22–26 Aug 2004, pp 574–577

Zhang H, Bevan A, Fuller D, Fang Y (2010) Archaeobotanical and GIS-based approaches to prehistoric agriculture in the upper Ying valley, Henan, China. J Archaeol Sci 37(7): 1480–1489

Chapter 3
Pattern Recognition and Classification Using VHR Data for Archaeological Research

Rosa Lasaponara and Nicola Masini

Abstract The extraction of the huge amount of information stored in the last generation of VHR satellite imagery, is a big challenge to be addressed. At the current state of the art, the available classification techniques are still inadequate for the analysis and classification of VHR data. This issue is much more critical in the field of archaeological applications being that the subtle signals, which generally characterize the archaeological features, cause a decrease in: (i) overall accuracy, (ii) generalization attitude and (iii) robustness. In this paper, we present the methods used up to now for the classification of VHR data in archaeology. It should be considered that: (i) pattern recognition and classification using satellite data is a quite recent research topic in the field of cultural heritage; (ii) early attempts have been mainly focused on monitoring and documentation much more than detection of unknown features. Finally, we discuss the expected improvements needed to fully exploit the increasing amount of VHR satellite data today available also free of charge as in the case of Google Earth.

Keywords Classification • Pattern recognition

R. Lasaponara (✉)
Institute of Methodologies for Environmental Analysis, CNR-IMAA,
C. da S. Loya, 85100 Tito Scalo, PZ, Italy
e-mail: lasaponara@imaa.cnr.it

N. Masini
Institute of Archaeological and Architectural Heritage, CNR-IBAM,
C. da S. Loya, 85050 Tito Scalo, PZ, Italy
e-mail: n.masini@ibam.cnr.it

R. Lasaponara and N. Masini (eds.), *Satellite Remote Sensing: A New Tool for Archaeology*, Remote Sensing and Digital Image Processing 16,
DOI 10.1007/978-90-481-8801-7_3, © Springer Science+Business Media B.V. 2012

3.1 Introduction

The VHR satellite data, today available at a spatial resolution less than 1 m, enables us to extract spectral/spatial pattern, geometric properties, along with other detailed information useful for a number of different application fields spanning from vegetation, mapping (Desclée et al. 2006; Lasaponara and Lanorte 2006), environmental monitoring (see for example Pulvirenti et al. 2011), urban expansion, cadastral mapping (see for example Pacifici et al. 2009; McFeeters 1996) and archaeology (see for example, Ciminale et al. 2009; De Laet et al. 2007).

In the field of archaeological investigations as for other applications, information retrieval and knowledge improvements are based on the extraction and analysis of what is generally hidden in the processed data sets. Moreover, informative content is not only stored in the diverse variables (i.e. pixel) but it generally spreads out all over the data sets and therefore can be unveiled by analysing spatial relationships between the single variables (and/or pixels) and their neighboring areas (spatial analysis).

To extract information from imagery we may use different methodological approaches and transformations which must not alter the informative content but it should make the information easier to manage for data analysis and clearer for interpretation.

Remote Sensing technologies acquire and provide measurements of electromagnetic energy from distant targets thus enabling the extraction of information about features, objects, and classes on the Earth's land surface. The interpretation of geospatial data is possible because objects made of diverse materials emit and/or reflect a different quantity of energy in diverse regions of the electromagnetic spectrum. Considering multispectral images, each pixel has a set of spectral values and therefore it can be represented as a vector in a multi-dimensional space whose axes correspond to the given image band in the multi-spectral image space.

Therefore, on the basis of spectral content we can identify and categorize the diverse surfaces (soil, vegetation, sea), materials (soil types, vegetation cover types, concrete) and objects (urban areas, archaeological feature) by classes or types, substance, and spatial distribution according to their specific characteristics (fresh snow, senescent vegetation, clear water, moisture content, grain size). The different spectral responses observed for diverse materials according to their characteristics, is generally known as spectral signatures.

Figure 3.1 shows some examples of spectral signatures observed for vegetation and soil and for different status of vegetation. Please note from the visible (around 0.50 nm) to short wave spectral range (to 2.50 nm) the diverse spectral responses which characterize and enable the discrimination between the vegetation and soil as well as between green and dry vegetation. Of course, the graph exhibits the expected spectral response for the various targets under observation, obviously what is observed in reality is quite similar but with some differences due to the atmospheric contamination (cloud, aerosol, etc), view angle geometry, etc. All the pre-processing steps tend to reduce the contamination and noise.

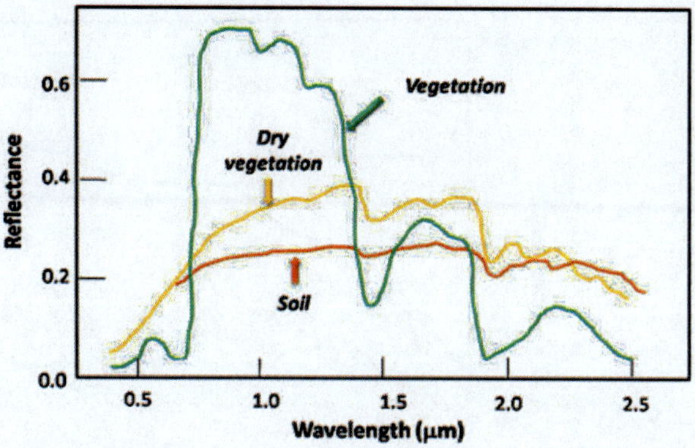

Fig. 3.1 Soil and vegetation spectral signatures

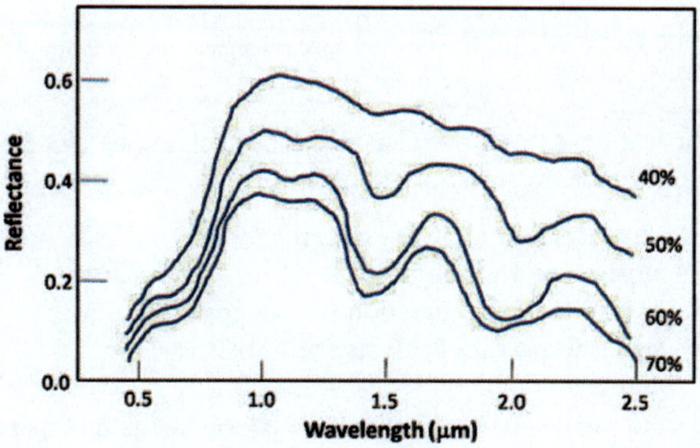

Fig. 3.2 Variations of vegetation spectral signatures according to different moisture contents

Figure 3.2 show the variations of spectral response of vegetation according to the variations in moisture content. Increase of moisture content generally produces a decrease in the spectral response due to the absorption of water. Figure 3.3 shows a similar behaviour also for silty soil, an increase in moisture induces a decrease in spectral response.

Pattern recognition and, in turn, image classification techniques have the overall objective to automatically or semi-automatically categorize all pixels of given scenes into known (pre-defined) or unknown (non predefined) classes or themes. The spectral pattern, or signature, of surface materials determines the assignment of given pixels to a specific class or category. One of the main purposes of satellite remote sensing is to carry out quantitative analysis to better interpret the observed data and classify features. Classification can be performed on pixel or object based using single or multiple image channels to identify and group pixels/object according to their spectral and /or shape characteristics. Over the years a number of different classification techniques have been devised, but the availability of VHR data has drawn considerable attention to the need of improving the capability

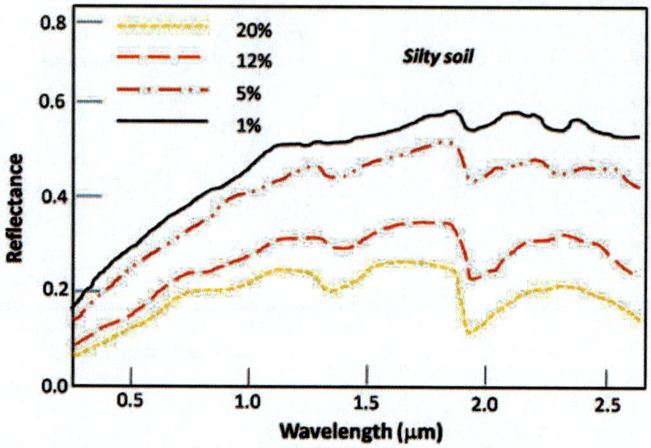

Fig. 3.3 Variations of silty soil spectral signatures according to different moisture contents

in feature extraction. Classification techniques can be grouped according to their specific characteristics as follows:

- unsupervised (self organizing);
- supervised (training);
- hybrid (self organization by categories);
- spectral mixture analysis (sub-pixel variations);
  and
- parametric classification, based on statistical parameters (mean & standard deviation);
- non-parametric classification approach, based on objects (polygons) in feature space;
- decision rules classification: it rules for sorting pixels into classes.

One of the most widely used categorizations is: (i) unsupervised (automatic data processing) and (ii) supervised (semi-automatic data processing). The unsupervised classification techniques are performed without any prior knowledge of the image. Pixels are grouped into a pre-defined number of classes, according to their reflectance features.

The main important difference between unsupervised and supervised classifications is the fact that the latter requires a prior knowledge of the considered classes and a stronger user interaction being a semiautomatic methodological approach.

Two are the basic steps in supervised classifications: (i) clustering or training which consists in providing known areas for each class, generally identified through in situ analysis, (ii) classification which is carried out by comparing the spectral signature to each pixel (under investigation) with the spectral signature of the training cluster.

A parametric classification is carried out on the basis of parametric signatures defined by statistical parameters (e.g., mean and covariance matrix) and attributes, such as, the number of spectral bands, mean, minimum and maximum value in each band, as well as the number of pixels and covariance matrix for each training cluster. Some examples of parametric classification tools are: (i) Minimum distance, (ii) Mahalanobis, (iii) Distance Maximum Likelihood.

A non-parametric classification is carried out using non-parametric signatures obtained in the n-dimensional feature space image considering *n* as the number of the spectral bands. A pixel is assigned to a given class according to its location, inside or outside the area in the feature space image. Among the non-parametric classification techniques we cite: (i) parallelepiped and (ii) feature space.

One more categorization is: (i) per pixel classification and (ii) object oriented classification (Bhaskaran et al. 2010; Hofmann 2001). In the latter, the spectral information is analysed jointly with various shape measurements, namely polygons, whose spectral and spatial attributes are the input of "traditional classifications". The object extraction can provide a greater number of meaningful features for the classification step and assures more flexibility and robustness. Object oriented classification is generally based on three steps: (i) the object extraction, (ii) segmentation and (iii) classification via a variety of classification techniques.

Nevertheless, it should be considered that pattern recognition and classification using satellite data is a quite recent research topic in the field of cultural heritage. Early attempts have been mainly focused on monitoring and documentation much more than detection of unknown features.

Trelogan (2000) classified Landsat images for urban sprawl monitoring using a semiautomatic method to assess the development of urban pattern close to an archaeological area.

Semiautomatic classifications of above ground archaeological remains have been carried out through pan-sharpened multispectral Ikonos images by De Laet et al. (2007). They used and evaluated the performance of different supervised classification methods, spanning from pixel based techniques, such as SAM, Parallelepiped, Minimum Distance, Maximum Likelihood, to the object based method implemented in the eCognition software (eCognition 2002). This procedure is based on three step analyses: (i) firstly, a segmentation stage provides spectral homogeneous regions, also considering shape and scale (Baatz and Schape 2000); (ii) secondly, the training stage is performed on the outputs from the segmentation, (iii) classification. In De Laet et al. (2007), the comparison of different methods was not objectively conducted, but only using visual analysis, which does not provide numerical results. The authors concluded that, as a whole, all the procedures they adopted were not capable to provide a unique class for archaeological structures on the site of Hisar (southwest Turkey). All the considered classification techniques were supervised methods and, therefore, not automatic but semiautomatic because all of them require a strong user-interaction mainly for the selection of training areas along with the definition of parameters for the classification step.

A similar approach based on the segmentation and classification also implemented in the *ecognition* and/or in similar commercial software (as ENVI) was applied by Jahjah and Ulivieri (2010) to the excavated archaeological remains of Nisa (Turkmenistan) and Babylon (Iraq) for a semiautomatic mapping of urban fabric. The authors highlighted that the results obtained from both *ecognition* and the similar approach from ENVI were strongly linked with the operator intervention and generally many archaeological elements were misclassified and not well defined in terms of accuracy. By contrast, results from the fully automatic methodology based on the Mathematical Morphology outperformed the other considered methods.

Good results from automatic classification addressed to crop and soil marks detection were obtained in Ciminale et al. (2009). The main aim of this study was the detection of circular features related to ditches and compounds of some Neolithic settlements in Apulia (southern Italy). The authors applied unsupervised classifications to IKONOS and SPOT previously processed by Global and Local geospatial analysis. Actually this was one of the first attempts addressed to a fully automatic identification of archaeological marks linked to buried cultural remains.

Aurdal et al. (2006) also focused on crop and soil marks classification but with quite unsatisfactory results.

3.2 Unsupervised Classification Algorithms

Unsupervised classification only requires a limited human intervention to have the foreknowledge of the classes. The importance of applying unsupervised classification in archaeological applications is that: (i) it is an automatic process, namely, it normally requires only a minimal amount of initial input, compared to supervised data set; (ii) classes do not have to be defined *a priori*; (iii) unknown classes may be discovered.

A number of unsupervised classification algorithms are commonly used in remote sensing, among them we outline (i) K-means clustering, (ii) ISODATA (Iterative Self-Organizing Data Analysis Technique), (iii) Migrating Means clustering Classifier and (iv) Mathematical Morphology based methods. ISODATA and K-means are quite similar algorithms. In both of them the user has only to indicate (i) the number of the predefined classes (clusters) and (ii) the iterations to be carried out. The only difference is that the K-means assumes that the number of clusters is known *a priori* whereas the ISODATA algorithm assigns "dynamically" the different number of clusters. These algorithms are iterative procedures, based on the following steps: (i) they first assign an arbitrary initial cluster vector, (ii) each pixel is classified to the closest cluster, (iii) new cluster mean vectors are calculated based on all the pixels in one cluster. The second and third steps are iteratively repeated until the "variations" between the iteration is small. Such variations can be computed and assessed in several different ways. For example, in the K-means algorithm, the cluster variability is optimized by minimizing the sums of square distances (errors) expressed by Eq. 3.1.

$$MSE = \frac{\sum [x - C(x)]^2}{(N - c)b} \tag{3.1}$$

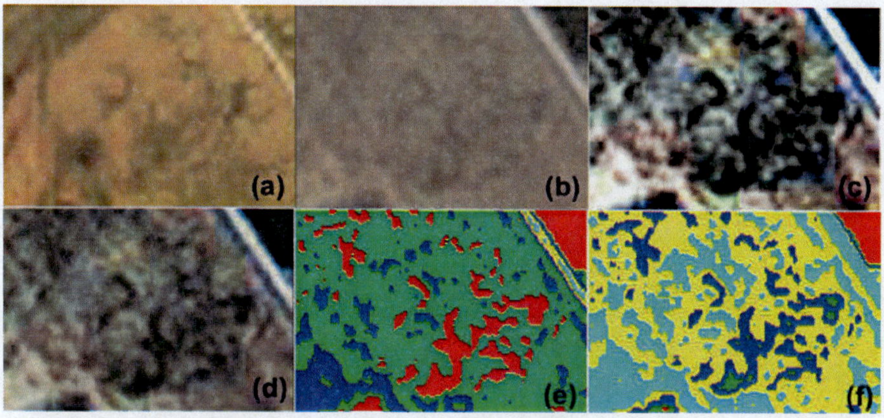

Fig. 3.4 Zoom of a test area selected in Palmori site: (**a**) orthorectified aerial image (**b**) QuickBird image, (**c**) satellite image enhanced by applying Gaussian (**c**) and equalization method (**d**). Results from ISODATA performed on the full scene (**e**) and on the subset (**f**)

where N is the number of pixels, c indicates the number of clusters, and b is the number of spectral bands, C(x) is the mean of the cluster that pixel x is assigned to.

Equation 3.1 clearly shows that the minimization of *MSE* implies that K-means works best for spherical clusters that have the same variance. This indicates that K-means algorithm tends to perform better for homogeneous surface/object as desert area, but it is quite unreliable for heterogeneous surfaces such as forest cover.

The ISODATA algorithm merges or splits clusters if, respectively the number pixels belonging to a given cluster is less than a certain threshold or if the centers of two clusters are within a threshold. ISODATA algorithm is considered more flexible compared to the K-means method, but it requires the empirical selection of many more parameters.

Both ISODATA and K-means algorithms may meet several problems. For example, for both of them the resulting classifications depend to a large degree on the arbitrary selection of initial parameters. This also leads to another inconvenience which is the so-called "reproducibility of classification". One more limitation is the fact that the number of clusters usually must be fixed *a priori*, and the problem is that it may not be known. Moreover, the spectral properties of specific targets, objects, classes are variable over seasons. Therefore, results cannot be directly transferred to other periods of the year or geographic areas. One more drawback of the unsupervised classifications is that some spectral clusters may be related to mixed classes of Earth surface materials and therefore physically meaningless. The analyst must know well enough the spectral behaviour of different targets to be able to identify specific information.

Figure 3.4 shows some results from the ISODATA classification obtained for Palmori which is a Neolithic archaeological area close to Foggia (Southern Italy). The typical circular and semicircular shapes of the compounds of the Neolithic sites are easily recognizable. Figure 3.4 shows a zoom of a test area selected in particular: (a) orthorectified aerial image (b) QuickBird image, (c) satellite image

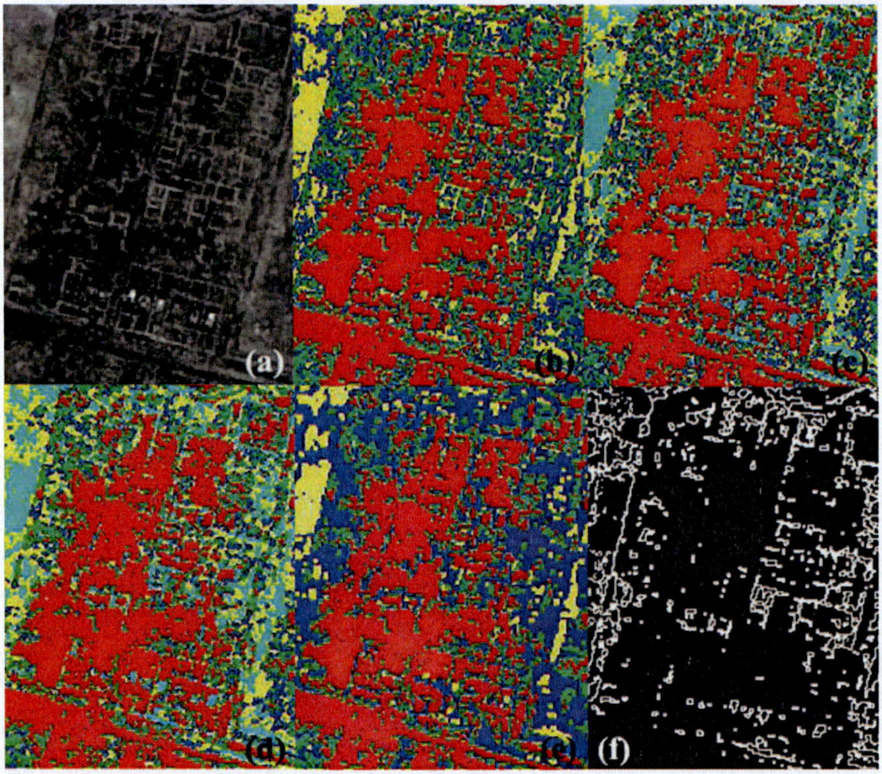

Fig. 3.5 (**a**) PAN; (**b**) K-means 6 classes; (**c**) K-means 5 classes; (**d**) post classification: major analysis; (**e**) post classification: clump; (**f**) vectorialization

enhanced by applying Gaussian (c) and equalization method (d). Figure 3.4e–f show the results from ISODATA performed on the full scene (e) and on a subset (f). The red class of Fig. 3.4e obtained from the ISODATA performed on the full scene is quite similar to the blue class of Fig. 3.4f obtained by only processing a subset of the whole scene.

Figures 3.5a–f show results obtained from K-means classification applied to Heraclea a Roman archaeological area in the Basilicata Region (Southern Italy). It is a Roman excavated town with above ground remains which clearly shows the typical roman urban layout composed of orthogonal streets (the so-called cardus and decumanus). The K-means classification was applied to a QuickBird panchromatic scene (Fig. 3.5a) considering both 6 and 5 classes (Fig. 3.5b–c, respectively). Before vectorialization step aimed at extracting the urban fabric, post classification procedures have been used to improve results from classification. Both majority analysis and clump were applied as implemented in the ENVI software (see Figs. 3.5d and e, respectively). The majority analysis enables us to manipulate the spurious pixels remaining in a given class by using a kernel. Whereas, the second post classification elaboration removes spatial incoherency, such as speckle or holes in classified areas by clumping adjacent similar classified areas together applying morphological operators by using a kernel of a given size. The morphological operators first perform a dilate operation and then an erode operation on the classified image. Note that, in this case the use of morphological operators is recommended to smooth these images instead of low pass filtering, whose application could contaminate the class information by adjacent class pixels. Finally, the vector (see Fig. 3.5f) is obtained from the clump outputs. The whole data processing from the classification to the vectorialization is completely automatic.

The most important implication of these promising results is the possibility of searching for archaeological features using an automatic processing tool. In such a way it should be possible to significantly reduce the amount of time and costs needed for field walking and ground detection. Vast geographical areas may be investigated using multispectral imagery to extract potential archaeological and palaeo-environmental features.

3.3 Spectral Separability Measures

It is possible to have a measure of the "spectral distance" (see for example Thomas et al. 1987) generally called separability between two or more spectral classes using specific numerical evaluation. There are a number of diverse separability tests useful to (i) determine the similarity/dissimilarity of two distributions, (ii) assess the type of distribution of data under investigation, (iii) discriminate the ability of an index or technique in separating and detecting distinct classes, (iv) evaluate if the separablity is statistically significant. Among the existing methods we list the following:

(i) Measures which look at the distance between class using means:

- Euclidean and Non-Euclidean Distances
- Divergence

(ii) Measures which look at both the differences between class means and the distribution of the values about those means (i.e. noise), such as:

- M statistic,
- JM Distance,
- Bhattacharyya Distance.

Some methods only work with one band at a time (e.g. Euclidean Distance, M statistic), while others can work on any number of bands (e.g. JM Distance).

The use of spectral separability can be very useful in archaeological application for many purposes, such as pre-processing to understand the performance of a given supervised classifier, to evaluate the capability in discriminating one class from the others (surface, shallow and buried archaeological features), to assess the performance of supervised classifiers.

A quantitative evaluation of spectral separability of archaeological marks and their surroundings was carried out by using one of the most widely used indices, the Jeffries–Matusita distance, by Lasaponara and Masini (2007), defined as follows:

- The Jeffries–Matusita (JM) distance is obtained from the Bhattacharyya distance (BD), shown in Eq. 3.2. Both JM and BD were devised to measure the statistical distance between two Gaussian distributions.

$$BD = \frac{1}{8}(\mu_b - \mu_{nb})^T \left[\frac{\Sigma_b + \Sigma_{nb}}{2}\right]^{-1} (\mu_b - \mu_{nb}) + \frac{1}{2}\ln \frac{\left[\left|\frac{\Sigma_b + \Sigma_{nb}}{2}\right|\right]}{\left[(|\Sigma_b|\Sigma_{nb}||)^{\frac{1}{2}}\right]} \tag{3.2}$$

Where m_b and m_{nb} are the mean of two classes (namely archaeological features and non archaeological features) and $\sum^b$ and $\sum^{nb}$ are the covariance matrix for the same classes.

Please, note that the first part of Eq. 3.2 represents the mean, whereas the second part is the covariance difference. BD should be as high as possible. A drawback of the Bhattacharya distance is that it does not provide any indication of threshold values for separability

$$JM = \sqrt{2(1 - e^{-BD})}. \tag{3.3}$$

where BD is Bhattacharya distance computed using formula 3.2

- The JM distance has an upper boundary of 1.41 (2), and a lower boundary of 0.
- The JM distance is asymptotic to the value 2 for increasing class separability. A value of 2 for JM distance would imply that the classification will be performed with 100% accuracy.
- When the calculated distance is zero, the signatures can be said to be totally inseparable.

Along with the Bhattacharrya (or Jeffries-Matusita) Distance there is also the Transformed Divergence which can be used to estimate the spectral separability between distributions. It is defined in formula 3.4

$$\mathrm{TD_{ij}} = 2000\left(1 - \mathrm{e}^{\left(\frac{-\mathrm{Dij}}{8}\right)}\right) \tag{3.4}$$

where i and j are the two signature classes being compared and Dij

$$D_{ij} = \frac{tr\left((Ci - Cj)\left(C_i^{-1} - C_j^{-1}\right)\right) + \left(\left(C_i^{-1} - C_j^{-1}\right)(\mu_i - \mu_j)(\mu_i - \mu_j)^T\right)}{2} \quad (3.5)$$

Where:

C_i is the covariance matrix of signature I, μ_i is the mean vector of signature I, tr is the trace function (matrix algebra), and finally T is the transposition function.

As for Bhattacharrya Distance also the Transformed Divergence ranges between 0 and 2, where 0 indicates the complete overlap between the signatures of two classes and 2 indicates a complete separation between the two classes.

Among the other distances we focus on the "statistically Mahalanobis Distance" different from the Euclidean distance which assumes that all the components of a given observation and/or spectral space contribute equally to the Euclidean distance of the observation from the center. Differently from the Euclidean distance, the Mahalanobis Distance also considers the variability of the given parameters and correlation between them.

In the Mahalanobis Distance (i) components with high variability receive less weight than components with low variability; (ii) being that correlation means that there are associations between the variables, it is considered by rotating the axes of ellipsoid. This yields the following general form for the statistical distance of two points.

$$MD(x, y) = \sqrt{(x - y)^t\, S^{-1}(x - y)} \quad (3.6)$$

where x and y are the two points

$$d_s(x, o) = \sqrt{x^t\, S^{-1} x} \quad \text{is the norm of x} \quad (3.7)$$

Within the classification process Mahalanobis Distance is used to evaluate the spectral separability among the areas considered as the training data set used for the learning process.

Lasaponara and Masini (2007) carried out statistical evaluation of spectral capability of satellite QuickBird data in detecting buried archaeological remains.

Cavalli et al. (2009) devised a spectral separability index specifically for archaeological features

$$SI = \left(1 - \frac{\int D_{marks}\, D_{background}\, dx}{\int D_{marks}^2\, dx \int D_{background}^2\, dx}\right) \times 100 \quad (3.8)$$

Where, D_{marks} represents the frequency distribution of the digital values of those pixels belonging to the archaeological spectral anomalies in all images, similarly for the $D_{background}$ corresponding to the frequency of pixels selected as background.

3.4 Supervised Classification Algorithms

The supervised classification algorithms require a preliminary knowledge necessary: (i) to generate representative parameters for each class of interest; and (ii) to carry out the training stage.

Over the years a number of algorithms have been developed for satellite data processing and, recently, the availability of VHR images has strongly pushed the implementation of new approaches to fit the complex need linked with the huge amount of geometrical and spectral information stored in VHR data.

Among the "traditional" supervised classifications the most common are:

(i) Maximum Likelihood Classifier (MLC), which is based on the evaluation of variance and co-variance for each class to assign a pixel to one of them according to the highest probability (see for example Kiema 2002).
(ii) Minimum-Distance to the Mean-Classifier, which is based on the evaluation of mean values for each class to assign a pixel to one of them, according the minimum values of Euclidian Distance
(iii) Parallelepiped Classifier which is based on the evaluation of a mean vector (instead of a single mean value) which contains an upper and lower threshold to assign a pixel to a given class or to disregard it as unclassified or null category.
(iv) Mahalanobis Distance classification which is based on the evaluation of the variance and co-variance similarly, but it considers all class covariances equal and all pixels are included in the closest class by threshold values, otherwise they may be unclassified.
(v) Spectral Angle Mapper algorithm (SAM) which is based on the spectral similarity measured by calculating the angle between the training and the under investigation spectra, considered as vectors in n-dimensional space, where n is the number of bands (Kruse and Lefkoff 1993).

In the following we will focus on MLC and SAM, because they are the most common for multispectral and hyperspectral data sets, respectively.

MLC is one of the most commonly used supervised classification algorithms. It assumes that each spectral class can be described by a multivariate normal distribution. The effectiveness of MLC depends on reasonably accurate estimation of the mean vector m and the covariance matrix for each spectral class data (Richards and Xiuping 1999).

The MLC (Lillesand and Kiefer 2000), as with other conventional hard classification techniques, assumes that all image pixels are pure. Nevertheless, this assumption is often untenable according also to the scale and spatial resolution of the investigated data sets. As an example, in mixed land cover compositions, as pixels increase in size, the proportion of mixed cover type distributed at pixel level will likewise increase and information at the sub-pixel level (Buck et al. 2003) will be of increasing interest. Consequently, in pixels made up of a mixture of features, conventional "hard" image classification techniques provide only a poor basis for the characterization and identification of features giving, in the best case, a compromised accuracy, or, in the worst case, a totally incorrect classification. In these conditions, the use of SMA can reduce the uncertainty in hard classification techniques since it is able to capture, rather than ignore, sub-pixel heterogeneity. The SMA allows for classifying the proportions of the different feature types (end-member classes) covered by each individual pixel. End-member classes can be taken from "pure" pixels within an image or from spectral libraries.

Over the years, different models of spectral mixtures have been proposed. Among the available models, the most widely used is the Mixture Tuned Matched Filtering (MTMF) by Ichku and Karnieli (1996) that is based on the assumption that the spectrum measured by a sensor is a linear combination of the spectra of all components within the pixels.

For archaeological applications some improvements may be expected in the case of Landsat TM or ASTER data (Abrams 2000), but the use of SMA for the VHR data generally should not exhibit significant variations in terms of accurracy and extracted details, apart the case of the scattering of pottery. In these conditions, geospatial analysis should be much more useful along with a pre-processing step oriented to sharpen the features to be extracted and to make them easier to identify.

Among the supervised classifications we briefly describe Spectral Angle Mapper, because it can be used for hyperspectral data processing. SAM compares image spectra to a known spectra or an endmember. The reference spectra can be directly obtained from field measurements, extracted from the image or taken from a spectral libraries. SAM treats both (training and unknown) spectra as vectors and calculates the spectral angle between them, using formula 3.9

$$u \cdot v = \sum_{i=1}^{n} (u_i v_i) = uv \cdot cos\alpha \tag{3.9}$$

where u_i e v_i are the components of the vectors u e v in in n-dimensional space. So the angle between the spectra is calculated with the following equation:

$$\alpha = \arccos \frac{u \cdot v}{uv} = \arccos \left(\frac{\sum_{i=1}^{n} (u_i v_i)}{\sqrt{\sum_{i=1}^{n} u_i^2} \sqrt{\sum_{i=1}^{n} v_i^2}} \right) \tag{3.10}$$

The α angle is in the range from zero to $\pi/2$.

High angle differences indicate high dissimilarity between the two spectra and vice versa low angle difference indicate low dissimilarity between the two spectra. Please note that SAM algorithm uses only the vector direction and not the vector length, thus assuring low sensitiveness to illumination conditions.

3.5 Classification Based on Supervised Learning Methods

3.5.1 *Artificial Neural Networks*

Today, the term neural network indicates a large class of classifications, regression and learning methods which exploit the basic idea of extracting linear combinations of the input layers (including multispectral images) to derive features and later use them to model the target as a nonlinear function of these features. Over the years a number of paper have been published in the topic. About the resurgence of Neural Networks in the mid 1980s we suggest the works of Werbos (1974) and Rumelhart et al. (1986).

The artificial neural networks (ANNs) act at the feature level mainly exploiting the pattern recognition capabilities of the ANN.

ANNs is a supervised classification process where the net is trained on a set of input regions of interest (ROIs). The neural networks gained popularity in the 1980s, and were extensively applied in many fields. In the early 1990s, the neural networks were applied to remotely sensed data since Werbos and Rumelhart developed a new learning scheme, based on the concept of a back-propagation algorithm.

From a theoretically point of view, ANNs can achieve an accurate result and, at the same time, assure a high generalization capability. The main advantages of the neural networks compared with the conventional classifiers are mainly linked to the fact that, they: (i) do not require previous knowledge about the distribution of the data; (ii) can adapt themselves to any type of data; (iii) can successfully deal with non-linear data patterns.

Nevertheless, the practical use of ANNs poses several problems. Firstly, a large variety of possible network architectures, setup parameters, and training algorithms. Secondly, possible bad choices of the above mentioned factors can provide erroneous results or affect subjectively the generalization capability.

The ANNs classifier approach adopted in this paper is fully described in Richards and Jia (2006) and implemented in ENVI routines. ANN is considered as a mapping device between an input set and an output set. The basic processing node in the ANNs is an element with many inputs and a single output.

The performed operation is described by relation 3.11:

$$0 = f(w'x + \theta) \tag{3.11}$$

Where θ is a threshold, w is a vector of weighting coefficients and x is the vector of inputs. The most common expression for the function f is that reported in formula 3.12.

$$f(z) = \frac{1}{1 + e^{-z/\theta_0}} \tag{3.12}$$

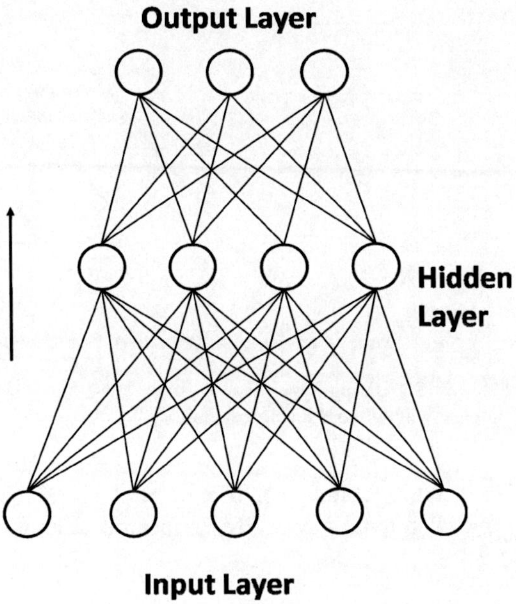

Fig. 3.6 Neural network is a two-stage classification model. The top level is defined according to the number of classes considered, as an example, in the case of K-classes, we have K units in the top level, with the kth unit modelling the probability of class k. There are K target measuments Yk, k = 1, 2, . . . ,K, each being coded as 0–1 variable for the kth class. Derived features (the hidden level) Zm are created from linear combinations of the inputs and then the target Yk is modelled as a function of linear combinations of the Zm

Where the argument z is $(w'x + \theta)$ and θ_0 is a constant, which is usually set at value 1. This leads 1 for z large and positive, and 0 for z large and negative. The threshold θ takes the place of the weighting coefficients. The outcome of the product $w'x$ is a scalar.

In the case of remote sensing data processing, the inputs are the satellite band images. The number of the input variables to a node will be defined by the network topology as well as data dimensionality. The ANNs used in the field of remote sensing data analysis can be schematized as:

(i) an input layer of nodes, which has the function of distributing the inputs to the processing elements of the next layer and scaling them if necessary;
(ii) an output layer from which the class labelling information is provided (Fig. 3.6).

One of the basic steps of the ANNs is the training process. For supervised learning, the Neural Net technique uses standard backpropagation. In order to minimize the difference between the output node activation and the desired output, learning is achieved by adjusting the weights in the node. The error is backpropagated through the network and weight adjustment is made using a recursive method.

The learning capability of ANN enables us to customize the image classification process. This can be time consuming and computationally complex compared with other standard classification techniques.

ANN-based fusion methods have more advantages especially when input data are obtained from multiple sensors, such as active and passive sensors (Radar and

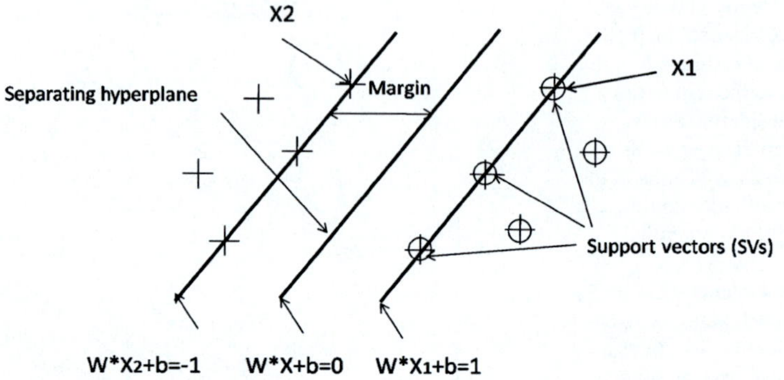

Fig. 3.7 Separating hyperplane is shown for a *linear* separation which enables us to start in the simplest way without losing generality

VHR satellite images). Moreover ANN is quite efficient for classifying high dimension data, such as hyper-spectral or long-term time-series data.

3.5.2 Support Vector Machines

In the last 10 years, SVMs (Support Vector Machines) have shown a great potential for data classification also in satellite data processing (Mountrakis et al. 2011). They are non parametric classifiers which show a great ability to optimize classification issues, minimizing the empirical classification errors while maximizing the class separations. The main objective is to generalise the problem avoiding overfitting. SVMs produce a model, based on the training data set, to predict the target values of the test data given only the test data attributes.

As in other supervised classifications the first step is the training, which involves the random separation of datasets into training and testing subsets. Please note that if there are categorical attributes, they must be first converted into numeric values. For example, in the case of a three-category attributes such as red, green, blue, the vectors for their numerical representation may be (0,0,1), (0,1,0), and (1,0,0). More in general, each example consists of a n number of data points (x1,. . . xn) followed by a label (or target). For two classes will be +1 or -1 representing, for example, archaeological or non-archaeological features. The given classes must be separated by an optimum hyperplane, as illustrated in Fig. 3.7. This hyperplane must minimize the distance between the closest +1 and -1 points (called support vectors) and maximise the margin between the two classes. This is known as the optimum separating hyperplane.

We can first consider a linear separation which enables us to start in the simplest way and to add complexity later by using kernel functions.

Considering Fig. 3.7, the separating hyperplane is defined as follows: (i) the normal vector w and (ii) – the offset b:

$$\text{hyperplane} = \{x\langle w, x\rangle + b\}$$

Where $\langle w,x \rangle$ is called inner product, scalar product or dot product.

In the training step w and b must be chosen from the labelled examples in the training set. The training must enable us to carry out the right prediction, namely on which side of the hyperplane a new point will lie. Considering Fig. 3.7, points lying in the right side are classified as positive, and viceversa points in the left direction are classified as negative. Note that the best hyperplane is "a fat plane", which separates the training set with maximal margin (see Fig. 3.7).

Support Vectors are the points nearest to the separating hyperplane, (which determine the position of the hyperplane), whereas all other points are non influent. From the mathematical point of view the weighted sum of the Support Vectors is the normal vector of the hyperplane.

In the case of non-separable training sets, SVM still considers linear separation, but admits training errors, which are measured as the distance to hyperplane multiplied by an error cost C. If it is not possible to reduce the penalty error, the separation may be easier considering higher dimensions. In this condition, a maximal margin hyperplane, there must be built as depend on inner products and, in the cases of very large dimensions, it cannot be manageable from the computational point of view. To overcome this drawback, the use of a kernel function enables us to maintain low dimensions, having performance (of an inner product) as in high dimensions. In this way, it is not necessary to know what the feature space really looks like, but it is enough to know outputs from the kernel function as a measure of similarity. Kernel methods received great attention in recent years. Initially they were used to face non-linearity within the SVM method. The idea behind them is very simple, but also very powerful. To classify two datasets which are not linearly separable it is possible to define a mapping function $f(x) : S^n \rightarrow S^m$ and work in the transformed space, at the cost of the transformation. SVMs (and many other linear algorithms) depend on data only through their inner products.

So if a method to evaluate directly the inner product of the mapped points can be found, without explicitly knowing the mapping, the problem becomes easier to solve. The use of Kernels makes this possible.

Diverse kernel functions may be adopted, such as

(i) nonlinear $k(p, q) = \langle p, q\rangle$,
(ii) polynomial $k(p, q) = (g\langle p, q\rangle + co)^d$,
(iii) radial basis function $k(p, q) = e^{-g\|p-q\|^2}$, etc.

As a whole, a Support Vector Machine is a maximal margin hyperplane in feature space built by using a kernel function in spectral space. The selection of SVM kernel classifier (along with kernel parameters) is considered as one of the most important steps in the implementation of SVM classifier. Lower the number of parameters to be defined, higher the robustness of the SVM implementation.

3.6 Hybrid Classifiers and Accuracy Evaluation

Archaeological features linked to buried settlements are really complex and traditional techniques, as a pixel-based classification, may be not effective. Traces of archaeological remains include different features which cannot be characterized by any specific color or tone of gray in the image, but rather by their heterogeneity. Archaeological marks (crop, soil, shadow) are very easy to extract in a visual photointerpretation process, but their heterogeneity makes their automatic or semi-automatic classification very difficult. To cope with this drawback, a pre-classification step may be necessary to make their classification easier. It is really important before running the classifier to make archaeological feature pattern easily recognizable, using spatial and/or spectral enhancement as, for example, feature extraction (PCA, TCT, spectral indices), spatial filtering, including mathematical morphology, whose outputs may be further processed using geospatial analysis to be later profitable classified (see for example Giada et al. 2003).

Mathematical morphology includes a big number of different operators, spanning from the simplest erosion or dilation, to the more complex, such as geodetic transformations or hit-and-miss transformations (see Chap. 2 and reference therein quoted).

Being that we firstly must know the shape to be recognized we may include the classifiers based on mathematical morphology inside the supervised classification category, but it must be noted that automatic procedures may run without any preliminary knowledge about the data set.

The use of a hybrid approach classification, that combines supervised and unsupervised, as well as pixels and objects, geometric shape and spectral feature characteristics, may provide improved performance also for scenes that contain a variety of subtle features.

Of course, diverse classifiers produce different results which must be evaluated using specific metrics.

In the case of unsupervised classifications it is not possible to evaluate accuracy and performance being that we have no training target to use as a basis data-set for numerical evaluations. Whereas, in the supervised classifications, we have a training data-set, sometimes called Region of Interest (ROI) which can be used to measure the achieved performance. First, we have to randomly divide the database into separate subsets: training and test samples. A common rule of thumb is to use 70% of the database for training and 30% for testing. It is absolutely important to measure the performance of a classifier using an independent balanced test set (number of samples in different classes very close to each other). When a data set is unbalanced the error rate of a classifier is not representative of the true performance of the classifier.

The accuracy of a classification process is generally carried out by comparing the classification results with ground truth or ROIs information.

A confusion matrix (contingency matrix) is generally considered and some "traditional" figures, below listed from (i) to (iv), are commonly used:

(i) The producer's accuracy is a measure indicating the probability that the classifier has correctly labelled an image pixel.
(ii) The user's accuracy is a measure indicating the probability that a pixel belongs to a given class and the classifier has labelled the pixel correctly into the same given class.
(iii) The overall accuracy is calculated by summing the number of pixels classified correctly and dividing by the total number of pixels.
(iv) Finally, the kappa statistics (K) can be also considered. It measures the increase in classification accuracy over that of pure chance by accounting for omission and commission error (Congalton and Green 1999). Overall accuracy is computed as the sum of the number of observations correctly classified (class1, as class 1, class 2 as class 2, etc.) divided by the total number of observations. This is equivalent to the "diagonal" of a square contingency table matrix divided by the total number of observations described in that contingency table.

Apart from the accuracy we can use other figure metrics to evaluate the performance of a classifier. As a whole, we can state that along with the accuracy one more widely used estimation is the robustness.

The classification accuracy also known as "predictive capability" refers to the ability of the classifier to correctly predict the class label of new or previously unseen data. The robustness is the ability of the classifier to model correctly and make right predictions also for noisy data or datasets with missing values. In the case of method based on machine learning, there are many specific metrics to evaluate performance, spanning from, log likelihood, mean squared error, 0/1 accuracy, and many others.

The reference quality as well as the optimization issues are domain-dependent and application-specific. In the case of archaeological application we can state that the quality of a pattern extraction methodology as well as the optimization of a given classifier is given by the possibility to reliably discriminate subtle archaeological features from their neighboring areas and this mainly lies in the capability to filter out and/or obscure background of archaeological features to enhance traces of past human activities still fossilized in the modern landscape.

References

Abrams M (2000) The Advanced Spaceborne Thermal Emission and Reflection Radiometer (ASTER): data products for the high spatial resolution imager on NASA's Terra platform. Int J Remote Sens 21(5):847–859

Aurdal L, Eikvil L, Koren H, Loska A (2006) Semi-automatic search for cultural heritage sites in satellite images. In: Proceedings of 'From Space to Place', 2nd international conference on remote sensing in archaeology, Rome, 4–7 Dec 2006. BAR International Series 1568, pp 1–6

Baatz M, Schape A (2000) Multiresolution segmentation: an optimization approach for high quality multi-scale image segmentation. In: Strobl J, Blaschke T (eds) Angewandte Geographische Informations verarbeitung XII. Wichmann, Heidelberg, pp 12–23

Bhaskaran S, Paramananda S, Ramnarayan M (2010) Per-pixel and object-oriented classification methods for mapping urban features using Ikonos satellite data. Appl Geogr 30(4):650–665

Buck PE, Sabol DE, Gillespie AR (2003) Sub-pixel artifact detection using remote sensing. J Archaeol Sci 30:973–989

Cavalli RM, Pascucci S, Pignatti S (2009) Optimal spectral domain selection for maximizing archaeological signatures: Italy case studies. Sensors 9:1754–1767

Ciminale M, Gallo D, Lasaponara R, Masini N (2009) A multiscale approach for reconstructing archaeological landscapes: applications in Northern Apulia (Italy). Archaeol Prospect 16:143–153

Congalton RG, Green K (1999) Assessing the accuracy of remotely sensed data: principles and practices. Lewis, Boca Raton

De Laet V, Paulissen E, Waelkens M (2007) Methods for the extraction of archaeological features from very high-resolution Ikonos-2 remote sensing imagery, Hisar (southwest Turkey). J Archaeol Sci 34:830–841

Desclée B, Bogaert P, Defourny P (2006) Forest change detection by statistical object-based method. Remote Sens Environ 102:1–11

eCognition User Guide (2002) http:\crwww.definiensimaging.com

Giada S, De Groeve T, Ehrlich D (2003) Information extraction from very high-resolution satellite imagery over Lukole refugee camp, Tanzania. Int J Remote Sens 24(22):4251–4266

Hofmann P (2001) Detecting informal settlements from Ikonos image data using methods of object oriented image analysisdan example from Cape Town (South Africa). In: Jurgens C (ed) Remote sensing of urban areas/Fernerkundung in urbanen Ra¨umen. Regensburger GeographischeSchriften, Regensburg, pp 107–118

Ichku C, Karnieli A (1996) A review of mixture modeling techniques for sub-pixel land cover estimation. Remote Sens Rev 13:161–186

Kiema JBK (2002) Texture analysis and data fusion in the extraction of topographic objects from satellite imagery. Int J Remote Sens 23(4):767–776

Kruse FA, Lefkoff AB (1993) Knowledge-based geologic mapping with imaging spectrometers. Remote sensing reviews, special issue on NASA Innovative Research Program (IRP) results, vol 8, pp 3–28

Lasaponara R, Lanorte A (2006) Multispectral fuel type characterization based on remote sensing data and Prometheus model. Forest Ecol Manag 234:S226

Lasaponara R, Masini N (2007) Statistical evaluation of spectral capability of satellite QuickBird data in detecting buried archaeological remains. In: Gomarasca M (ed) GeoInformation in Europe. Millpress, Rotterdam. ISBN 9789059660618657 663

Lillesand TM, Kiefer RW (2000) Remote sensing and image interpretation. Wiley, New York

McFeeters SK (1996) The use of Normalized Difference Water Index (NDWI) in the delineation of open water features. Int J Remote Sens 17(7):1425–1432

Mountrakis G, Im J, Ogole C (2011) Support vector machines in remote sensing: a review. ISPRS J Photogramm 66(3):247–259

Pacifici F, Chini M, Emery WJ (2009) A neural network approach using multi-scale textural metrics from very high-resolution panchromatic imagery for urban land-use classification. Remote Sens Environ 113(6):1276–1292

PCI Geomatics (1998) OrthoEngine reference manual. PCI Geomatics, Richmond Hill

Pulvirenti L, Chini M, Pierdicca N, Guerriero L, Ferrazzoli P (2011) Flood monitoring using multi-temporal COSMO-SkyMed data: image segmentation and signature interpretation. Remote Sens Environ 115(4):990–1002

Richards JA, Jia X (2006) Remote sensing digital image analysis -hardback, 4th edn. Springer, Berlin/Heidelberg, 476 p

Richards JA, Xiuping J (1999) Remote sensing digital image analysis an introduction. Springer, New York

Rumelhart D, Hinton G, Williams R (1986) Learning internal representations by error propagation. In: McClelland JL, Rumelhart DE, The PDP Research Group (eds) Parallel distributed processing: explorations in the microstructure of cognition. The MIT Press, Cambridge, vol 1, pp 318–362

Thomas IL, Ching NP, Benning VM, D'Aguanno A (1987) A review of multichannel indices of class separability. Int J Remote Sens 3:331–350

Trelogan J (2000) Remote sensing and GIS in the Chora of Chersonesos. In: The Study of ancient territories. Chersonesos and Metaponto. 2000 annual report. Institute of Classical Archaeology, The University of Texas, Austin, pp 25–31

Werbos P (1974) Beyond regression: new tools for prediction and analysis in the behavioral sciences. PhD dissertation, Harvard University, Cambridge

Chapter 4
Pan-Sharpening Techniques to Enhance Archaeological Marks: An Overview

Rosa Lasaponara and Nicola Masini

Abstract The application of pan-sharpening techniques to very high resolution (VHR) satellite data can fruitfully improve the enhancement of archaeological marks and facilitate their detection. Nevertheless, the quantitative evaluation of the quality of the fused images is one the most crucial aspects in the context of data fusion. This is because (i) data fusion application is a rather recent technique applied to archaeology; (ii) the criteria generally adopted for the data fusion evaluation can not fit the needs of this type of application. This chapter provides an overview of pan-sharpening techniques and quantitative evaluation of their capability in (i) preserving spectral fidelity and (ii) sharpening spatial and textural content.

Keywords Pan-sharpening • Assessment criteria • Very high resolution • Archaeological marks

4.1 Introduction

The use of pan-sharpening techniques enables the integration of the complementary information acquired from the panchromatic and multispectral (MS) imaging sensors. The higher spatial resolution of the panchromatic can be suitably merged with the spectral capability of multispectral channels.

R. Lasaponara (✉)
Institute of Methodologies for Environmental Analysis, CNR-IMAA,
C. da S. Loya, 85100 Tito Scalo, PZ, Italy
e-mail: lasaponara@imaa.cnr.it

N. Masini
Institute of Archaeological and Architectural Heritage, CNR-IBAM,
C. da S. Loya, 85050 Tito Scalo, PZ, Italy
e-mail: n.masini@ibam.cnr.it

R. Lasaponara and N. Masini (eds.), *Satellite Remote Sensing: A New Tool for Archaeology*, Remote Sensing and Digital Image Processing 16,
DOI 10.1007/978-90-481-8801-7_4,

Over the years, a number of algorithms have been developed for data fusion, among them we will focus on the most widely used: Intensity-Hue-Saturation (IHS), Principal Component Analysis (PCA), different arithmetic combination (e.g., Brovey transform), Zhang algorithm, high-pass filtering, Ehlers algorithm, multi-resolution based methods (e.g., pyramid algorithm, wavelet transform), and Artificial Neural Networks (ANNs).

The evaluation and numerical comparisons of the diverse pan-sharpening algorithms is quite complex. Several statistical indicators have been designed for assessing the performance of image fusion algorithms in terms of preserving both: (i) spectral fidelity and (ii) spatial and textural content. In this chapter we focus on the specific requirements necessary for the enhancement of archaeological features through pan-sharpening techniques as well as on the evaluation of their performance.

4.2 Data Fusion in Remote Sensing for Archaeology

The use of data fusion techniques can fruitfully improve the enhancement of archaeological marks and make their detection easier by exploiting jointly the higher spatial resolution of the panchromatic image and the multispectral properties of the spectral channels (Fig. 4.1). Moreover, another advantage of using data fusion products is that the increased spatial resolution can fruitfully provide a

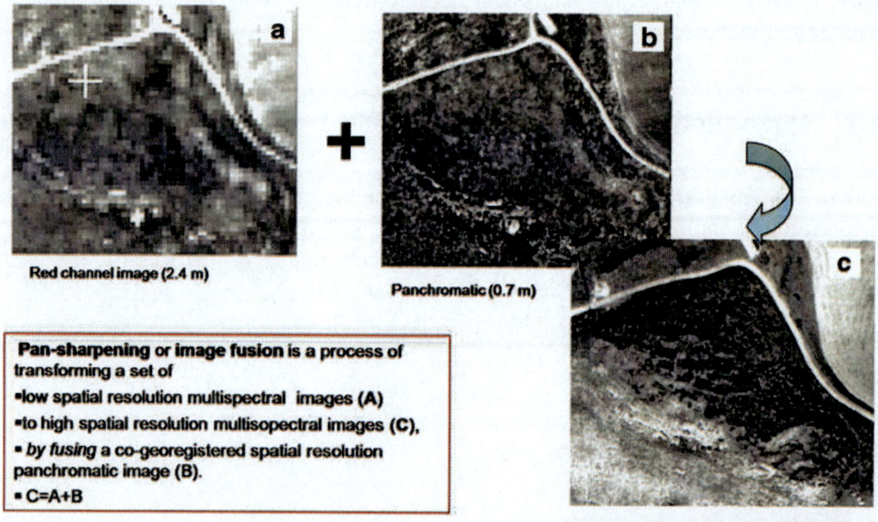

Fig. 4.1 Pan-sharpening allows us to exploit the higher spatial resolution of the panchromatic image and the multispectral properties, thus improving the enhancement of archaeological features

more accurate localization of the archaeological features. This more accurate localization, from the initial spatial resolution of multispectral data around meter (2.4 m for QuikBird, 2 m for GeoEye) to the sub-meters spatial resolution of panchromatic (0.6 m for QuickBird or 0.5 m for GeoEye) can be very helpful during in situ survey, such as GPS (Global Position System) campaigns, geophysical prospection or excavation trials.

Nevertheless, in order to take advantages from data fusion techniques, it is mandatory to quantitatively evaluate the benefits of different algorithms and approaches (Alparone et al. 2007). The quantitative evaluation of the quality of the fused images is yet one of the most crucial aspects in the context of data fusion. This issue is particularly relevant in the case of the identification of archaeological marks, because (i) data fusion application is a rather recent topic approached in the field of remote sensing of archaeology (Lasaponara and Masini 2007; Aiazzi et al. 2008; Lasaponara et al. 2008); (ii) the criteria generally adopted for the data fusion evaluation cannot fit the needs of remote sensing archaeology that are mainly focused on the identification of small features, that can be easily obscured by noise.

The best results from data fusion is that the multispectral set of fused images should be as identical as possible to the set of multispectral images that the corresponding sensor (Alparone et al. 2007; Aiazzi et al. 2008; Lasaponara et al. 2008) would observe with the high spatial resolution of panchromatic.

As no multispectral reference images are available at the requested higher spatial resolution, the assessment of the quality of the fused products is not obvious. Several score indices or figure metrics have been designed over the years (see Thomas and Wald 2004) to evaluate the performances of the fused images. Both intra-band indices and inter-band indices have been set up in order to measure respectively, spatial distortions (radiometric and geometric distortions) and spectral distortions (colour distortions).

In order to assess the performance of data fusion algorithms, three properties should be verified as expressed by Wald et al. (1997): (1) the data fusion products, once degraded to their original resolution, should be equal to the original; (2) the data fusion image should be as identical as possible to the MS image that would be acquired by the corresponding sensor with the high spatial resolution of the Pan sensor; (3) the MS set of fused images should be as identical as possible to the set of MS images that would be acquired by the corresponding sensor with the high spatial resolution of Pan.

As no multispectral reference images are available at the requested higher spatial resolution, the verification of the second and the third property is not obvious. In order to overcome this drawback, diverse methodological approaches can be used: (i) the Wald protocol (Wald et al. 1997), (ii) the Zhou protocol (Zhou et al. 1998), and, finally, (iii) the QNR (Quality with No Reference) index devised by Alparone et al. (2007). In Sect. 3 we provide an overview of the currently available pan-sharpening techniques and in Sect. 4 detail for their numerical evaluation.

4.3 Overview on Pan-Sharpening Techniques

Pan-sharpening is a branch of image fusion that is receiving ever increasing attention from the remote sensing community for multiple applications in different fields such as pattern recognition, visual enhancement, classification, change detection, object detection and area surveillance (Pohl and Van Genderen 1998).

New-generation space-borne imaging sensors operating in a variety of ground scales and spectral bands provide huge volumes of data having complementary spatial and spectral resolutions. Constraints on the signal to noise ratio (SNR) impose that the spatial resolution must be lower, if the desired spectral resolution is larger. Conversely, the highest spatial resolution is obtained whenever no spectral diversity is required. The trade-off of spectral and spatial resolution makes it desirable to perform a spatial enhancement of the lower resolution multi-spectral (MS) data or, equivalently, to increase the spectral resolution of the data-set having a higher ground resolution, but a lower spectral resolution; as a limit case, constituted by a unique panchromatic image (Pan) bearing no spectral information.

According to Jiang Dong et al. 2009, pan-sharpening may be obtained at different levels, which are categorized as follow: (i) signal level, (ii) pixel level, (iii) feature level, and (iv) decision level as follows (see Fig. 4.2):

(1) Signal level fusion algorithms elaborate signals acquired from different sensors to create a new signal characterized by an improved signal-to noise ratio respect to that of the original signals.
(2) Pixel level fusion algorithms elaborate information on a pixel-by-pixel basis to create a fused image from a set of pixels in original images to improve features visibility and results from image processing tasks such as segmentation, classification, etc
(3) Feature level fusion algorithms elaborate information on feature level after the recognition and extraction of objects from pixel intensities, edges or textures in the original data sources.
(4) Decision-level fusion algorithms elaborate information at a higher level of abstraction, namely information extraction is carried out using results from multiple algorithms which are then combined using given decision rules.

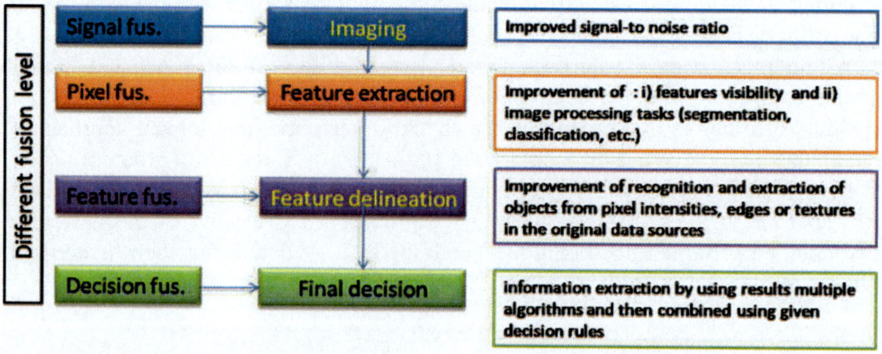

Fig. 4.2 Pan-sharpening categorization

Over the years numerous image fusion techniques have been developed. Among them we will focus on the most widely used: intensity-hue-saturation (IHS), principal component analysis (PCA), different arithmetic combination (e.g. Brovey transform), Zhang, high-pass filtering, Ehlers, multi-resolution analysis-based methods (e.g., pyramid algorithm, wavelet transform), and Artificial Neural Networks (ANNs).

These methods are based on the following general protocol: the high-resolution geometrical information of the Pan scene is extracted and incorporated into the low-resolution MS images, by a proper modelling of the relationships between the Pan and MS bands.

In general, the image fusion methods can be divided into these three main classes depending on how the spatial details are extracted from the Pan image:

(i) techniques based on arithmetic combinations of multispectral images resampled at the higher spatial resolution of Pan;
(ii) component substitution (CS) techniques based on a spectral transformation of the MS data followed by replacement of the first transformed component with the Pan image. Later a reverse transformation is carry out to yield back the sharpened MS bands;
(iii) techniques that employ multi-resolution analysis (MRA) to extract the geometrical information that will be added to the MS bands, from the Pan image.

4.4 Pan-Sharpening Techniques Based on Arithmetic Combinations

The simplest approach for fusing the Pan and MS images is based on arithmetic combinations of different scenes performed at the pixel level. This approach is generally obtained through two computation steps:

(I) Re-sampling of the selected spectral bands to the panchromatic spatial resolution;
(II) selection and application of the arithmetic combinations, which span from the simple multiplication, to Brovey transformation, Synthetic Variable Ratio (SVR), and Ratio Enhancement

4.4.1 Multiplicative Approach Fusion

This algorithm is one of the simplest pan-sharpening algorithms being only based on the multiplication of each multispectral band with the panchromatic image. The image fusion is carried out as follows:

(i) selection of spectral bands and resampling to panchromatic spatial resolution;
(ii) application of multiplication to obtain the transformation

$$Fusion_k = Multi_k \times Pan \tag{4.1}$$

$Fusion_k$ is the kth fused multispectral band, $Multi_k$ is the kth original MS image resampled to the Pan spatial resolution.

The advantages of the multiplicative approach are the following: (i) it does alter the spectral characteristics of the original image data, (ii) it is simple and quite fast

4.4.2 Brovey Transformation

Brovey equation is designed on the basis of the assumption that spectral range of the panchromatic image is the same as that covered by the multispectral data. The transformation is defined by the following equation:

$$Fusion_k = \frac{Multi_k}{Multi_{sum}} \times Pan \tag{4.2}$$

$Fusion_k$ is the kth fused MS band, $Multi_k$ is the kth original MS image, $Multi_{sum}$ is a synthetically panchromatic image obtained by the sum of MS bands at the same resolution as the original MS images.

The image fusion is carried out as follows:

(i) selection of spectral bands and resampling to panchromatic spatial resolution;
(ii) application of Brovey transformation.

Brovey transformation is implemented in the most popular image processing software such as PCI ENVI, ERDAS, etc.

4.4.3 Synthetic Variable Ratio (SVR) and Improved-SVR

The SVR pan-sharpening method (Munechika et al. 1993) is based on Eq. 4.3

$$Fusion_k = \frac{Multi_k}{Pan_{syn}} \times Pan \tag{4.3}$$

$Fusion_k$ is the kth fused multispectral band, $Multi_k$ is the kth original MS image, Pan_{syn} is a synthetically panchromatic image obtained at the same resolution as the original MS images by Suits et al. (1988) equation:

$$Pan_{syn} = \sum_{i=1}^{4} \varphi_i X\, S_{Li} \tag{4.4}$$

The parameters φ_i are only computed for some typical land cover types: (i) urban, (ii) soil, (iii) water, (iv) trees and (v) grass using a regression analysis between

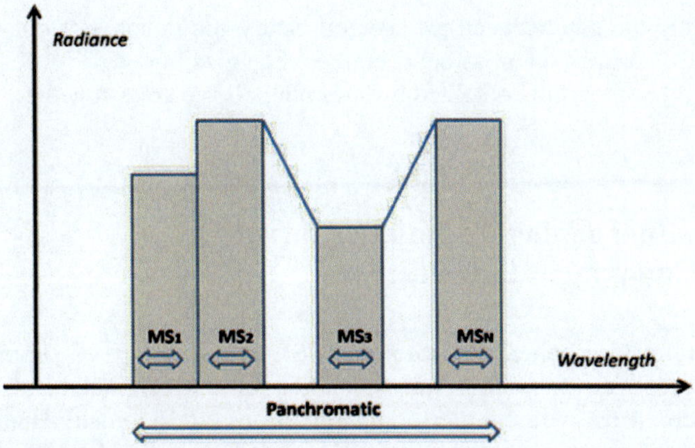

Fig. 4.3 Computation scheme of Wang et al. 2008 algorithm

the values simulated through a given atmospheric model and those measured. The histogram matching was used to force the original Pan image to the PanSyn.

Zhang (2004) developed a data fusion algorithm specifically devised for VHR satellite images and mainly based on the computation of the parameters φ_i using a multiple regression analysis instead of only consider the five main land covers listed before (urban, soil, water, trees and grass).

According to Zhang (2004), this statistics-based fusion technique may solve the two major problems in image fusion: (i) color distortion and (ii) operator (or dataset) dependency the following ways:

(1) the reduction of the color distortion is obtained through least squares technique to find the best fit between the grey values of the image bands being fused and to adjust the contribution of individual bands to the fusion result;
(2) the elimination of dataset dependency is carry out by employing a set of statistic approaches to estimate the grey value relationship between all the input bands

This algorithm is available in a PCI-Geomatics routine (PANSHARP). In the PANSHARP routine, if the original MS and Pan images are geo-referenced, the resampling process can also be accomplished together with the fusion within one step. All the MS bands can be fused at one time. The fusion can also be performed solely on user-specified MS bands.

Wang et al. 2008, developed an improved synthetic variable ratio method for image fusion, mainly based on the assumption that the radiance of the synthesized panchromatic band can be obtained integrating the radiance of MS bands under the hypothesis that the panchromatic band covers exactly the same range as the MS bands to be processed. Therefore, in these conditions, the radiance of the synthesized panchromatic images can be obtained as the integral. The computation is arranged in this way (see Fig. 4.3):

(i) the radiance of each MS band is assumed as equal to the rectangular area;

(ii) regarding the gap between the spectral bands, the radiance is obtained as the area of the trapezoid areas between the rectangles;
(iii) finally, the integral is obtained by summing all the areas together.

4.5 Pan-Sharpening Techniques Based on Component Substitution

Pan-sharpening Component Substitution (CS) is a typology of simple and fast technique based on a spectral transformation of the original bands in a new vector space. Most widely used transformations are Intensity-Hue-Saturation (IHS), Principal Component Analysis (PCA; see Richards and Jia 2006 and Sect. 2.3.4.2 in this book), and Gram-Schmidt orthogonalisation procedure (Laben et al. 2000; Aiazzi et al. 2007).

The rationale of CS fusion is that one of the transformed components (usually the first component or *intensity*, I_L) is *substituted* by the high-resolution Pan image, *P*, before the inverse transformation is applied.

4.5.1 Principal Component Analysis Pan-Sharpening

In pan-sharpening based on PCA (Chavez et al. 1991), the PAN image replaces the first principal component. Prior to the substitution, the PAN image is stretched such that its mean and variance are similar to that of the first principal component (Fig. 4.4).

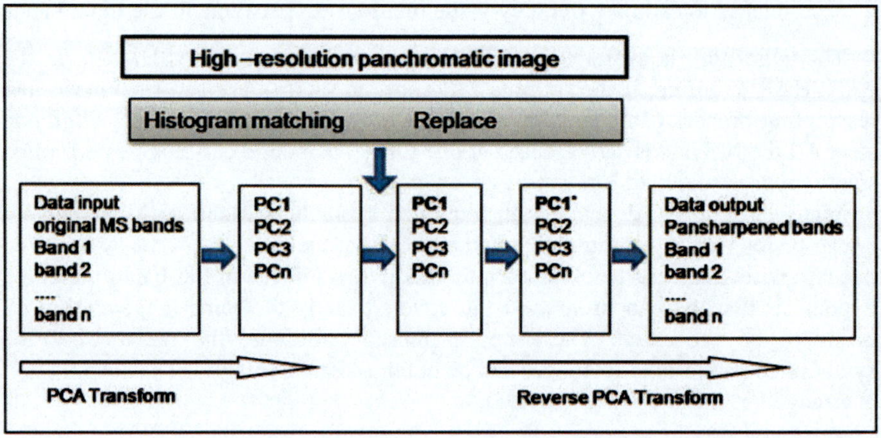

Fig. 4.4 PCA pan-sharpening scheme

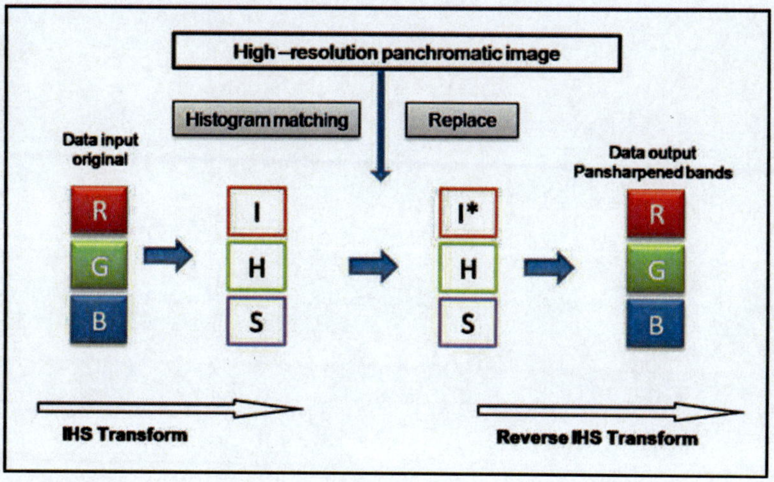

Fig. 4.5 IHS scheme

4.5.2 Intensity Hue Saturation (IHS) Pan-Sharpening

IHS fusion technique, originally defined for three bands only, has been extended to an arbitrary number of spectral bands (Tu et al. 2004).

The IHS transform converts a color MS image from the RGB space into the IHS color space. Because the intensity (I) band resembles a panchromatic image, it is replaced by a high-resolution Pan image in the fusion. A reverse IHS transform is then performed on the Pan together with the hue (H) and saturation (S) bands, resulting in an IHS fused image.

As for the case of pan-sharpening based on PCA, to ensure a *global* preservation of radiometry, the Pan band is histogram-matched to *Intensity*, in such a way that the two images exhibit same *global* mean and variance.

However, since the histogram-matched Pan image and *I* may not have the same *local* radiometry, spectral distortion, may occur and appear as local colour changes in a composition of three bands at a time. To mitigate local spectral distortion, *I* may be taken as a linear combination of the MS bands with weighted coefficients. These are adjusted to the overlap between the spectral response of each MS channel and that of the Pan image. In principle, if the low-pass approximation of the Pan image synthesised by combining the spectral channels exactly matches the low-resolution version of Pan, spectral distortion does not occur (Tu et al. 2004; Aiazzi et al. 2007) (Fig. 4.5).

4.5.3 Gram Schmidt

The Gram Schmidt (GS) pan-sharpening (Kodak / RSI patent) is based on the Gram Schmidt transformation. As PCA, this transformation rotates the original axes applying orthogonalization process to obtain a new reference system with less correlated components.

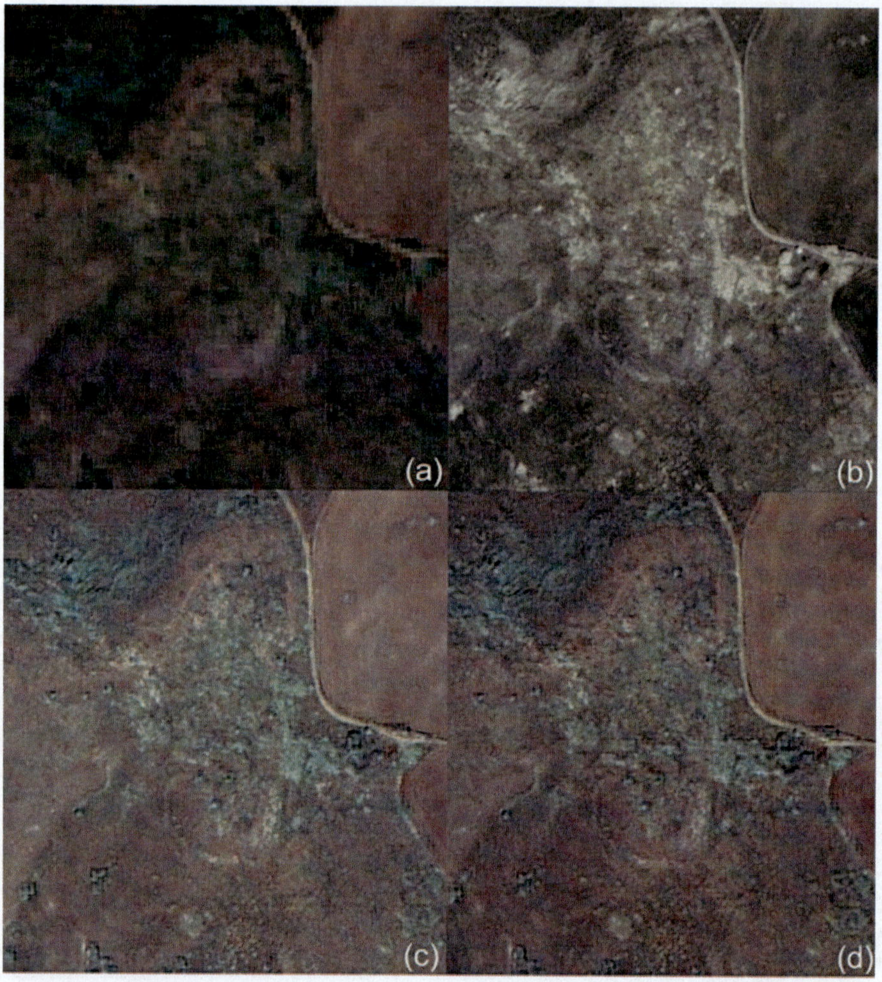

Fig. 4.6 Pan-sharpening of QuickBird multispectral images: Yrsum (Irsina, Southern Italy). (**a**) RGB of multispectral images; (**b**) Pan; (**c**) RGB of pan-sharpened channels by PCA method; (**d**) RGB of pan-sharpened channels by Grahm-Schmidt method. Grahm-Schmidt method allows us to discriminate the archaeological features better than PCA method

The GS approach is based on the following steps:

(i) Pan simulation by averaging the multispectral bands;
(ii) GS transformation using simulated Pan, assumed as the first band, and the MS bands;
(iii) Pan replaces the first GS component;
(iv) inverse GS transform to obtain the pan-sharpened products

Compared to PCA pan-sharpening, GS fusion products are less dependent on the dataset under processing.

Figure 4.6 shows the pan-sharpened RGB image of an archaeological site obtained by PCA e Gram Schmidt method. The latter allows us to discriminate the archaeological features better than PCA method.

4.6 Pan-Sharpening Techniques Based on Fourier Transform

The pan-sharpening techniques based on the Fourier Transform (Chavez et al. 1991) are based on the following computational steps:

(I) extraction of the high frequency components from the PAN image using a high pass filter;
(II) injection of the high frequency components into low frequency components of the multispectral images, to generate the corresponding sharpened images.

Examples of this technique are: (i) Ehlers fusion (Ehlers 2004) method which is implemented in commercial image processing ERDAS; (ii) high pass filtering techniques available as routine in the most popular image processing software such as PCI, ENVI, ERDAS.

4.6.1 *Ehlers Fusion Technique*

The Ehlers (Ehlers 2004) fusion technique is based on the following steps:

I. IHS transform is used to separate the spatial and color information
II. the Fourier transform, low-pass (LP) filter, is applied to intensity (obtained from the IHS transform)
III. the Fourier transform, high-pass (HP) filter, is applied to Pan
IV. the inverse Fast Fourier transform (FFT^{-1}), is applied both to the filtered images are converted back into the spatial domain and summed to merge the low and high frequency information derived from the coarse Intensity channel and Pan, respectively
V. the final fused images are obtained applying the inverse IHS transformation (IHS^{-1}) (Fig. 4.7).

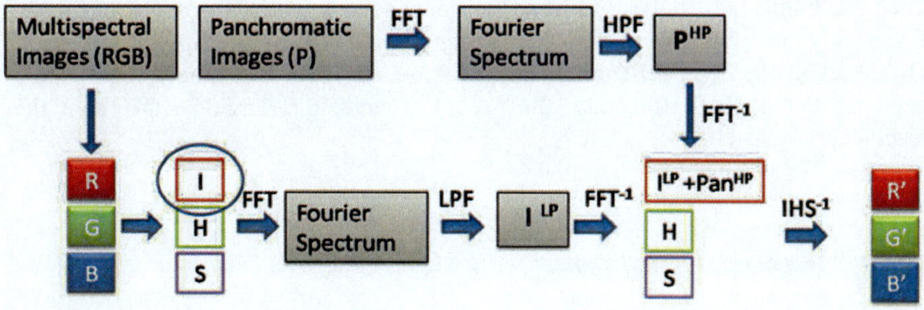

Fig. 4.7 Ehlers fusion technique scheme

4.6.2 High Pass Filtering (HPF) Approach

High pass filtering (HPF) pan-sharpening uses a high pass convolution filter kernel whose size is generally equal to the ratio between the spatial resolution of the Pan and MS images. The pan-sharpening products are obtained as follows:

(i) Pan is filtered using HPF, weighted (weights equal to kernel size) and added to each multispectral band.
(ii) A linear stretch is applied to pan-sharpened products to match the mean and standard deviation of MS images

The process may be also expressed by Eq. 4.5.

$$MS_k = \{W_{high} \cdot HF[PAN]\} \cdot \{W_{low} \cdot LF[MS'_k]\} \tag{4.5}$$

Where MS_k is the kth pan-sharpened image, PAN is the panchromatic image, MS'_k is the original multispectral image, HF and LF correspond to the high and low pass filter operators, respectively

High frequency components are related to the spatial detail, whereas low frequency components contain spectral information.

The performance of these techniques depends on filtering, Kernel types, window size, and weights ($W_{\text{high}}$and W_{low}.) which determine the amount of frequency components to be combined. Generally high pass filter provide satisfactory results also for multitemporal and multisensor data sets.

4.7 Multi-scale Pan-Sharpening Techniques

Pan-sharpening techniques based on a multiscale or multiresolution approach substantially split the spatial information of the MS bands and Pan image into a series of band-pass spatial frequency channels. The high frequency channels are inserted into the corresponding channels of the interpolated MS bands. The sharpened MS bands are synthesised from their new sets of spatial frequency channels. Over the years a number of pan-sharpening methods based on Laplacian pyramid (Núñez et al. 1999; Garzelli and Soldati 2001; Aiazzi et al. 2002). have been widely used since the 1980. Recently, the use of wavelet has largely substituted the Laplacian approach.

4.7.1 Wavelet Transforms

Recently, wavelet data analysis has become one of the fastest growing research areas, being used for a number of different fields and applications, such as signal imaging, numerical analysis, biomedical data analysis, and image processing, including image compression and reconstruction, image segmentation, pattern recognition, and satellite image fusion.

Wavelet methods transform a signal from time domain into time-frequency domain. It is evident that signal wavelet decomposition using Discrete Wavelet Transform (DWT) provides an alternative to the Discrete Fourier Transform (DFT). The main ability of DWT is the multi-resolution time-scale analysis, which is also the main advantage of DWT respect to DFT.

In the field of image processing, DWT acts in this way: an image is decomposed, with each level corresponding to a coarser resolution band.

The general scheme of the wavelet-based pan-sharpening highlights the fact that detail spatial information is extracted from the Pan image using wavelet transforms and injected into the MS image through the following three steps:

(i) the Pan image is decomposed into a set of low-resolution images characterized by corresponding wavelet coefficients (which provide spatial details for each level);
(ii) a MS image replaces the low-resolution Pan at the resolution level of the original MS image
(iii) Pan spatial detail is injected into each MS band by carrying out a reverse wavelet transform using the corresponding wavelet coefficients.

Compared to other standard pan-sharpening techniques, wavelet generally performs better minimizing both color distortion and de-noising effects; nevertheless, spectral content of small objects can often be lost in the fused images. Some other general limitations are (i) the computational complexity; and (ii) the need to set up threshold values for certain parameters thus limiting the exportability to different applications, geographic regions, land cover and land use, and surface characteristics.

Pan-sharpening techniques based on DWT are implemented in many commercial image processing softwares, such as ENVI, PCI, ERDAS.

Some improved DWT, such as Ridgelet, Curvelet, and Contourlet transformation could provide better performance and results, but they have a greater computation complexity and still require setting of threshold values for given parameters (Fig. 4.8).

4.8 Artificial Neural Network

In the pan-sharpening based on Artificial Neural Networks (ANNs; additional information are in Chap. 2 Rumelhart et al. 1986; Werbos 1990) the general schematic approach can be summarized as follows:

(i) two co-registered images are decomposed into several blocks with size of M and N;
(ii) features of the corresponding blocks are extracted from the original Pan and MS images to obtain the normalized feature vectors;

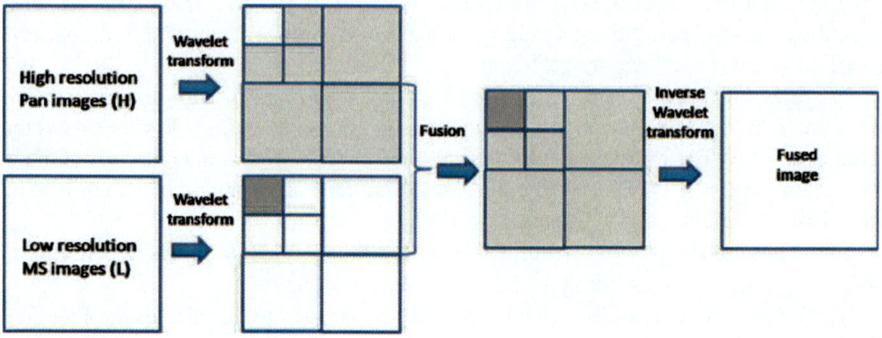

Fig. 4.8 Wavelet data fusion scheme

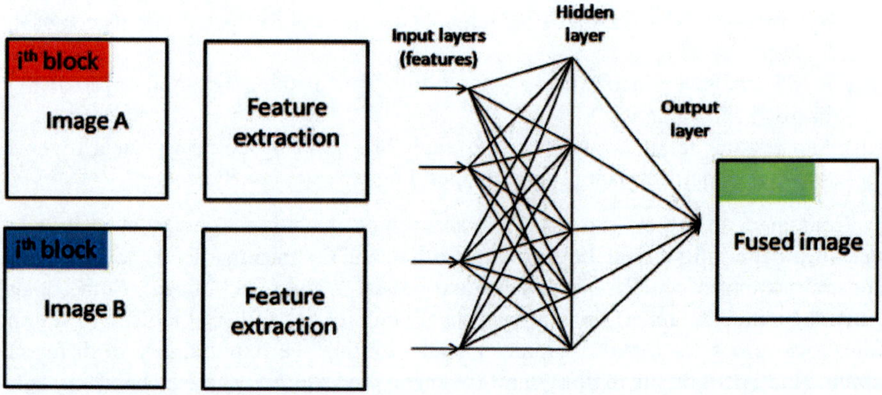

Fig. 4.9 ANN pan-sharpening scheme

(iii) training step is carried out from selected vector samples (Fig. 4.9).

The ANN pan-sharpening methods are quite efficient for high dimension data, such as hyper-spectral or long-term time-series data, as well as for data from multiple sensors, such as active and passive sensors, Radar and VHR satellite images. The learning step of ANN can be time consuming and computationally complex.

4.9 Integrated Data Fusion Approaches

The most difficult aspect in data fusion is the integration of multisource remote sensing data as in the case of active and passive satellite sensors. As an example, fusion of SAR and optical imagery is much more difficult than the ordinary pan-sharpening because the SAR data do not correlate with multispectral imagery. Therefore, classical approaches of multisensor image fusion such as techniques based on multiplication (such as Brovey transform), or component substitution (PCA, IHS) are not suitable for fusing together the spectral detail of optical data with the spatial and texture features of SAR image.

To overcome these drawbacks, feature and decision level fusion or multisource classifiers can be used (Schistad Solberg et al. 1994; Schistad Solberg et al. 1996).

In this context, wavelet multiresolution analysis has been applied successfully (Sveinsson et al. 2001; Cakir et al. 1999; Chibani and Houacine 2000; Alparone et al. 2004). For example, Alparone et al. (2004) is based on the following three steps:

(i) extraction of SAR texture through rationing the (despeckled) SAR image with its low pass approximation,
(ii) extraction of details from the Pan image using á trous wavelet
(iii) high pass filtering based on modulation (HPFM) fusion method.

Wavelet transforms are one of the most widely used methods for extracting feature from multisource remote sensing data. They are also used in conjunction with other methods, such as neural network classifier.

One more example of image fusion method of optical (Pan and MS) and SAR data is in Garzelli (2002). The proposed approach is mainly based on the wavelet analysis and provides an integrated map in which selected information/features/object from SAR data are injected into the optical imagery.

4.10 Data Fusion Performance Evaluation

The assessment of the quality of the fused products is quite difficult, because no multispectral reference images are available at the requested higher spatial resolution. Over the years, several score indices or figure metrics have been devised (see, for example Thomas and Wald 2004) to evaluate the performances of the fused images. Spatial distortions (radiometric and geometric distortions) and spectral distortions (colour distortions) are generally evaluated using both intra-band indices and inter-band indices.

Wald et al. (1997) suggested that to assess the performance of data fusion algorithms, the three following properties should be fulfilled:

1. The data fusion products, once degraded to their original resolution, should be equal to the original.
2. The data fusion image should be as identical as possible to the MS image that would be acquired by the corresponding sensor with the high spatial resolution of the Pan sensor.
3. The MS set of fused images should be as identical as possible to the set of MS images that would be acquired by the corresponding sensor with the high spatial resolution of Pan.

As no multispectral reference images are available at the requested higher spatial resolution, the verification of the second and the third property is not obvious.

In order to overcome this drawback, these following methodological approaches can be used: (i) the Wald protocol, (ii) the Zhou protocol, and, finally, (iii) the QNR (Quality with No Reference) index devised by Alparone et al. (2007).

4.10.1 Wald Protocol

In order to solve the problems linked to the unavailability of the multispectral reference images, Wald et al. (1997) suggested a protocol to be applied in order to evaluate the quality of data fusion products. Such a protocol is based on the following three steps:

1. spatial degradation of both the Pan and MS images by the same factor,
2. fusing the MS images at the degraded scale;
3. comparing the fused MS images with the original reference MS images.

The Wald protocol assumes a scale invariance behaviour.

This means that performances of fusion methods are supposed to be invariant when fusion algorithms are applied to the full spatial resolution. Nevertheless, in the context of remote sensing of archaeology, the small features, which represent a large amount of the archaeological heritage, can be lost after degrading both the Pan and MS. In these situations, the archaeological features will be missed, and, therefore, the evaluation of data fusion results could not be performed over the targets of interest.

4.10.2 Zhou Protocol

As an alternative to the Wald's protocol, the problem of measuring the fusion quality may be approached at the full spatial scale without any degradation by applying Zhou's Protocol (Zhou et al. 1998). Such a protocol is based on the following three criteria:

(1) Both the spectral and spatial qualities are evaluated, but by using separate scores from the available data: the first from the low resolution MS bands and the second one from the high resolution Pan image.
(2) The evaluation of spectral quality is performed for each band by computing an absolute cumulative difference between the fused and the input MS images.
(3) The evaluation of spatial quality is obtained as the correlation coefficient (CC) computed between the spatial details of the Pan image and of each of the fused MS bands.

Such spatial details are extracted by using a Laplacian filter. Unfortunately, some problems can arise by using Zhou's Protocol (Alparone et al. 2007). Firstly, the two quality measures follow opposite trends. Secondly, at degraded scale, the obtained results cannot be in agreement with objective quality indices.

4.10.3 QNR Index

The QNR (Quality with No Reference) index devised by Alparone et al. (2007) is a "blind" index capable of jointly measuring the spectral and spatial quality at the full scale. This index should allow to overcome the drawbacks that can arise when using Zhou's protocol.

The QNR computes both the spatial and spectral distortion from the quality index (Q) by Wang and Bovik (2002).

This index combines the correlation coefficient with luminance and contrast distortion. It was devised for image fusion to assess the quality of output image, as well as for evaluating image processing systems and algorithms. Given an image *X* and its reference image *Y*, the quality index proposed by Wang and Bovik (2002) is calculated as:

$$Q = \frac{(2\mu_x + \mu_y + C1)(2\sigma_{xy} + C2)}{\left(\mu_x^2 + \mu_y^2 + C1\right)\left(\sigma_x^2 + \sigma\mu_y^2 + C2\right)} \tag{4.6}$$

Where $C_1 = (k_1 L)$ and $C_2 = (k_2 L)$, μ_x and $\mu_y \mu x$ indicate the mean of the two images X and its reference image Y, σx and σy are the standard deviation, $\sigma x, y$ represents the covariance between the two images, and L is the dynamic range for the image pixel values, $k_1 << 1$ and $k_2 << 1$ are two constants chosen equal to 0.01 and 0.03, respectively.

Although the values selected for k_1 and k_2 are arbitrary, it was experienced that the quality index is insensitive to variations of k_1 and k_2. Note that C_1 and C_2 are solely introduced to stabilize the measure. In other word, just to avoid the denominator approaches zero values for flat regions.

To measure the overall image quality the mean quality index can be rewritten as a three factor product, that can be regarded are relatively independent.

$$\begin{aligned} Q(x, y) &= f(l(x, y), c(x, y), s(x, y)) \\ &= \frac{(\sigma_{xy} + C_3)}{(\sigma_x \sigma_y + C_3)} \frac{(2\mu_x \mu_y + C_1)}{(\mu_x^2 + \mu_y^2 + C_1)} \frac{(2\sigma_x \sigma_y + C_2)}{(\sigma_x^2 + \sigma_y^2 + C_2)} \end{aligned} \tag{4.7}$$

where C_3 is a small positive constants as C_1 and C_2.

In particular, among the three factor of Eq. 4.2, the first (varying between –1 and 1) represents the correlation coefficient between the two image x and y; the second (varying between 0 and 1) measures the similarity between the mean luminance values of x and Y, and finally, the third (varying between 0 and 1) measures the contrast similarity.

The rationale The QNR (Quality with No Reference) index devised by Alparone et al. (2007) is that the Q index calculated between any two spectral bands and between each band and the Pan image should be unchanged after fusion. In order to obtain a single index, both the spectral and spatial distortion indices are complemented and combined together to obtain a single index that measures the global quality of the fused image.

In detail, the spectral distortion is computed as follows:

The spectral distortion is obtained by computing the difference of Q values from the fused MS bands and the input MS bands, re-sampled at the same spatial resolution as the Pan image

The Q is calculated for each couple of bands of the fused and re-sampled MS data to form two matrices with main diagonal equal to 1

The measure of spectral distortion Dl is computed by using a value proportional to the *p-norm* of the difference of the two matrices.

$$D_\lambda = \sqrt[p]{\frac{1}{L^2 - L} \sum_{l=1}^{L} \sum_{\substack{r=1 \\ r \neq l}}^{L} \left| Q(\hat{G}_e, \hat{G}_r) - Q(\tilde{G}_e, \tilde{G}_r) \right|^2} \tag{4.8}$$

where L is the number of the spectral bands processed, and $Q(\hat{G}_e, \hat{G}_r)$ denotes that Q is calculated for each couple of bands of the fused and resampled MS data.

The spatial distortion is computed two times: (1) between each fused MS band and the Pan image; and than (2) between each input MS band and the spatially degraded Pan image. The spatial distortions Ds are calculated by a value proportional to the *q-norm* of the differences

$$D_S = \sqrt[q]{\frac{1}{L} \sum_{l=1}^{L} \left| Q_n(\hat{G}_l, P) - Q_n(\tilde{G}_l, \tilde{P}) \right|^q} . \tag{4.9}$$

where L is the number of the spectral bands processed, and $\hat{G}_l, P$ denotes the Q is calculated between each fused MS band and the Pan image, and $Q_n(\hat{G}_l, P)$ denotes the Q is calculated between each input MS band and the spatially degraded Pan image.

Just as an example, we refer to an application related to an archaeological site in Southern Italy (Lasaponara et al. 2007). It is Metaponto which was before a Greek colony (700 BC–200 BC), then it was frequented by Romans (200 BC–400 AD).

QuickBird imagery has been used to identify and study linear features which have been thought to ancient field divisions or other features of archaeological interest. Such features are better visible from NIR channel respect to other channels. To improve the details pan-sharpening has been carried out by using Gram Schmidt, PCA and Wavelet Transform methods.

The NIR pan-sharpened products (Fig. 4.10) have been compared by visual inspection by about ten colleagues including experts of remote sensing, archaeologists and geophysicists. According the involved researchers Wavelet pan-sharpening provides the best result in terms of visibility of archaeological features.

Fig. 4.10 Visualization of NIR pan-sharpened images obtained by using Gram Schmidt (*upper*), PCA (*medium*) and Wavelet transform (*lower*) methods

Table 4.1 Evaluation of three pan-sharpening NIR channels by using the Wald and Alparone protocols

Pan-sharpening method	Wald protocol Bovik score index				Alparone protocol
	I component	II component	III component	Q	QNR score index
Gram Schmidt	0.508	0.999	0.506	0.257	0.632
PCA	0.727	0.886	0.640	0.413	0.673
Wavelet	0.773	0.999	0.769	0.595	0.863

Then the same pan-sharpened dataset has been evaluated in a quantitative way by applying the Wald and Alparone protocols. For both of them the best performance was clearly achieved by using the wavelet-based data fusion (see Table 4.1).

4.11 Final Remarks

This chapter is composed of two parts, the first provided an overview of diverse pan-sharpening techniques; whereas, the second part focused on the protocols available for the numerical evaluation of performance.

The most common used methods are described in detail. In particular, we focused the different methodological approaches, such as pan-sharpening techniques based on: (i) arithmetic combinations (among them Multiplicative approach, Brovey transformation, SVR); (ii) component Substitution (among them HIS, PCA and Gram-Schmidt); (iii) Fourier Transform (Ehlers, HPF); (iv) Multiscale/multiresolution (wavelet, Laplacian); (v) Artificial Neural Networks.

The simplest approach is the arithmetic combination of different scenes performed at the pixel level. The CS fusion method is based on diverse types of transformations which are applied to synthetise a Pan image, later substituted by the real pan and, finally, the application of the reverse transformation provides the fused products. The pan-sharpening techniques based on the Fourier Transform extract the high frequency components from the PAN image (using a high pass filter) and inject them into low frequency components of the multispectral images. Multiscale/multiresolution approaches split the spatial information of the MS bands and Pan image into a series of band-pass spatial frequency channels. The high frequency channels are inserted into the corresponding channels of the interpolated MS bands. In the pan-sharpening based on Artificial Neural Networks the images are firstly decomposed into several blocks, later features are extracted from the original Pan and MS scene and finally the training step is carried out.

Moreover we also highlighted the protocols in use for the numerical comparison and evaluation of different approaches, in terms of spectral fidelity and spatial resolution performance.

The visual inspection is not enough to compare results obtained from different algorithms but automatic quality assessment is highly desirable to evaluate performance and benefits of Pan-sharpening techniques. Considering that archaeological features are characterized by quite subtle signals, It clearly shows that the quantitative evaluation of performance from pan-sharpening methods is a critical issue for archaeological applications.

Nowadays, several mathematical methods have been adopted to evaluate the quality of merged imagery (i) Specific protocols for quality measures such as well as (ii) Statistical indices, such as cross entropy, mean square error, and signal-to-noise ratio. Nevertheless, up to now analytical studies of these measures are still lacking. New performance assessment criteria and automatic quality assessment methods will be addressed in future research.

Moreover, specific context-adaptive indices need to be developed, because outputs from the existing evaluation procedures cannot fit the specific requirement of cultural features sharpening. Although the application of pan-sharpening techniques for enhancing cultural features is a quite recent topic, we would like to point out their potentiality for improving the identification of unknown archaeological buried structures.

References

Aiazzi B, Alparone L, Baronti S, Garzelli A (2002) Context-driven fusion of high spatial and spectral resolution data based on oversampled multiresolution analysis. IEEE Trans Geosci Remote Sens 40(10):2300–2312

Aiazzi B, Baronti S, Selva M (2007) Improving component substitution pan-sharpening through multivariate regression of MS + PAN data. IEEE Trans Geosci Remote Sens 45(10): 3230–3239

Aiazzi B, Baronti S, Alparone L, Lasaponara R, Masini N (2008) Data fusion techniques for supporting and improving satellite-based archaeological research. In: Lasaponara R, Masini N (eds) Advances in remote sensing for archaeology and cultural heritage management. Aracne, Roma, pp 31–36

Alparone L, Baronti S, Garzelli A, Nencini F (2004) Landsat ETM + and SAR image fusion based on generalized intensity modulation. IEEE Trans Geosci Remote Sens 42:2832–2839

Alparone L, Aiazzi B, Baronti S, Selva M, Garzelli A, Nencini F (2007) Quality assessment without reference of pan-sharpened MS images. In: Proceedings of EARSel symposium 5–7 June 2007, Bozen

Cakir HJ, Khorram S, Dai XL, de Fraipont P (1999) Merging SPOT XS and SAR imagery using the wavelet transform method to improve classification accuracy. In: Geoscience and remote sensing symposium. IGARSS '99 proceedings, IEEE 1999 International, 28 June–2 July 1999

Chavez PS Jr, Sides SC, Anderson JA (1991) Comparison of three different methods to merge multiresolution and multispectral data: Landsat TM and SPOT panchromatic. Photogramm Eng Remote Sens 57(3):295–303

Chibani Y, Houacine A (2000) On the use of the redundant wavelet transform for multisensor image fusion. In: Proceedings of the 7th IEEE international conference on electronics, circuits and systems, Lebanon, 17–20 Dec 2000. ICECS 2000, vol 1, pp 442–445

Dong J, Zhuang D, Huang Y, Jingying Fu (2009) Advances in multi-sensor data fusion: algorithms and applications. Sensors 9:7771–7784

Ehlers M (2004) Spectral characteristics preserving image fusion based on Fourier domain filtering. In: Ehlers M, Kaufmann HJ, Michel U (eds) Remote sensing for environmental monitoring, GIS applications, and geology IV. Proceedings of SPIE, Bellingham, pp 1–13

Garzelli A (2002) Possibilities and limitations of the use of wavelets in image fusion. In: Proceedings of the IEEE international geoscience and remote sensing symposium, Toronto, 24–28 June 2002. IGARRS 02, vol 1, pp 66–68.

Garzelli A, Soldati F (2001) Context-driven image fusion of multispectral and panchromatic data based on a redundant wavelet representation. In: Proceedings of the IEEE/ISPRS joint workshop on remote sensing and data fusion over urban areas, Rome, 8–9 Nov 2001, pp 122–126

Laben CA, Bernard V, Brower W (2000) Process for enhancing the spatial resolution of multispectral imagery using pan-sharpening. US Patent 6,011,875

Lasaponara R, Masini N (2007) Detection of archaeological crop marks by using satellite QuickBird multispectral imagery. J Archaeol Sci 34:214–221

Lasaponara R, Lanorte A, Coluzzi R, Masini N (2007) Performance evaluation of data fusion techniques for archaeological prospection based on satellite data. In: Proceedings of SPIE, vol 6749, 67492W. doi: 10.1117/12.738204

Lasaponara R, Masini N, Aiazzi B, Alparone L, Baronti S (2008) Satellite-based enhancement of archaeological marks through data fusion techniques. In: Proceedings of the SPIE, vol 7110, 711024. doi:10.1117/12.801569

Munechika CK, Warnick JS, Salvaggio C, Schott JR (1993) Resolution enhancement of multispectral image data to improve classification accuracy. Photogramm Eng Remote Sens 59(1): 67–72

Núñez E, Otazu X, Fors O, Prades A, Palà V, Arbiol R (1999) Multiresolution-based image fusion with adaptive wavelet decomposition. IEEE Trans Geosci Remote Sens 37(3):1204–1211

Pohl C, Van Genderen JL (1998) Multisensor image fusion in remote sensing: concepts, methods and applications. Int J Remote Sens 19:823–854

Richards JA, Jia X (2006) Remote sensing digital image analysis. Hardback, 4th edn. Springer, Berlin/Heidelberg, 476 p

Rumelhart D, Hinton G, Williams R (1986) Learning internal representations by error propagation. In: McClelland JL, Rumelhart DE, The PDP Research Group (eds) Parallel distributed processing: explorations in the microstructure of cognition. The MIT Press, Cambridge, vol 1, pp 318–362

Schistad Solberg AH, Jain AK, Taxt T (1994) Multisource classification of remotely sensed data: fusion of Landsat TM and SAR images. IEEE Trans Geosci Remote Sens 32(4):768–778

Schistad Solberg AH, Taxt T, Jain AK (1996) A markov random field model for classification of multisource satellite imagery. IEEE Trans Geosci Remote Sens 34(1):100–113

Suits G, Malila W, Weller T (1988) Procedures for using signals from one sensor as substitutes for signals of another. Remote Sens Environ 25:395–408

Sveinsson JR, Ulfarsson MO, Benediktsson JA (2001) Cluster-based feature extraction and data fusion in the wavelet domain. In: Geoscience and remote sensing symposium, 2001. IGARSS '01. IEEE 2001 International, vol 2, 9–13 July 2001, pp 867–869

Thomas C, Wald L (2004) Assessment of the quality offused products. In: Oluic M (ed) Proceedings of 24th EARSeL Symposium on New Strategies for European Remote Sensing, Dubrovnik, 25–27 May 2004. Balkema, Rotterdam, pp 317–325

Tu TM, Huang PS, Hung CL, Chang C-P (2004) A fast intensity-hue-saturation fusion technique with spectral adjustment for IKONOS imagery. IEEE Trans Geosci Remote Sens 1(4):309–312

Wald L, Ranchin T, Mangolini M (1997) Fusion of satellite images of different spatial resolutions: assessing the quality of resulting images. Photogramm Eng Remote Sens 63(6):691–699

Wang Z, Bovik AC (2002) A universal image quality index. IEEE Signal Proc Lett 9(3):81–84

Wang L, Cao X, Chen J (2008) ISVR: an improved synthetic variable ratio method for image fusion. Geocarto Int 23(2):155–165

Werbos PJ (1990) Backpropagation through time: what it does and how to do it. Proc IEEE 78:1550–1560
Zhang Y (2004) Understanding image fusion. Photogramm Eng Rem Sens 70:657–661
Zhou J, Civco DL, Silander JA (1998) A wavelet transform method to merge Landsat TM and SPOT panchromatic data. Int J Remote Sens 19(4):743–757

Part II
Satellite Remote Sensing for Cultural Heritage Documentation and Management

Chapter 5
Remote Sensing and Integration with Other Geomatic Techniques in Archaeology

Gabriele Bitelli

Abstract It's a fact that today archaeologists can interact ever more closely with the modern sciences of surveying, which as a whole may fall under the name of Geomatics. A scientific approach to Archaeology may be in fact reflected in the first place in the adoption of methods, both objective and based on rigorous criteria and tools, for knowing and describing the goods. Remote Sensing, as a technique for surveying a site from a distance by means of special detectors and using management methods based on mathematical and physical models, is by nature a practical expression of this phenomenon, whether it is meant in the most common meaning of use of satellite data or extending its acceptation to aerial photogrammetry. The paper will review some of the issues related to the use of Remote Sensing in the archaeological context, but particularly in relation to other disciplines of Geomatics; Remote Sensing is not, as noted, a technique that is "self contained" and restricted in its scientific and technical areas independently, but it relies on the wealth of other techniques and scientific fields, and they must fit perfectly, especially for applications in Archaeology which have by nature a highly multidisciplinary vocation.

Keywords Geomatics • Surveying • Remote Sensing • Data integration

5.1 Introduction

Geomatic techniques that can be taken into account today in Archaeology are numerous, and almost always refer to digital data: the latter feature is especially important if you want to integrate different data, for example in the framework of an analysis within a Geographical Information System (GIS).

G. Bitelli (✉)
Department of Civil, Environmental and Materials Engineering, University of Bologna, Viale Risorgimento, 2, Bologna, Italy
e-mail: gabriele.bitelli@unibo.it

R. Lasaponara and N. Masini (eds.), *Satellite Remote Sensing: A New Tool for Archaeology*, Remote Sensing and Digital Image Processing 16,
DOI 10.1007/978-90-481-8801-7_5,

Remote Sensing is certainly among the techniques of Geomatics one of the most interesting and promising for a number of activities in Archaeology, not only in support of prospection and mapping but to support complex spatio-temporal analysis; the scientific community is rapidly realizing its importance and its use is increasing within the archaeological projects (Campana and Forte 2006; Lasaponara and Masini 2008; Scollar et al. 2009; Wiseman and El-Baz 2008). The trend is fostered by the development of new sensors in the optical field with increasing spatial resolution and radiometric and spectral characteristics (with new bands that can make an important contribution in archaeological prospecting), and also by the spreading diffusion in the use of radar imagery.

This situation has been certainly facilitated by different factors, and not secondarily by the familiarity and skill with satellite imagery due to the emergence of the so-called Earth Viewers systems, or Virtual Globes, such as Google Earth (since 2005), which made available, to a wider user community, large databases of georeferenced images that were hitherto only used by experts (with very high costs for acquisition and management). This phenomenon has further led archaeologists towards synoptic analysis of the territory, following on the other hand a scientific and cultural trend; it has been considered very positive and useful for the research and field activity, and in a few years has produced results of considerable interest, allowing, among others, the support and guide for a large number of new researches and also the discovery of sites.

In order to better define the matter, it should be noted however that the availability of this vast corpus of information is not always accompanied by real knowledge of what it represents and what the limits are of its use. It is for example known that these Earth Viewers systems, increasingly sophisticated and suppliers of very high resolution (VHR) images for large regions, although in many parts of the world provide good quality results (Cheng and Chaapel 2010), cannot generally replace the cartographic data for rigorous mapping applications, either for the treatment that the images have undergone, or because the level of accuracy of their geo-referencing is not the same everywhere. Moreover, different regions are provided with images acquired by satellite platforms and sensors with different characteristics (e.g. geometrical resolution); also restricting the analysis to a single type of images, the quality of georeferencing does not always guarantee everywhere constant outcomes.

A similar problem apply for three-dimensional data (Digital Terrain Models, DTM) coming from global elevation models: this information is certainly very useful for a more expressive and communicative representation of geographical information, but the archaeologist must be aware that they can't be directly used for some rigorous geospatial analysis, both because of the level of resolution and intrinsic quality of the global DTMs and for the processing required to refer to a vertical geodetic datum.

Besides these precautions, it is nonetheless true that for large regions lacking cartographic data of sufficient quality, the images provided by Earth Viewers systems are one of the few opportunities to access a georeferenced dataset and are now standard on exchanging geographical contents. It should also be noted that a system like Google Earth has developed in recent years strong collaborative activities with the user community that have helped expand the database with their own multimedia content (e.g. photographic images and text) and three-dimensional models. In this sense, the archaeological community could further explore the possibility of a greater use of these platforms to perform multidisciplinary researches.

Faced with an increasingly larger access to satellite images in Archaeology, it takes good understanding and knowledge of the image properties by the user, and what are the geomatic techniques and disciplines that may be useful in working with these data, not only for their processing but also to enrich them for the creation of integrated information systems.

This paper tries to briefly review some ways for applying and integrating the new geomatic techniques in support of using Remote Sensing as a new powerful tool for Archaeology. Priority will be given to clarifying how the adoption of these techniques can modify the concept of mapping in the archaeological field: an important example in this sense is the move from classical surveying approaches made by using local reference systems, not shareable and sometimes not well documented or maintainable, to global absolute systems that can instead enable integration into GIS and updating over time.

5.2 Geomatic Techniques in Support of Remote Sensing Use in Archaeology

A growing experience in different international projects highlights the potentiality of using together some digital surveying and mapping techniques in archaeological applications, and specifically for support or integration of satellite imagery: topographical surveying, photogrammetry, Global Navigation Satellite Systems (GNSS), aerial laserscanning (Lidar).

These techniques can be taken in account for a multiscale research, as each of them can be better employed at different investigation scales, as shown in Fig. 5.1. In this context, the characteristics of each method can be summarized in the following Table 5.1. Many of these techniques are already being used as standard methods on archaeological sites, others are becoming common. In a number of cases an effective integration of different techniques and data, acquired with various accuracy/scale and using different reference systems, is needed. The geometrical coherence among all the information must be taken into account and controlled for errors. The establishment of shared reference systems, or the insertion of local systems into absolute systems by highly accurate measurements is a crucial point to be addressed from the beginning of each multisource and multitechnique surveying activity, and good documentation by metadata records is crucial. For these purposes the main technical requirements are related to data processing (e.g. GNSS in its manifold methods) and geometric mathematical

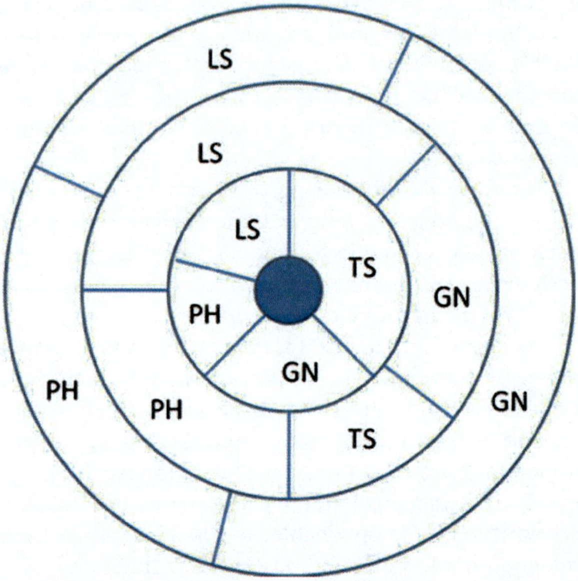

Fig. 5.1 Employment of four main geomatic techniques (*TS* topographical surveying, *GN* GNSS, *PH* photogrammetry, *LS* laserscanning) in integration with Remote Sensing at different scales: regional (*exterior ring*), archaeological site and its neighborhood (*intermediate ring*), structures (*interior ring*)

transformations; a good knowledge of local geodetic infrastructures is in general worthwhile. The same concepts for systematic, coherent and accurate georeferencing can be carried out through monitoring procedures on sites or objects, realized over the years or decades for maintenance and control.

As regarding Remote Sensing, it can be used for qualitative or for quantitative analysis in Archaeology, and data requirements are often very different. What methods and tools should support the satellite imagery analyst in the two cases?

The first case refers for example to a land cover classification made with multispectral or hyperspectral data, to know whether the required conditions are met to further investigate the existence or the characteristics of a site, also underground; this investigation is often integrated today in Archaeology with geophysical prospecting. In this situation it is essential on the one hand to have good knowledge of physical phenomena but also the availability of appropriate image processing tools, both to perform a radiometric calibration – reducing noise or atmospheric effects on the images – and to carry out an accurate classification according to one of many available techniques.

As for classification techniques, along with the well-established pixel-oriented procedures, it is becoming increasingly common the adoption of an object-oriented classification approach, mainly due to the spread of high resolution imagery and advanced segmentation algorithms.

Figure 5.2 shows a risk management application in an area of great cultural interest. On December 26, 2003 a major earthquake struck a large part of the Arge-Bam Citadel, the largest adobe building in the world, located in Bam, a city in the Kerman province of south-eastern Iran. The Citadel is listed by UNESCO as World Heritage Site of "Bam and its Cultural Landscape". While a 15 m ground resolution ASTER after-event image can't highlight the damage (Fig. 5.2a), a couple

Table 5.1 Main characteristics of various geomatic techniques applicable in association with satellite Remote Sensing in a multiscale integrated analysis for archaeological applications

Technique	Product	Applications	Coverage	Precision accuracy	Notes
Topographical surveying by total stations and levels	1D, 2D, 3D coordinates of significant points	Survey of sites for large scale applications, ground control points (GCP) determination, control and monitoring surveys	Small, medium	Very high	Time consuming, reduced possibility for automatic operations
Photogrammetry	Vector cartographical database, raster products (photoplanes, orthophotos), surface models (DTM, DSM) by image matching procedures	Mapping, monitoring over time of sites and environment, landscape analysis, feature identification	Medium or large (aerial photogrammetric blocks)	Depending on the photographic scale and the flight height	Use of balloons, kite systems or more sophisticated Unmanned Aerial Vehicles (UAV) for low altitude acquisition. Growing use of close-range digital photogrammetric systems, possibility of integration with aerial derived data
GNSS	3D coordinates of single points or tracks (acquired by continuous kinematic surveys)	Mapping, map updating, acquisition of Ground Control Points (GCP), training sites for image classification, accurate morphological 3D description of areas (DSM)	Small, Medium	High, depending on the surveying technique adopted	Growing availability of networks of permanent stations around the world. Real time kinematic methods with high productivity
Lidar	Dense point clouds, DSM, DTM	3D models for sites and structures, feature localization, landscape analysis, high accuracy orthophoto production	Medium, large	High/medium, depending on the instrument and the flight height	Complex data processing (e.g. feature extraction, 3D modeling). Spreading of terrestrial systems (TLS): possibility of integration of terrestrial and aerial datasets

Fig. 5.2 (**a**) Near infrared ASTER nadir image, 15 m resolution, of city of Bam after the 2003 Earthquake (courtesy of LP DAAC-USGS), with the Arge-Bam Citadel area evidenced; (**b**) pre- and (**c**) after-event pan-sharpened QuickBird imagery of the Citadel, 0.60 m resolution (courtesy Digital Globe); (**d**) Building damage map for the city of Bam, based on object-oriented analysis of QuickBird imagery captured before and after the event (Gusella et al. 2005). Collapsed structures are shown in red and non-collapsed in light-blue. The inset shows a detail of the building classification

of pre- and post- event pan-sharpened QuickBird images clearly show the effects of the earthquake on the structures and support a damage assessment evaluation (Fig. 5.2b).

Starting from high resolution imagery, and using an object-oriented approach, a citywide reclassification was performed (Fig. 5.2c), where a nominal 40% of objects were classified as collapsed (Gusella et al. 2005).

An application like this one, using two pre- and post-event images, requires a careful process of co-registration between the two images, as indeed happens in all cases of change detection. In these applications a good absolute georeferencing is therefore not essential rather than a relative one, of an image over the other. Working with high resolution, while it facilitates the understanding and the visual inspection of the phenomenon, in many cases some problems arise, for example due to different off-nadir angles for the two images that make it difficult to match procedure and segmentation; it is enough to mention the presence of different shadowing effects.

On the other hand, in many circumstances we have to describe in rigorous and metric terms an existing phenomenon, clearly detectable by the images, for example to produce a cartographic product. It is therefore essential to use a good geometric processing of the data, both for the sensor-to-Earth coupling and with regard to the georeferencing problem.

As to the latter question, there has been a significant change in the methodology of archaeologists. If in the past they worked mainly with local reference systems, confined to the areas of excavation, being satisfied by a coarse absolute georeferencing of the site on a small scale map, now it is possible to overcome this disconnection and operate directly in a absolute reference system adoptable for cartographic mapping, with high precision and with all the benefits that entails, especially for the opportunity to easily integrate other data acquired simultaneously or that will become available over time, detected on the ground or from space. This is an essential condition to have single management of all the data coming from an archaeological multidisciplinary investigation within a GIS system. A GIS is in fact the essential tool to maintain and manage in a systematic and coherent way a complex database containing all the information needed to make geospatial and geostatistical analysis, with different possibilities to represent the archaeological site and the territory in which it is located. Many GIS applications use today the satellite imagery as base layer and maintain 2D and 3D data georeferenced with the same criteria.

For these reasons it is therefore critical to define precisely the cartographic-geodetic datum in which the images have to be georeferenced: global systems are normally used today (e.g. UTM-WGS84) but in many cases attention must be paid also to systems developed at national level, often far back in the past. In the last case the parameters of the transformation needed to interface the two systems have to be carefully determined, for instance if we want to make use of information derived from old representations (historical maps, surveys and measurements performed in the past).

Aspects of image georeferencing are very important and very sensitive, because a failure at this stage undermines the work and produces unusable results. There are several problems depending on the sensor, the platform used and the level of image processing purchased: for high resolution images you can use georeferencing methods based on rigorous physical models, provided by the supplier of the images, or non-parametric methods, based on general mathematical models whose parameters no longer have a direct physical meaning (e.g. rational polynomial functions). Especially in the latter case it is important to have at one's disposal the coordinates of an appropriate number of ground control points (GCP) clearly visible on the image, and they should normally be in the cartographic-geodetic system where one wants the final products. For this application satellite positioning techniques provided by GNSS are of course of fundamental importance: available systems are today the American Navstar Global Positioning System (GPS), the Russian Glonass, the Chinese Compass, the European Galileo (in advanced phase of development). GNSS, and more commonly GPS, are now quite popular among archaeologists but often for absolute positioning of sites and objects with rather limited accuracy requirements, and hence are usually not suitable for orientation of high and very high resolution satellite images useful for mapping of significant features of a site. This involves to dispose of the more suitable instrumentation and surveying methodologies (static, rapid static, post-processed or real-time kinematic, undifferenced approach by Precise Point Positioning), that today can supply results with very good accuracy at lower costs (Vittuari 2008).

Adopting GNSS technique for absolute positioning applications can benefit to the availability of networks of permanent stations, now available in many areas of the globe with a density of points variable in different regions. Where these networks are more dense, you can directly use data from permanent stations for differential correction of acquired measurements in the area of excavation, thereby reducing the efforts on the ground in terms of surveyors and work, obtaining the final coordinates already included in a well-defined geodetic-cartographic frame worldwide. As mentioned before, the latter fact can be considered as one of the major changes occurred in these years in archaeological surveying, both in strictly technical terms and as an evolution of the overall approach to the subject of surveying an archaeological site: we moved from a work suitable only for local purposes, on a reference system not shared or not shareable to others, to surveys that produce absolute coordinates in a global geodetic system (WGS84), known and usable by anyone. It was a sort of revolution for mapping, which can be today addressed through a universal system, not limited by territorial or geopolitical boundaries. It is clear that this is in particular beneficial for an archaeologist who works in different areas, sometimes in buffer zones between states with different cartographic conventions, for multinational collaborative projects and for researchers who make use of the synoptic data provided by the satellite imagery over large regions.

Paradoxically, archaeologists are now operating on the site with coordinates that have global scope in the world but which can have an accuracy comparable to, or greater than, those determined in the past with traditional surveying in a narrow local context; often the absolute positioning is determined with an accuracy of one order of magnitude greater than what was estimated for the archaeological activity in the traditional sense. And, of course, also image classification procedures from multispectral datasets can benefit from a higher accuracy in the definition of training sites.

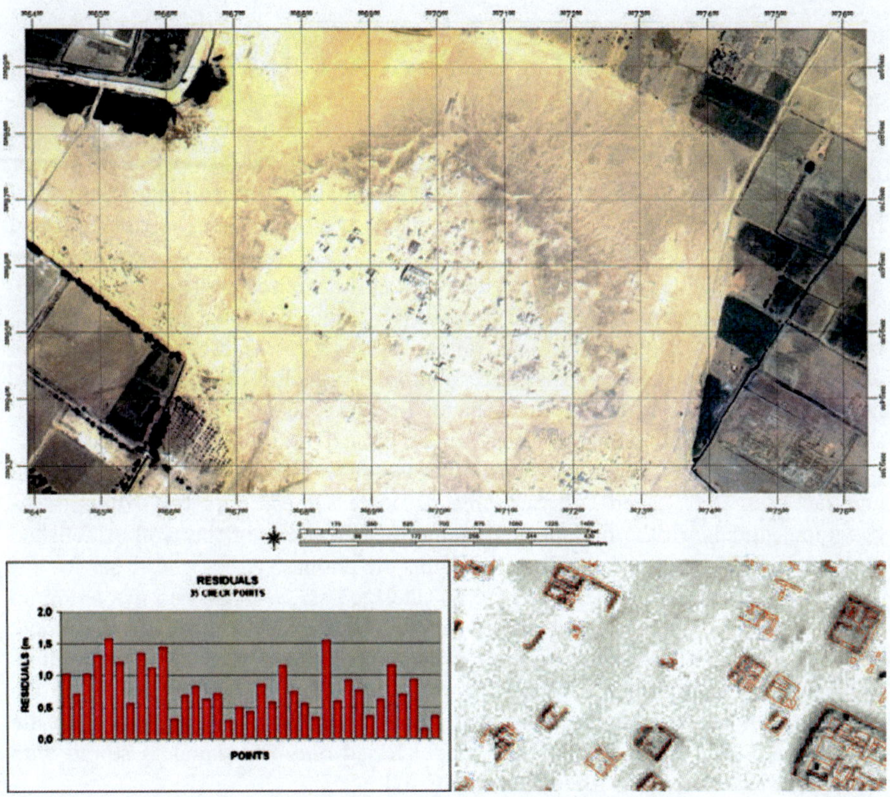

Fig. 5.3 (*top*) Pan-sharpened orthophoto from Ikonos imagery; (*bottom left*) absolute value of the residuals on the check points after RPF application using a single GCP; (*bottom right*) The orthoimagery fits very well with the basemap layer of the buildings surveyed by topographical equipment in the same reference frame

Another stage in the analysis of satellite imagery that requires other external geomatic information is the generation of orthophotos or some spatial analysis procedures performed in a GIS environment (Menze and Sherratt 2006; Poli et al. 2004), where a DTM or a DSM (Digital Surface Model) is needed. For applications over large areas at a national, regional or basin scale, global models such as GTOPO30 (30 arc sec resolution) or preferably SRTM – Shuttle Radar Topography Mission (1–3 arc sec resolution) and the recent GDEM – ASTER Global Digital Elevation Model (30 m resolution) can be used, while for large-scale projects more precise terrain models are required, obtained by aerial surveys (Lidar or photogrammetry), by digitizing contour lines of appropriate map scale or by ground surveys (GNSS in kinematic mode).

In the case study of Fig. 5.3, a DEM derived from accurate ground measurements (total station and GPS) was used for the creation of an orthophoto (Fig. 5.3a) for the archaeological site of Bakchias (Egypt) using a Geo Ortho Kit Ikonos image (1 m geometric resolution of panchromatic) in the UTM-WGS84 system (Bitelli and Zanni 2005). The quality assessment of the geocoding and orthorectification process, thanks also the quality of DTM obtained through terrestrial surveys, showed that in this experiment a comparable accuracy value was reached either by a rigorous model (Toutin 2001) and by a non-parametric Rational Polynomial Function (RPF) model, although applied using only one GCP and the rational coefficients derived from metadata. In the latter case the residuals on 35 check points had a mean value of 0.80 m with a standard deviation of 0.38 m (Fig. 5.3b). The orthophoto is a pan-sharpened product, obtained by a data fusion procedure between panchromatic and multispectral bands; it is a known fact that this solution allows to combine the high spatial resolution of panchromatic images with the radiometric resolution of the spectral bands (in this case RGB).

There is also an increasing number of optical and radar satellite systems providing stereo-imagery at medium-high resolution (Jacobsen 2005; Buyuksalih and Jacobsen 2007), and software able to process these data for extraction of elevation models; in the optical field, some modifications of the well-established photogrammetric image-matching algorithms are adopted.

Inexpensive but very interesting stereo satellite data, with ground resolution up to 15 m, come from ASTER sensor (Abrams et al. 2002) in near infrared band, where along-track nadir and backward images are acquired on the same area. To produce a DEM from ASTER imagery, there are two basic approaches: the generation of an absolute model, when some ground control points (GCPs) are known, or the creation of a relative model based only on metadata stored with the image. The process is illustrated in Fig. 5.4.

Figure 5.5 shows the results of a test carried out on an area of archaeological interest in Southern Italy, 2.5 km × 2.5 km wide, with heights varying between 250 and 1,000 m AMSL (Bitelli et al. 2005): an absolute DEM obtained by ASTER imagery using 4 GCPs was compared against a reference model derived by photogrammetry (stereoplotting scale 1:31,000) and against SRTM.

In some tests, ASTER derived absolute DEMs show a better accuracy in respect to SRTM, with the important addition of the information coming from ASTER multispectral coverage, useful for a large number of scientific applications (Hirano et al. 2003; Poli et al. 2004).

Orthophotos from high resolution multispectral satellite imagery, either in panchromatic or in pan-sharpened, certainly constitute a product of major interest for archaeological investigation (Lasaponara and Masini 2006), being a suitable base for GIS establishment, to conduce metric analysis and cartographic restitution on the site and to study the anomalies of interest observed on large areas (e.g. soil, crops and shadow marks). In many cases it was found that the data fusion procedure that provides more information on anomalies is obtained by considering the near-infrared band, provided by the main high-resolution sensors currently used.

Frequently, however, the interest is not only related to the current state of an area but may also be directed to its change along the last decades, especially in the fairly frequent case of major changes occurred in the territory, due mainly to strong

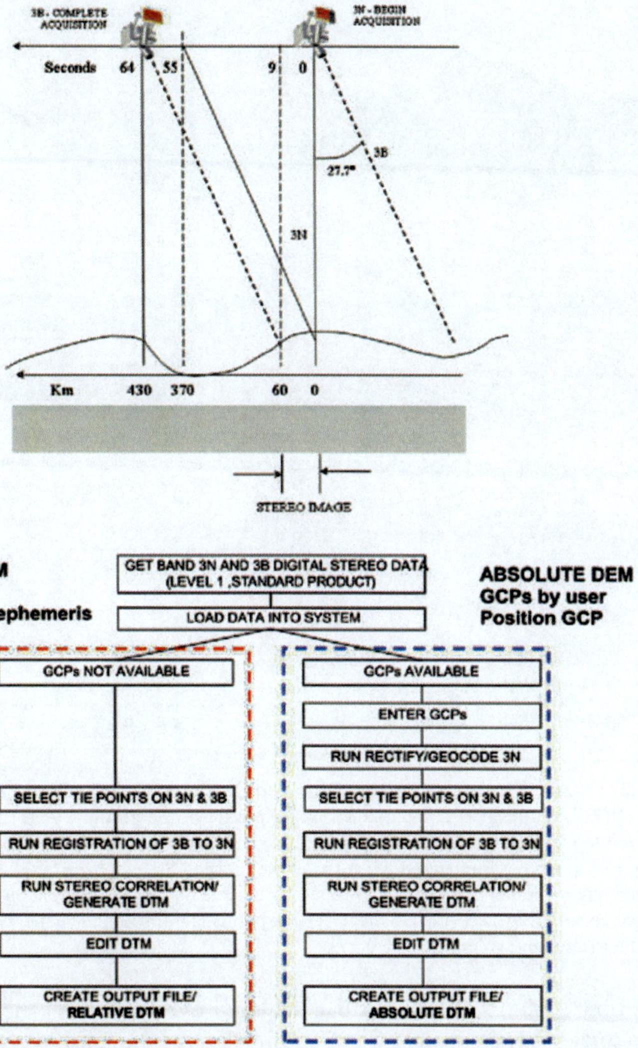

Fig. 5.4 Deriving a Digital Terrain Model from stereo ASTER satellite imagery: configuration of a stereocouple provided by 3N and 3B near infrared bands (*up*) and processing stages (*down*) (From Lang and Welch 1999)

anthropogenic activity (urbanization, transformation of agricultural land, new hydraulics systems, etc.). For this purpose, when it is important to study a site and the landscape in detail, and in a wider span time than that covered by the satellite platforms still operating, an interesting possibility is constituted by the recovery of old satellite images declassified from the military to civilian use.

An example of great interest to archaeologists is given by the images from Corona, the first US operational space photo reconnaissance project operating in

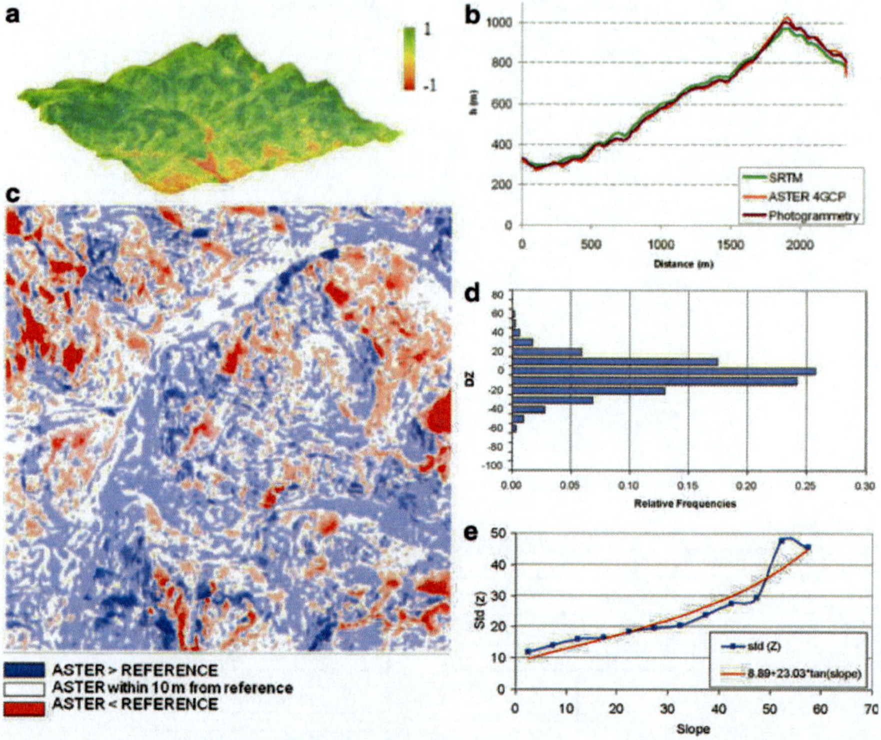

Fig. 5.5 (**a**) ASTER derived DEM, with superimposition of Normalized Difference Vegetation Index (NDVI) values; (**b**) comparison along a profile of the absolute ASTER DEM with SRTM and a reference photogrammetric model; (**c, d**) considering the overall area, 50% of the zone is within ±10 m from reference DEM; (**e**) relation between StDev of the heights and slope values in the study area (the population was grouped by classes of five degrees). Comparable results were obtained considering a set of well chosen check points, confirming the dependency of the precision from the local slope values

the period 1960–1972. Corona had acquired over 860,000 panchromatic images, with a total coverage from 1.5 to 2 million km^2 of the Earth's surface. The images are now provided in digital form by USGS at different scan resolutions; as the original film was of a very large format, each image is provided in four single files with a reduced longitudinal overlap.

These images, characterized by a panoramic geometry, indeed represent an important historical source for change analysis in large regions and for the detection of ancient ruins, roads, irrigation networks, soil mark sites, morphological artefacts, vanished landscapes and associated environmental features dating back to over four decades ago. Unfortunately, the metrical use of these particular kind of imagery is difficult, due to severe distortions affecting the acquisition (Fig. 5.6a) together with the problems of finding technical data related to the missions; their direct use in a GIS environment following conventional procedures is therefore not possible.

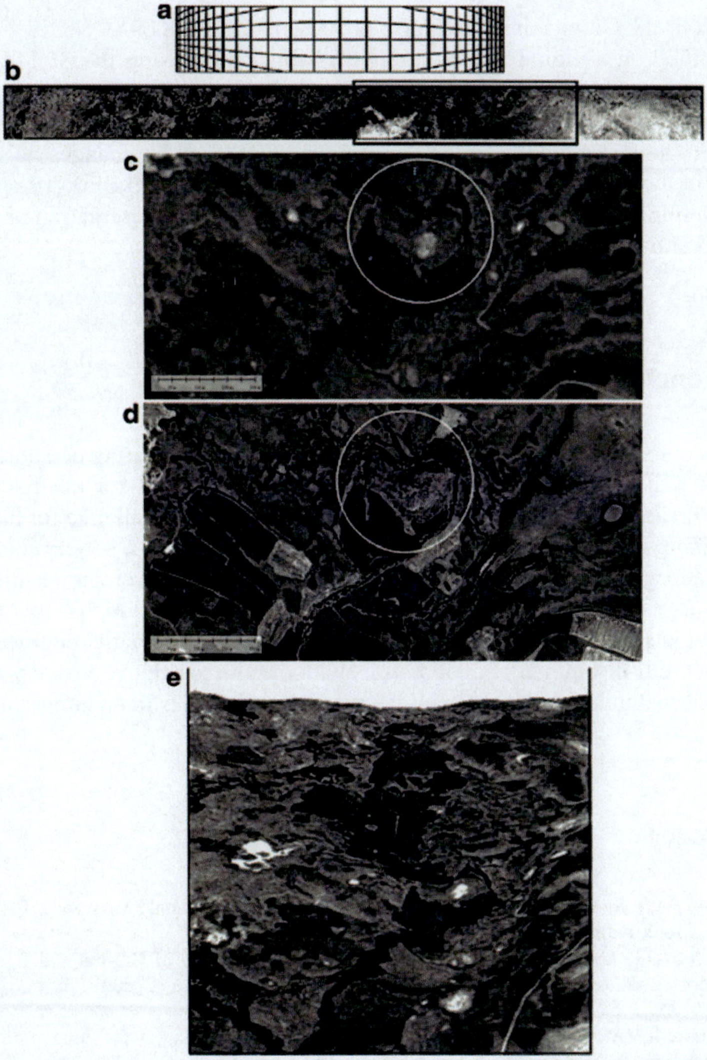

Fig. 5.6 (**a**) digital version of the original film (5.54 × 75.69 cm): evidenced the quarter containing the Tilmen Höyük site. (**b**) The film distortion for a panoramic camera; Comparison between (**c**) Corona (1968) and (**d**) QuickBird (2003) orthoimages for the area surrounding the archaeological site; (**e**) Corona orthoimage superimposed on a terrain model obtained from ASTER stereoscopic imagery

Some experiences were anyway carried out by different authors (e.g. Beck et al. 2007; Galiatsatos et al. 2008) trying to solve this problem. A test was performed in the framework of a multidisciplinary research project carried out by the University of Bologna, in collaboration with Istanbul University and the Gaziantep archaeological museum, at the archaeological site of Tilmen Höyük (south-eastern Turkey).

A March 1968 Corona image (Fig. 5.6b) belonging to KH-4A series has been orthorectified on a digital photogrammetric workstation using the SRTM dataset as terrain model; the results (Bitelli and Girelli 2009) show better accuracy in the area of the archaeological site, where good GCPs are available as obtained by GPS static and fast-static measurements, and clearly highlight the changes occurred in the area in the last 40 years (Fig. 5.6c, comparison with QuickBird orthophoto of 2003). Draping the old image on a DTM can improve the representation of the past territorial situation (Fig. 5.6d).

5.3 Conclusions

The archaeologist is today faced with the promising and fascinating perspective of a spreading use of satellite imagery as a common instrument for documentation, prospection and monitoring of sites and entire regions. The challenge for the future is not principally related to the availability of new sensors with better radiometric, spectral and geometric characteristics, but instead to the development of diffuse skills in data management, data processing and data integration. To this purpose archaeologists can greatly benefit from the contribution of geomatics engineering, in a modern multi-disciplinary collaboration, with the aim to achieve a selection of the best suited methods for operating all the surveying activities in an integrated way.

References

Abrams M, Hook S, Ramachandran B (2002) ASTER users handbook, version 2. EROS Data Center, Sioux Falls

Beck A, Philip G, Abdulkarim M, Donoghue D (2007) Evaluation of Corona and Ikonos high resolution satellite imagery for archaeological prospection in western Syria. Antiquity 81:161–175

Bitelli G, Girelli VA (2009) Metrical use of declassified satellite imagery for an area of archaeological interest in Turkey. J Cult Herit 10S:35–40

Bitelli G, Zanni M (2005) Use of Ikonos high resolution imagery in archaeology: the example of the ancient city of Bakchias in Egypt. In: Oluic M (ed) New strategies for European Remote Sensing. Millpress, Rotterdam, pp 23–30

Bitelli G, Carrara A, Gusella L (2005) Evaluation and use of DEMs derived from ASTER imagery. In: Proceedings workshop Italy – Canada 2005 "3D digital imaging and modeling: applications of heritage industry medicine and land", CD-Rom, Padova, 8 pp

Buyuksalih G, Jacobsen K (2007) Comparison of DEM generation by very high resolution optical satellites. In: Bochenek Z (ed) New developments and challenges in remote sensing. Millpress, Rotterdam, pp 627–637

Campana S, Forte M (2006) In: Proceedings of the 2nd international conference on remote sensing in archaeology: from space to place, British Archaeological Reports, Rome

Cheng P, Chaapel C (2010) WorldView-2 satellite – Pan-sharpening and geometric correction. GeoInformatics 13(4):30–33

Galiatsatos N, Donoghue DNM, Philip G (2008) High resolution elevation data derived from stereoscopic CORONA imagery with minimal ground control: an approach using Ikonos and SRTM data. Photogramm Eng Rem S 74(9):1093–1106

Gusella L, Adams BJ, Bitelli G, Huyck CK, Mognol A (2005) Object oriented image understanding and post-earthquake damage assessment for the 2003 Bam, Iran, Earthquake. Earthq Spectra 21(S1):225–238

Hirano A, Welch R, Lang H (2003) Mapping from ASTER stereo image data: DEM validation and accuracy assessment. ISPRS J Photogramm 57:356–370

Jacobsen K (2005) Analysis of digital elevation models based on space information. In: Oluic M (ed) New strategies for European Remote Sensing. Millpress, Rotterdam, pp 439–451

Lang HR, Welch R (1999) Algorithm theoretical basis document for ASTER digital elevation models, version 3.0. (http://eospso.gsfc.nasa.gov)

Lasaponara R, Masini N (2006) On the potential of Quickbird data for archaeological prospection. Int J Remote Sens 27(15–16):3607–3614

Lasaponara R, Masini N (2008) Advances on remote sensing for archaeology and cultural heritage management. In: Proceedings of the 1st International EARSeL workshop Aracne, Rome

Menze BH, Sherratt AG (2006) Detection of ancient settlement mounds: archaeological survey based on the SRTM terrain model. Photogramm Eng Rem S 72:321–327

Poli D, Remondino F, Dolci C (2004) Use of satellite imagery for DTM extraction, landscape modeling and GIS applications. In: Proceedings of the international workshop on processing and visualization using high-resolution imagery, Pitsanulok, Thailand

Scollar I, Tabbagh A, Hesse A, Herzog I (2009) Archaeological prospecting and remote sensing. Cambridge University Press, Cambridge

Toutin T (2001) Geometric processing of Ikonos Geo images with DEM. In: ISPRS joint workshop "High resolution mapping from space", Hanover, Germany

Vittuari L (2008) Precise global georeferencing of sites and geodetic techniques for morphological surveys within a common reference frame. In: ARCHAIA. Guidelines on research planning, characterisation, conservation and management in archaeological sites, British Archaeological Reports International Series, BAR S1877, 59–65, Archaeopress, Oxford

Wiseman J, El-Baz F (2008) Remote sensing in archaeology. Springer, New York

Chapter 6
Integrated Methodologies for the Archaeological Map of an Ancient City and Its Territory: The Case of Hierapolis in Phrygia

Giuseppe Scardozzi

Abstract The chapter concerns the research activity conducted by the Institute for Archaeological and Monumental Heritage of the Italian National Research Council in the ancient city and in the territory of Hierapolis in Phrygia (Pamukkale, south-west Turkey), in cooperation with the Italian Archaeological Mission. The research, conducted during the 2003–2010 campaigns, aimed to the reconstruction of the layout of the city and to the study of the ancient topography of its territory. It is based on the integration of different non-destructive study methods and technologies: systematic archaeological and topographical surveys, geophysical prospecting and processing, analysis and interpretation of aerial photos and multi-temporal high resolution satellite images (in particular, space photos of 1960s and 1970s taken by USA reconnaissance satellites; panchromatic and multispectral images taken by QuickBird-2 and Ikonos-2 between 2002 and 2009). During the research, satellite images were not only used to replace and integrate the aerial photographs in the finding and location of traces of archaeological and paleo-environmental elements, but they were accurately corrected in their geometric distortions and employed to integrate available cartography, as base-maps for the field works and in the GIS and webGIS of Hierapolis and its territory.

Keywords Declassified satellite images • Ancient topography • Multitemporal • Very high resolution satellite images • Hierapolis • Archaeological map

G. Scardozzi (✉)
Institute of Archaeological and Monumental Heritage, CNR-IBAM,
Prov.le Lecce-Monteroni, 73100 Lecce, Italy
e-mail: g.scardozzi@ibam.cnr.it

R. Lasaponara and N. Masini (eds.), *Satellite Remote Sensing: A New Tool for Archaeology*, Remote Sensing and Digital Image Processing 16,
DOI 10.1007/978-90-481-8801-7_6, © Springer Science+Business Media B.V. 2012

6.1 Introduction

The research in ancient topography offers an important contribution to the expansion of knowledge of archaeological and monumental heritage because it is aimed at (i) the study of single settlements and monumental archaeological complexes, (ii) the location of new archaeological evidences, (iii) the reconstruction of the ancient settlement system of a specific context and (iv) the analysis of its historic development.

The archaeological maps derived from this research, in digital format and inserted in specific GIS, are not only a tool in historic and archaeological research, but are also fundamental for the agencies that overlook the protection, management and valorisation of cultural heritage, and its fruition within correct urban and territorial planning policies.

Today, traditional methods, such as aerial photography (historic and recent, vertical and oblique), direct archaeological surveys, literary and epigraphic sources, archive documents, historic cartographies, have been enriched by satellite imagery whose use is quickly spreading, especially due the availability of images with high spatial resolutions, nowadays between 1 and 0.5 m. Specifically, we refer to the images taken by optic sensors which are particularly apt to be used in archaeology, not only for the level of detail reached, but also for their spectral and radiometric characteristics and for the temporal resolution. Radar imagery has by now reached the same geometric resolutions and is becoming an effective investigative tool for those applications that require elevated detailed DEMs.

Generally satellite images are a complementary documentation, though not alternative, to remote-sensed aerial images, particularly useful when other tools of archaeological research (for example, base-maps) are missing, as in the case of archaeological research carried out in Turkey for several years by the Institute for Archaeological and Monumental Heritage of the Italian National Research Council in collaboration with the Salento University.

6.2 The Hierapolis of Phrygia Survey Project

Since 2003, the CNR-IBAM has conducted archaeological research in the ancient city and in the territory of Hierapolis in Phrygia (Pamukkale, south-west Turkey), in cooperation with the Italian Archaeological Mission (directed by Prof. Francesco D'Andria), finalized with the reconstruction of the layout of the city and the study of the ancient topography of its territory (Fig. 6.1). The investigations are based on the integration of different non-destructive methods and technologies: systematic archaeological and topographical surveys, geophysical prospecting and processing, analysis and interpretation of multi-temporal high resolution satellite images and oblique aerial photos (Scardozzi 2007a, b, 2008a; Castrianni et al. 2008a).

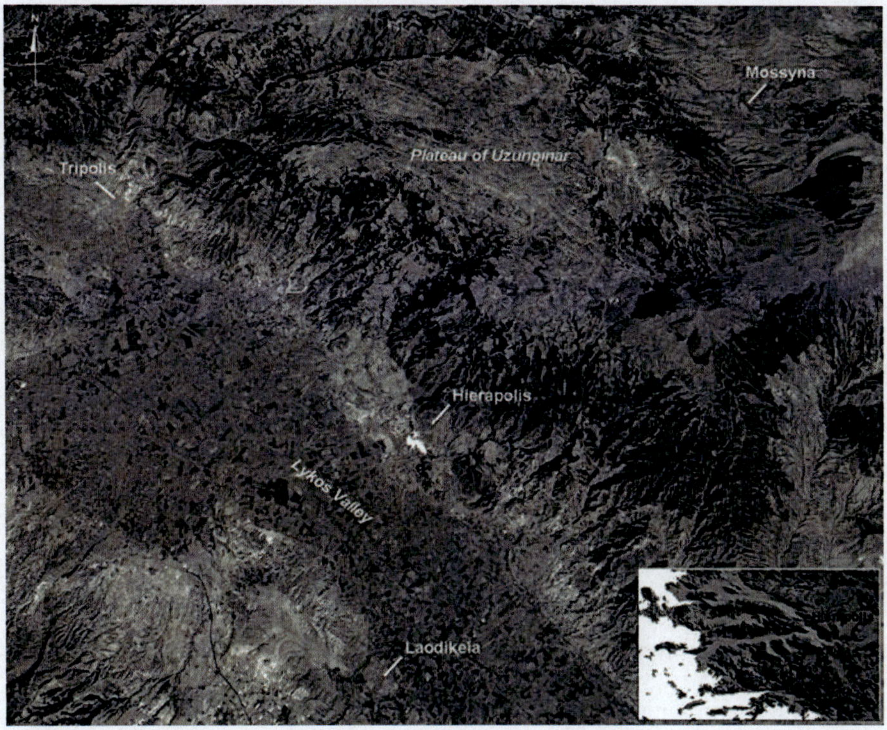

Fig. 6.1 The territory of Hierapolis in Phrygia in an SRTM DEM on which a Landsat ETM + image was georeferenced

Satellite images have complemented the limited availability of vertical aerial photos and up-to-date maps on an adequate scale for field work. Only for Hierapolis, oblique aerial photos taken during different surveys were available, together with a 1:1,000 scale map recently completed by the Faculty of Architecture of the Politechnic of Turin (D'Andria et al. 2008; Scardozzi and Spanò 2008).

Therefore, some satellite images acquired by QuickBird-2 and Ikonos-2 in different years (from 2002 to 2009), were processed and used during field work to find traces linked to buried archaeological structures. Thanks to their high resolution in ground (re-sampled to 0.60 and 1 m in panchromatic mode, and 2.40 and 4 m in multispectral mode), radiometric (11 bit) and spectral domain, they are particularly suitable in detailed archaeological surveys. For these reasons, they have been used in each phase of the research: from the activities on the field to data elaboration and management, up to the presentation of the results, providing for the first time planimetric synchronic images of all the city and large sectors of its territory. These images have thus been used in place of aerial photography for the discovery of new archaeological features, including those of small dimensions, and for their contextualization in relation to those already known.

Satellite images represented a fundamental support for research carried out on the field, supplying large amounts of data and indications for archaeological excavations and survey activities. Thus allowing detailed analysis of single sectors and monumental complexes of the urban area, the necropolises and the surrounding territory. The systematic archaeological surveys, together with the constant examination of satellite and aerial imagery and ground verification of the traces, have thus allowed us to recover a considerable amount of data concerning the urban fabric of Hierapolis.

The geo-referencing of available satellite images, coupled with ground measurements with differential GPS, allowed systematic investigations not previously carried out, along with the vectorization of all the remains and traces related to the site. During the research, satellite images were also employed to map evidence recorded in different years and integrated into the archaeological map of Hierapolis in a GIS environment. Maps of some parts of the territory were produced, orthorectifying QuickBird satellite images with the use of available DTMs, DEMs, and DSMs, of Ground Control Points and Check Points surveyed with differential GPS systems and of Rational Polynomial Coefficients.

For the urban area, in which numerous control points are present, it was also possible to use the 2005 QuickBird ortho-image for the up-date of modern topography, for the reconstruction of the urban layout and as a base of several sheets in a 1:1,000 scale of the archaeological map of the city. In the Hierapolis territory, instead, different solutions were adopted and different degrees of precision were obtained, according to the possibility of orthorectifying the remote sensed images only through terrain models and RPCs or also measuring GCPs and CPs.

Regarding two areas with particular concentrations of archaeological features (the immediate suburb of Hierapolis and an area located around 2 km south of the city), we are currently extracting cartographic themes from clippings of an Ikonos stereo-pair (see above), orthorectified through 3D ground models elaborated on the basis of SRTM and ASTER data and thanks to the use of GCPs surveyed on the field, in order to produce topographic maps aimed for archaeological research in a 1:10,000 scale.

On these digital maps, organized in informative layers, all the archaeological evidence located on the ground during surveys is georeferenced; within each single layer, the different entities, both cartographic and archaeological, are filed through unique codes. Particular attention was then taken in the rendering of the altimetric data, obtained through comparison of the various different methods of generation of the contour lines: vectorization of the contour lines with a 10 m offset of the Turkish maps in a 1:25,000 scale; extraction of the lines with a 1 and 5 m offset from DSMs and DEMs obtained from Ikonos and ASTER stereo-pairs and from SRTM data. Furthermore, in the urban area the archaeological and topographical surveys and the applications with aerial photographs and high-resolution satellite images have also been integrated with geophysical surveys (geo-magnetic and Ground Penetrating Radar), carried out in 2007–2008 and 2010 in some parts of the urban area of Hierapolis (Fig. 6.2), characterized by a deep layer of alluvial and colluvial sediments or by thick limestone deposits (up to 4 m high) that cover the ancient remains in the western sector of the city.

Fig. 6.2 Different methods of surface exploration integrated with archaeological surveys and satellite remote sensing: *above*, geophysical prospecting (Magnetometry and Ground Penetrating Radar); *down*, balloon for low altitude aerial photography

All the data gathered from these diverse and integrated methods of systematic surface investigation have been inserted in the digital cartography of Hierapolis, where the whole urban area and the necropolises are represented, with the structures brought to light and surveyed during the 50 years of research of the Italian Archaeological Mission.

6.3 Multitemporal Satellite Data for Archaeological Research

In the Hierapolis of Phrygia Survey Project, satellite images are utilized for the identification and the spatial characterization of several archaeological features (Fig. 6.3) (D'Andria et al. 2008; Scardozzi 2008a). Images have also allowed the

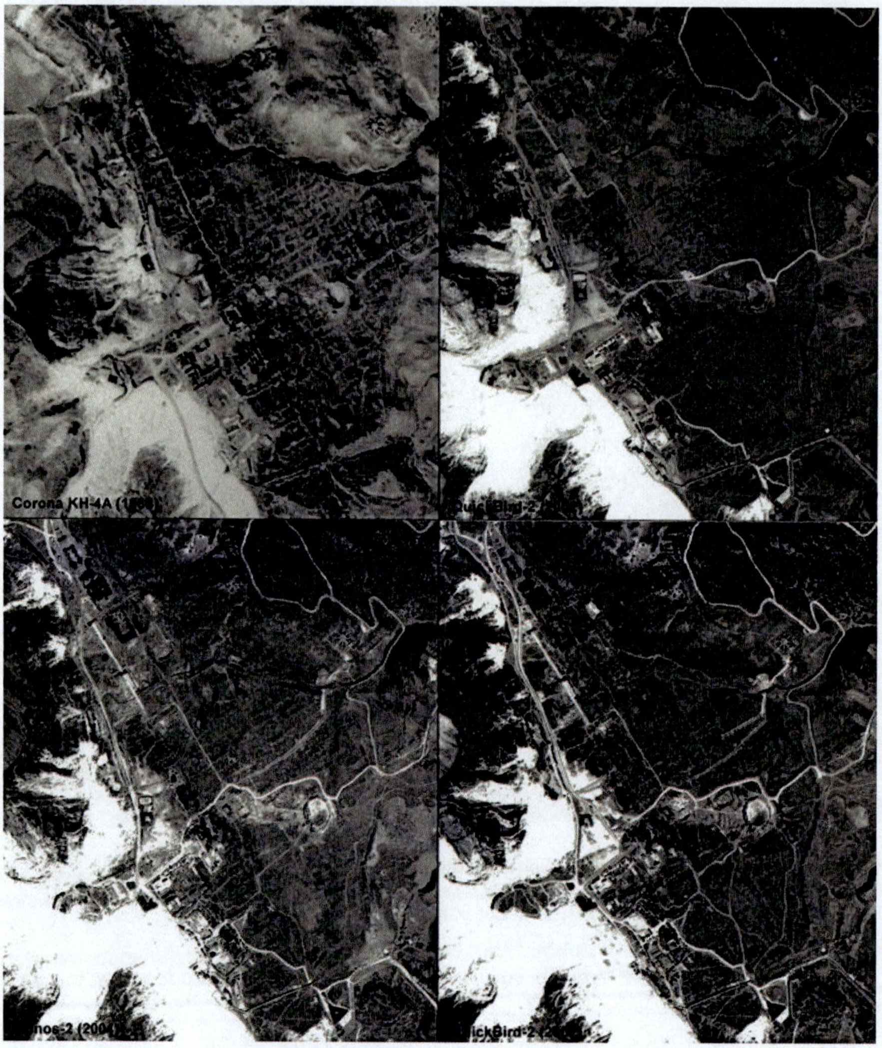

Fig. 6.3 Hierapolis in Phrygia: example of satellite multi-temporal documentation

detection of paleo-environmental elements and contributed to the reconstruction of the ancient landscape. Regarding excavated or partially emerging monuments or archaeological sites, images have been used in order to document and contextualize them.

In some cases, the processing of multispectral satellite images made it possible to better locate archaeological and paleo-environmental features (Lasaponara et al. 2008a, b, 2010). Some elaborations and data fusion between panchromatic and multispectral data were tested, in order to highlight different archaeological anomalies and traces and to recognize elements that can provide a reconstruction of the ancient landscape. So, different enhancement techniques were applied and multispectral elaborations of the satellite images were used, like the colour composite RGB of the original bands (3-2-1, 4-3-2, 4-3-1, 4-2-1) and the calculation of spectral indexes. They allowed us to show the presence, the kind and the status of the vegetation and its different growth connected with the buried evidence: in particular the Normalized Difference Vegetation Index (NDVI) and the Tasselled Cap Transformation (TCT) have been used. Moreover, a "fusion" between the panchromatic datum and the multispectral information has been made, in order to integrate the high spatial resolution of the first kind of image and the spectral information acquired in the various bands with a lower spatial resolution. To this aim some techniques of data fusion have been tested, like HSV Sharpening, Colour Normalized (Brovey) Sharpening, Principal Components Spectral Sharpening, getting new pan-sharpened images in real colours or false colour, with a high spatial resolution (0.70 m). The Gram-Schmidt Spectral Sharpening and Zhang's algorithm have emerged as the two mostly accurate methods with a much higher relation between high spectral quality and maintenance of the high spatial resolution.

In particular, considering the different kinds of information that can be obtained by the various bands and the geological characteristics of the territory of Hierapolis, panchromatic images are better for the detection of damp-marks, while red and infra-red channels are better for the detection of crop and soil-marks (Lasaponara and Masini 2007; Masini and Lasaponara 2007).

During field work, a constant verification of the geo-referenced traces and anomalies found in all the satellite images was carried out, in order to specify their real pertinence to archaeological elements, the interpretation and, if possible, also the chronology, avoiding misunderstandings and mistakes. During the survey, precise topographical measurements with the use of Total Station and differential GPS (in static, RTK and cinematic modes) allowed us to record all the remains preserved in situ and visible on the surface. Using differential GPS systems, very high resolution Digital Elevation Models of limited areas of the city were also realized. Results obtained by the comparison and integration of collected data with multitemporal aerial photos (particularly oblique from the helicopter and an aerostatic balloon: Fig. 6.2) were very important, because they were useful for the documentation of the archaeological areas. They also helped us to understand some topographical situations and the articulation of monumental complexes. During the last years, in particular, aerial photos surveys of the ancient city with a balloon are replicated. They are finalized to the documentation of (i) the excavation areas and (ii) all the sectors of Hierapolis investigated by systematic archaeological intra-site surveys (Figs. 6.4 and 6.5).

In the QuickBird satellite images used for the research project, a poor visibility of the archaeological traces highlighted by the micro-relief (shadow sites) was noticed. Even if we used images acquired in the morning, with the sun still low on the horizon (so that, the micro-relief should be highlighted by long shadows), it is not possible to appreciate the slightest altimetric variations. Initially, in order to avoid this kind of difficulty, some anaglyphs were processed experimentally;

Fig. 6.4 QuickBird-2 image in infra-red false colours (RGB 4-3-1) and aerial photo from a balloon: numerous remains and traces of the regular urban layout of Hierapolis in the northern sector of the city are visible

in fact, they were obtained by duplicating the same satellite image, with an optical artifice that allows a quick perception of the third dimension, without achieving a real stereoscopy. Moreover, digital models of the ground, with a low-middle resolution, have been processed, based on both radar and optic satellite data: specifically, DEMs are based on SRTM data and DSMs processed from a

Fig. 6.5 QuickBird-2 panchromatic image and aerial photo from a balloon: numerous remains and traces of the regular urban layout of Hierapolis in the southern sector of the city are visible

stereo-pair taken by Terra ASTER sensor in 2004 (Fig. 6.6). On these digital terrain models, both the panchromatic and the pan-sharpened images have been georeferenced, in order to associate archaeological traces to the morphology of the terrain. In this way, it was possible to study the territory in the third dimension.

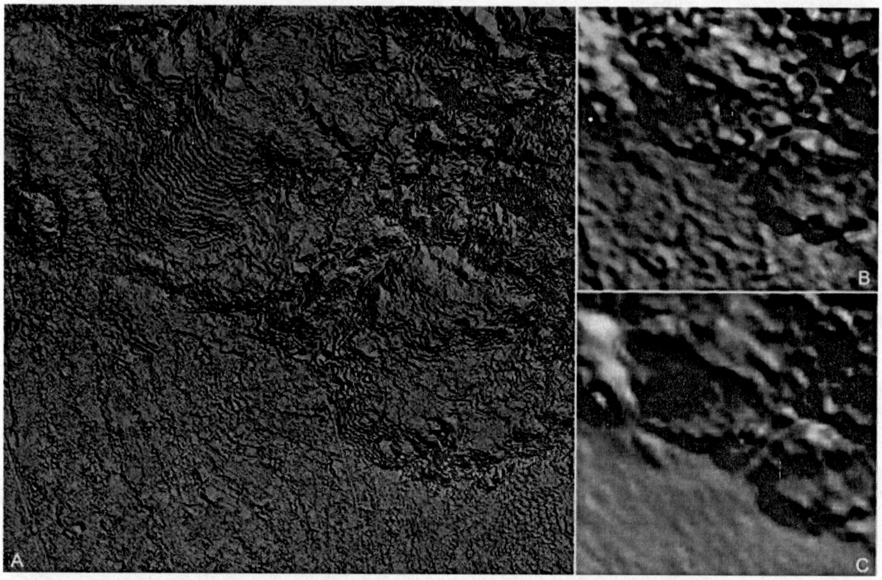

Fig. 6.6 Comparison between a DSM extracted from an Ikonos-2 stereo-pair (**a**), a DSM extracted from an ASTER stereo-pair (**b**) and a DEM based on SRTM data (**c**)

All the data were also integrated with the multitemporal QuickBird images from Google Earth, that covered all the study area in plan and 3D view. A stereo-pair taken by the satellite Ikonos in 2004 has also been used; it regards the urban area and a sector of the surrounding territory. They have been processed to get stereoscopic epipolar panchromatic images, from which, using automatic techniques of auto-correlation, typical of digital photogrammetry, a high resolution DSM was extracted. It was overlayed with the panchromatic satellite images and with real colour and false infra-red colour pan-sharpened images. Moreover, it was possible to display an anaglyph of the epipolar images that provided a real stereoscopy view in which the micro-relief seems to be strongly exalted and measured. The 3D models of the ground are particularly useful, not only for the study of the territory, but also for the presentation of research results both with static images and flight simulations.

For a contextualization of Hierapolis in its territory and for a general study of the latter, satellite images were employed, with medium-low spatial resolution, i.e. the above-mentioned ASTER stereo-pair acquired in the near infrared, with a 15 m ground resolution, SPOT-1 images from 1986, with a 10 m maximum geometric resolution in the panchromatic, and those acquired by Landsat 4 and 5 TM and Landsat 7 ETM + satellites between 1990 and 2003 (GeoCover 1990 and 2000), available in NASA World Wind; they are pan-sharpened images with 30 and 15 m of geometric resolution, in different RGB visualizations (bands 7-4-2, 5-4-2 and 3-2-1).

Apart from recovering a multi-resolution documentation (medium-low, high and very high), the availability of multi-temporal images enabled the processing and interpretation of "historic" changes. This was particularly important for the study of territories strongly affected by transformations which destroyed archaeological evidence in recent decades, as evident by the declassified USA reconnaissance space photos taken in the 1960s and 1970s. So, our data set was enriched with images from Corona KH-2 and KH-3 of 1961, from Corona KH-4A of 1968, and from Hexagon KH-9 of 1973, 1976, 1979 and 1980, with a geometric resolution between 2.74 and 9 m (Scardozzi 2008b). Although often characterized by geometric distortions, this type of satellite image represents an important documentation especially when historic aerial photos are rare or missing.

The analysis and interpretation of these historical images, properly geo-referenced, were very useful because they provided a valuable data source for the reconstruction of the urban layout and of the ancient landscape. In fact, they show Hierapolis and the surrounding territory before a few important transformations of the last decades, such as the construction of buildings (for example, six hotels inside the ancient city) that have destroyed ancient remains, the creation of infrastructures, the expansion of modern villages, and the diffusion of mechanized agriculture. Traces and anomalies found in older satellite images were constantly confronted with the modern situation documented by the recent ones of QuickBird (of the years 2002, 2003, 2005, 2006, 2007 and 2009) and Ikonos (2004). For example, in the Corona KH-4B photos of 1968 several traces of the ancient roads and water channel networks are clearly visible, more than today, especially in the south-western area of the city.

The same historical images also show clearly the situation before the construction of modern structures at the tip of the North Necropolis, and in the area between the *Frontinus* South Gate and the South Necropolis: the traces and remains of the main roads and water channels northward and southward are visible. Furthermore, the landscape around the ancient city has been transformed in the last 40 years: for example, in the Corona images the white calcareous formations covering the slope that from the terrace of Hierapolis descends towards the *Lykos* valley and the village of Pamukkale appear less extensive than today. Moreover, some roads and water channels for the irrigation, that climbed along this slope and now are obliterated, are clearly visible. Moreover, the hills east of the ancient city in the 1960s were still not covered by coniferous woods and in the images some ancient quarries of travertine are visible, along with the remains of the old road that from Hierapolis climbed to the large plateau north of the city.

The USA declassified images also provided an important contribution for research in the territory of the city. Particularly, satellite images Corona and Hexagon have supplied important data about ancient orthogonal land divisions present in the northern part of the plateau north of Hierapolis. Thus showing ancient limits better preserved than today, when some of them were destroyed because of new buildings and agricultural works. The images of the 1960s and 1970s also show the traces and the remains of the main ancient roads leading out of the city, some ancient quarries of travertine and alabaster north-west of Hierapolis, now partially destroyed by the revival of extractive activities, and many other archaeological presences are documented in a situation better preserved than nowadays.

6.4 High Resolution Satellite Images for the Production of Base-Maps and Cartographies for Archaeological Research

During the research in the territory of Hierapolis, satellite images replaced aerial photos and also integrated cartographies used during surveys and in the GIS and webGIS environments (Scardozzi 2009; Castrianni et al. 2010a). For the urban area and the surrounding necropolises a digital cartography on a large scale (1:1,000) is available (elaborated in cooperation with the Architecture Faculty of the Polytechnic Institute in Turin) whereas for the rest of the territory (that in ancient times was under the control of Hierapolis) only raster Turkish maps on a scale 1:25,000 (Harita Genel Komutanlığı – General Command of Mapping of Turkey), drawn in the 1970s and updated in the 1990s, are available. These maps are poorly detailed, therefore scarcely useful particularly during archaeological and topographical surveys. Consequently, space-maps of some sectors of the territory have been produced, with tolerances of cartographies in a scale between 1:10,000 and 1:5,000. GCPs and CPs were collected through a differential GPS to obtain a high degree of accuracy, necessary for the elimination of distortions typical of satellite images and to "link" them to the ground by means of real coordinates. To obtain good ortho-rectifications it was also necessary to take into consideration the position of satellites during acquisition in relation to Earth. These parameters, commonly named Rational Polynomial Coefficients (RPCs), are released in association with each image as metadata.

We carried out the geometric correction of satellite images by means of (i) the GCPs and CPs, (ii) the RPCs, (iii) the DTMs obtained through vectorization of contour lines photoprojected in the cartography of the General Command of Mapping of Turkey, (iv) the DEMs based on SRTM data and (v) the DSMs extracted from a stereo-pair acquired in NIR by Terra satellite with ASTER sensor in 2004. From the satellite ortho-images were processed space-maps to be used during archaeological surveys and for the positioning of archaeological evidence in the GIS of Hierapolis. For small sectors of the territory of the city, cartographies specifically aimed at archaeological research were also obtained from satellite images.

During the last two decades the level of specialization in the development of cartographies for archaeological purposes, aimed at compensating for the technical incompatibilities caused by the different uses for which commercial maps were made, has progressively increased (Piccarreta and Ceraudo 2000; Piccarreta 2003). In their most frequent use, in fact, archaeological maps are assembled by importing detailed archaeological elements, almost always resulting from direct observation and survey, on a generic topographic base, which usually lacks upgrades and is often inadequate in scale. The result is a composite product that shows deep deficiencies since the archaeological data is inserted in an often too approximate topographic context. Consequently the necessity for a specifically aimed photogrammetry emerged, in which the possibility is given to choose both the scale to be adopted for the represented objects and which elements need to be emphasized or require more accuracy in the details. For example, one of the most neglected aspects in commercial cartography is the rendering of altimetry, only represented as a tendency; instead, a correct rendering of the contour lines is fundamental for the reading of the topographic context and for the graphic representation of the archaeological elements that must be, as much as possible, real and never symbolic. When the aerophotographic base allows it, that is when the scale of the available stereoscopic images is large enough, the cartography obtained can boast remarkable detail: this means topographic maps 1:1,000 and 1:2,000 in scale, in which the contour lines can be plotted with an interval up to 0.50 m.

It is evident how the cartographies aimed to archaeological research will result more easily achievable if they are produced by or in presence of an archaeologist, who has direct knowledge of the requirements for representation and is familiar with photographic interpretation of landscape morphology and of traces referable to buried archaeological remains. Each represented element in the cartography will have its own alphanumeric and colour code according to the information layer to which it belongs: Planimetry, Hydrology, Geomorphology, Vegetation, Administration Limits and Toponomy, Reference Points and Topographic Landmarks, Archaeology. The latter layer, on the basis of encoding layers developed in the Ancient Topography and Photogrammetry Laboratory of the University of Salento, is further divided in nine functional areas that allow a precise identification of all represented objects.

This branch of research can be connected, conceptually and methodologically, to that of the possibilities of cartography production and update offered by high resolution satellites ortho-images, particularly in study contexts, as the territory of Hierapolis, where there is no adequate cartography available and it is difficult or impossible to retrieve stereoscopic aerophotographic coverage (Holland and Marshall 2003; Eisenbeiss et al. 2004; Gianinetto and Scaioni 2004; Toutin 2004; Jacobsen 2006).

In the Hierapolis of Phrygia Survey Project this methodology was applied to limited sectors of the territory located about 2 km south-east of the ancient city, near the modern villages of Yeniköy and Küçükdereköy, characterized by a concentration of archaeological evidence dating from the Hellenistic period to the Ottoman age (Fig. 6.7).

The systematic archeological surveys have relieved the presence of a lot of architectural materials in white marble and travertine re-used in the village of Küçükdereköy; they pertain to a big village of the Hellenistic and Roman ages (that seems to life also in Byzantine times), on which now lies the modern settlement. The ancient settlement, perhaps under the control of Hierapolis, is situated to the slopes of the hills that climb steeply toward the Küçükçokelz mountain, in correspondence of a natural road of penetration (the close valley of the "Küçük Dere" stream) that allowed to goes up from the *Lykos* valley to the plateau that threshold to the north the valley itself (the plateau of Uzunpınar: see below). On the hills immediately to the south-east and above all to west, in direction of Yeniköy, are situated the necropolises of the ancient village, dated to the

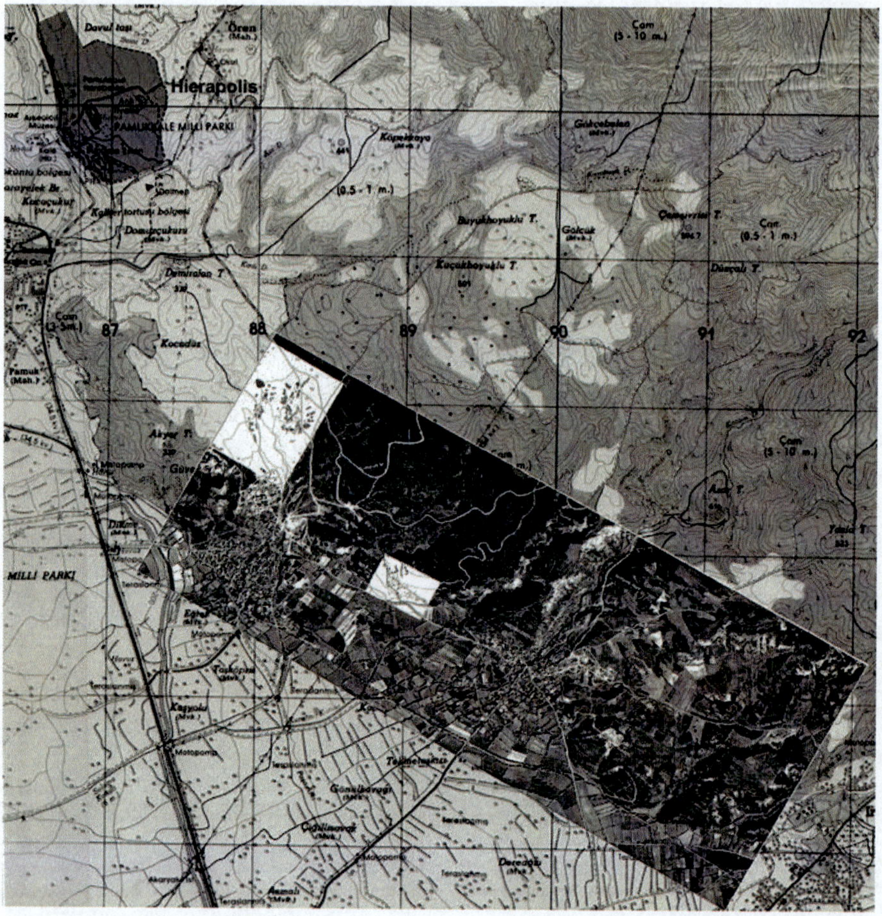

Fig. 6.7 The territory south-east of Hierapolis in the Turkish map (scale 1:25,000): a space-map of the area between Yeniköy and Küçükdereköy and two topographic maps of areas with particular concentrations of archaeological evidence (with extraction of cartographic and archaeological themes from satellite ortho-images) were realized

Hellenistic and Imperial periods, the main of which is c. 500 m north-west of Küçükdereköy, along a slope southwards (Fig. 6.8). It appears widely disturbed by digs of looters; there are c. 130 tombs, about the 80% of which are graves pit, excavated in the rocky bench. The remaining tombs are travertine sarcophagi placed on the rocky bench or on simplex bases in calcareous blocks; in some cases they are placed upon more elegant bases (with a square plan and moulded seat) or on *hyposoria*. The sarcophagi have elementary coffins (sometimes not completely detached from the rock bank) provided with a lower moulding, like type II of the North Necropolis of Hierapolis (Vanhaverbeke and Waelkens 2002). The main number of lids are very simplex: there are examples with or without

Fig. 6.8 Topographic map and images of the necropolis north-west of Küçükdereköy

corner *acroteria* and moulded edges, like types I-III of the North Necropolis of Hierapolis. Also a *tumulus*, obviously referable to an aristocratic family outstanding in the settlement, is present in the central area of the necropolis: it was surrounded by cylindrical drum formed of squared blocks, and a *dromos* gives access to the quadrangular funerary chamber. The central position taken by the *tumulus* of the necropolis hints at a centralizing function exerted by the funerary monument on the tombs immediately surrounding it, that show an orientation similar to that of the funerary chamber of the *tumulus*.

Generally the tombs are characterized by the presence at once of more orientations, but a coherent layout can be discerned within the single groups of two or three burials located in different parts of the necropolis, possibly referable to family groups.

The tombs orientation is conditioned on the one side by the ground's orography, while on the other by viability. A main road has in fact been located, with a north-east/south-west direction, in the lower part of the slope, that must have connected the ancient settlement at Küçükdereköy to Hierapolis; this route is connected a service road, east-west oriented, that runs across the central and eastern parts of the necropolis. Several quarry fronts found in the area immediately north of the necropolis confirm the local origin of the material in which the sarcophagi and funerary monuments were made, together with the nearby settlement's structures. To this regard, it is interesting to observe the presence of unfinished sarcophagi, with corner *actroteria* only roughly outlined, that attest for the condition in which these were transported from the nearby quarries, without the necessary finishing work, carried out directly on site.

On the hill immediately to the north of the modern village of Yeniköy (in Turkish "new village"), the ruins of an old Turkish settlement ("Eskiköy", in Turkish) were also found, almost completely destroyed perhaps from one of the many earthquakes that have struck the area, such as in the seventeenth, eighteenth or nineteenth century (Ambraseys and Finkel 1995; Altunel 2000). The ruins of the houses (in which architectural ancient elements were also reemployed) cover a surface of about 25 ha, and they are poorly perceivable on the ground, where are scarcely remains of stonewalls and traces of micro-relief linked to buried structures; ruins and traces were well visible in the satellite images (QuickBird-2 e Ikonos-2), that allowed the vectorial restitution of many features. Over 400 structures of different size, square or rectangular in plan, were individuated; they are isolated or grouped and also some ancient roads through the buildings were found. Near the old Turkish village, some enclosures and a cemetery with graves pit are also present.

For the archaeological research in the area between Yeniköy and Küçükdereköy, only a Turkish map 1:25,000 in scale and updated in 1990s was available, absolutely inappropriate to positioning of the single archaeological evidences surveyed; furthermore, the aerophotographic coverage of the area is not available. For this work the stereo-pair captured in 2004 by Ikonos-2 was used (Di Giacomo et al. 2008a). The satellite panchromatic image with a smaller off-nadir angle has been ortho-rectified by a DEM and a DSM, elaborated respectively on SRTM data and an ASTER stereo-pair, and by GCPs ant CPs collected during field work with a differential GPS system. The Ikonos-2 stereo-pair was also processed to create a high resolution DSM; nevertheless, in this phase we experimented on the limit of the algorithms in the shaded areas, in which the 3D surface model resulted very disturbed and not very useful for the reconstruction of the terrain's morphology, requiring many interpolations. The equivalent points identified in both the stereoscopic images also represented the starting point for their processing to obtain an anaglyph, very useful to highlight the archaeological data from the analysis of micro-relief.

From the Ikonos-2 ortho-image was produced a space-map of the entire area of Yeniköy and Küçükdereköy in the tolerance of a 1:10,000 scale. For the area of the old Turkish village and of the necropolis described, from the same ortho-image many thematisms were also extracted with the aim to create new topographical maps: modern urbanization and topography, hydrology, archaeological features and traces, etc. During the restitution and vectorization phases, the most feasible, correct and detailed way to represent morphology of the territory has been researched, so that information, in particular regarding altimetry, would be legible. Different methods have been compared for the creation of the contour lines: the vectorization of that ones with 10 m interval of the Turkish cartography in a 1:25,000 scale, and the extraction of contour lines with 1 and 5 m interval from the high resolution DSM extracted from the Ikonos-2 stereo-pair, and from DEM and DSM based respectively on SRTM and ASTER data. The contour lines deriving from the Turkish cartography are obviously the only ones resulting from real measurements (aero-photogrammetric restitution), while the others derive from an automatic process of extraction, though they come close to the real aspect of the studied territory.

The topographical maps obtained (in the tolerance of a 1:10,000 scale) are in accordance with the standard themes of cartographic production: they are set in layers and the different homogeneous elements of the map follow a logical affinity. On these maps, all the archaeological remains detected on the ground during archaeological and topographical surveys have been geo-referenced using a high precision GPS; moreover, numerous archaeological traces were detected and geo-referenced in the new maps also thanks to the analysis of a QuickBird-2 image of the study area taken in 2005. The data acquired through satellite images processing and analysis, ever having been validated by ground survey, are systematically inserted in the GIS of Hierapolis, contributing to the enrichment of the archaeological map of the city's territory.

During research in the territory of Hierapolis, these new topographical maps aimed at archaeological research, QuickBird-2 and Ikonos-2 ortho-images, and space-maps (within the Turkish cartography available) were also used during field work in a system, called *Ulixes*, developed to provide support to archaeological survey, speeding the positioning and documentation of archaeological presences. The system allows to navigate on cartographies (vector and raster), vertical aerial photos and high resolution satellite images visualized on a Tablet PC linked to a GPS receiver with metric precision (Di Giacomo et al. 2008b). In *Ulixes*, the images and cartographies are transferred to a software written specifically for the system and reduced to an indexed mosaic of 500 m^2 images with known coordinates; the system is capable of locating the position of the user and recalling in real time the portion of the image that is being scanned. In the occurrence of an archaeological find, the operator can associate the location of the ancient evidence to a record designed especially for *Ulixes*, equipped with extremely flexible vocabularies, that can be updated and implemented directly from the field. The system allows not only to quicken and optimise recording and documentation of archaeological evidence during the survey, but also a more effective management of the data collected which, after verification and validation, update in real time the Hierapolis GIS. The navigation on high resolution satellite images furthermore allow a quick and accurate confirmation on the field of traces and anomalies visible (or to easily relocate them) and the monitoring of the ancient evidence previously located.

Finally, satellite images were also very useful for the disclosure of data in the webGIS for the monuments and necropolises of Hierapolis in Phrygia (http://antares.ibam.cnr.it/atlante-hierapolis; Castrianni et al. 2008b); the organisation in layers enables navigation on the 2005 QuickBird satellite ortho-image (suitably "lightened" for web exploration) and – should the user require it – on the archaeological map superimposed on the image. This satellite image, which provides a fairly up-to-date and detailed overhead view of the city and its surroundings, may be replaced in future with a more recent image or accompanied by other layers, such as further satellite images acquired in different years and/or by different platforms. The layer with the archaeological information is linked to a database that allows the users to make simple or complex queries, integrating spatial and chronological data. The final users is thus provided with cartographic data, the satellite image and historic and archaeological data, for a highly dynamic approach to the knowledge of the ancient city.

6.5 Geophysical Prospecting in the Hierapolis of Phrygia Survey Project

In order to gather further knowledge about the less well-known sectors of the urban area, archaeological surveys and remote sensing data have been integrated with geo-magnetic and georadar surveys carried out by S. Giese and C. Hübner of the GGH (Giese, Grubert & Hübner Gbr.) of Freiburg in 2007–2008, and by N. Masini and R. Persico of CNR-IBAM in 2010 (D'Andria et al. 2008; Scardozzi 2008a, 2010a). The integration of different remote sensing technologies, was particularly necessary in some areas of the city, where colluvial and alluvial sediments and a thick layer of limestone, which covers the western side of Hierapolis from the Medieval age (generated by the calcareous water from the thermal springs along the seismic fault that run into the ancient urban area), prevented the observation of the articulation of the urban layout.

Geo-physical surveys tested in various points with different geo-morphological characteristics allowed us to identify and locate traces of features invisible on the surface, such as paved streets, squares and remains of buildings. They therefore contributed to the reconstruction of the original plan of the city and were merged in the numeric cartography of Hierapolis, enriching the heterogeneous archaeological thematic data. In particular, we worked in several areas of the city, for an overall surface of 6 ha investigated with geo-magnetic surveys and one of 1 ha with GPR. In the North *Agorà*, covered by thick colluvial and alluvial deposits, geo-magnetic surveys have confirmed the absence of structures at the centre of the large square and showed only the possible remains in a building of uncertain function and chronology in the north-western corner, facing the North *Stoà*.

Other anomalies related to Byzantine wall structures were also identified, the north-east corner, near the structure of the same date previously brought to light by excavations. The presence of slab paving was not identified, thus confirming that the beaten earth surface of the square, characterized by multifunctionality, could have been also used for athletic and gladiator games.

Fig. 6.9 Results of geo-magnetic prospecting in the central area of Hierapolis, georeferenced on a QuickBird-2 image (processing GGH): (**a**) the area north-west the Civil *Agorà*; (**b**) *insulae* 61 and 70; (**c**) the area east of the Church with Pillars

The investigations included part of the *Stoà*-Basilica, that lines the square itself on the eastern side, and the area included between the *Agorà* and the North Theatre, where several elements for the reconstruction of the ancient topography of the area were collected. Further research by means of GPR was aimed at the analysis of the inner structure of the Basilica (for the study of which a section was created through the application of electric tomography) and of the scene of the Theatre, both covered with colluvial deposits.

In the city's central area (Figs. 6.9 and 6.10), the integration of geo-magnetic surveys and investigations with the GPR allowed us to recover important elements

Fig. 6.10 Results of GPR prospecting in central area of Hierapolis, georeferenced on a QuickBird-2 image (processing GGH): (**a**) the area north-west the Civil *Agorà*; (**b**) *insulae* 61 and 70; (**c**) the area east of the Church with Pillars

on some *insulae* of the urban layout (61, 70 and 79), covered by colluvial deposits, where anomalies referable to houses are evident, and on the northern, southern and eastern sides of the Civil *Agorà* (the remains of which are today partially covered by the modern structures of the Pamukkale Termal), inserted in a small lake that was formed along the Hierapolis seismic fault and covered by limestone deposits up to 3–4 m thick. Despite the strong disturbances created by buried modern elements (mostly drainage), we also identified traces in the north-western corner of the square, linked to the collapsed remains of a large building north of the so-called Porta Bella and stretches of several *stenopoi* flanked by other buildings.

The data that emerged in the area in front of the Nymphaeum of the Sanctuary of Apollo and immediately east of the so-called Church with Pillars are also quite interesting. In particular, in the last area where no evidence was visible on the surface (characterized by earth layers alternating with recently formed limestone deposits), several anomalies were evident. These are referable to a building located on the eastern side of the *plateia*, oriented following the urban layout and, according to its position in the central part of the city, most likely of public function: the peripheral walls and the roads along them are identifiable, and their existence was also verified by menas of short archaeological excavations.

In the south-western part of the city, lastly, between the area covered by the modern structures of the Palmiye Otel, now demolished, the *plateia*, the Gymnasium and the Byzantine walls, geo-magnetic surveys allowed to reconstruct part of the urban layout, characterized in this area by *insulae* with houses and large public complexes. Our knowledge of the area is strongly limited by the presence of limestone deposits formed from canals that in medieval and modern times ran across this part of the city, by layers of colluvial accumulation and by interventions carried out beginning from the 1960s for the construction of the hotel; for this reason, wall remains are poorly visible from the surface and most of the archaeological evidences are obliterated. Lastly, at the southern end of the *plateia*, completely covered by calcareous layers, GPR measurements highlighted the buried remains of the road flanked by *tabernae*.

6.6 Archaeological Results

Integration between different methodologies and technologies was fundamental in the Hierapolis of Phrygia Survey Project for the knowledge of city plan and its transformations in the main historic periods (Hellenistic, Roman, Byzantine, Ottoman). The systematic archaeological surveys and geo-physical prospecting, together with the constant analysis of the satellite images, space and aerial photos, and the traditional study of ancient maps, allowed us to recover abundant data on the urban layout of Hierapolis, previously not systematically studied.

Sometimes the new data acquired from archaeological surveys and from satellite images processing had directed the strategies of the archaeological excavations. The discovery of the processional route to the *Martyrion* of St. Philip is surely the main example of this research methodology (Scardozzi 2007b); the traces and anomalies linked to the route were initially recognized in the satellite 2002 QuickBird-2 image and later verified on the ground during the campaign of 2003; finally, the ancient structures were excavated. Cleary visible in the image (integrated with oblique aerial photographs) were traces of the ancient road network, generally consisting of semi-emerging walls delimiting the edges of the streets, or of piles of stones lying on the roads' surface, which are the remains of collapsed houses of late construction that were built on the streets themselves, or materials that were removed from adjacent areas in the Ottoman period, so that they could be used for agricultural purposes. Within this regular network (a rectangular grid), barely perceptible on the ground, an anomaly was discovered, corresponding to the north-east end of the *stenopos* 13, which was oriented differently to that of the rest of the network, with a bend heading towards the *Martyrion* of St. Philip. The verification on the ground and the excavation which followed this discovery clarified many aspects of the structure of Hierapolis in the Early-Byzantine period; in fact, before these researches the Sanctuary of St. Philip, excavated between 1957 and the early 1970s, stood isolated on a hill to the north-east of Hierapolis, just outside the Early-Byzantine city walls. For the first time the excavations showed how it was possible to reach the area of the *Martyrion* of St. Philip from the *stenopos* 13, which initially followed most of the same route as it did in the imperial Roman period, before bending away near the end. The new road passed through the city walls via a monumental gate and passed over a seasonal watercourse with a wide bridge. At the foot of the hill on which the *Martyrion* stood, in 2003 a building with an octagonal plan was found; it is a bath and belonged to the same monumental phase of building of the *Martyrion*, at the beginning of the fifth century AD, and is currently still being excavated; from here the pilgrims completed the ascent to the Sanctuary of St. Philip by a long flight of steps.

The complete urban layout of Hierapolis – previously studied only in short excavations conducted between 1969 and 1971, which brought to light some stretches of paved roads and numerous crossroads – has been systematically examined in 2004 and 2008, thanks the integration of data from the analysis of satellite and aerial images, from systematic archaeological and topographical surveys, and from geo-physical prospecting. It was thus possible to study the characteristics of the layout, to record its regularities and anomalies due to the geo-morphology of the terrain, to study its conservation in the Early-Byzantine period and subsequent disintegration, and to identify sectors of the city in the eastern part of the urban area lying outside the city walls constructed at the beginning of the fifth century AD. The main roads coming out of the city were also examined and reconstructed, in relation to the necropolises to the north, east and south. All the data collected were inserted in the new archaeological map of the city (D'Andria et al. 2008); from 2008 it is in updating with research activities regarding in particular the city walls (Castrianni et al. 2010b), the urban layout and road network, the monuments of the central area (Sanctuary of Springs and surroundings), the hill of St. Philip, the North Theatre and the Roman and Byzantine water canals network in the town and in the area immediately surrounding.

In the 2005–2007 campaigns, archaeological surveys were extended to include the territory of Hierapolis, first of all in the eastern sector of the *Lykos* valley and subsequently in the plateau north of the city (Scardozzi 2007a; Castrianni et al. 2008a). The research aims to reconstruct the ancient settlement pattern of the area and its development during time; the surveys allowed the identification of numerous archaeological evidences, which contributed to the detailed reconstruction of the historical development of the territory, from prehistoric times to the modern period, and particularly from the Hellenistic age until the Seljuk period, i.e. for the entire period in which Hierapolis was in life.

Regarding the area immediately surrounding the city, in a radius of about 3 km numerous ancient quarries were identified, used extensively for the buildings of the urban area and the necropolises from the Hellenistic to the Byzantine times. There are many travertine quarries to the north, east and south-east of Hierapolis, even quite close to the city, in some cases also still in use. A little to the north of the urban area, at the foot of the mountains located just south-east of the village of Karahayıt, some ancient quarries have been identified from which a local polychrome *breccia* with reddish matrix was extracted, used in various monuments in Hierapolis. Quarries of white marble were founded to the north of the northern necropolis and further north on the mountain known as Marmar Tepe (i.e. "Mountain of Marble"); the stone, which has grey marbling and is medium-to-large grained, was used in Hierapolis above all as a construction material, whereas for sculptured decorations imported white marbles were used, brought both from Thiounta, in the northern area of its own territory (see below), or from the quarries of Laodikeia *ad Lykum*, and from further away, such as the quarries of Docimium, near Afyon, in the Akdagi massif, or those of Aphrodisias (on the western side of the Babadağ). Immediately to the north and north-west of the city, numerous quarries of alabaster *fiorito* and *listato* were also found (Scardozzi 2010b), which have been identified with the "marble of Hierapolis" mentioned in the Roman and Byzantine literary sources and extensively exported.

Again in the area immediately surrounding Hierapolis, evidence of intensive occupation of the territory has been found, with rural settlements of medium size, identifiable as farms, spanning a period from the Late Hellenistic – Early Imperial ages to the Early-Byzantine period; some have nearby small travertine quarries, from which blocks for their construction were extracted. The structures discovered often consist of remains of emerging walls and in many of them large monolithic crushing basins for olives and stone elements of olive oil and wine presses have been found; sometimes the visible remains on the surface are negligible, but the multispectral processing of the satellite images in some cases makes it possible to highlight the traces of the buried walls (Lasaponara et al. 2008a, b, 2010).

The aqueducts, which brought water to Hierapolis along three main routes, from north, north-east and east, have been identified, documented and mapped (Scardozzi 2007c). Between 6 and 13 km long, they consisted of terracotta pipes of medium size (diameters between 20 and 40 cm; lengths between 52 and 62 cm), in some cases simply buried in the earth, in others laid in hollows carved out of the rock or carved stone supports; on some stretches traces have been found of rebuilding (including small variations in the route) or of two parallel pipes, next to which there was also in some cases an open channel. This system for supplying water to the city, built in the Roman period and still in use at least up until the Early-Byzantine period, took drinking water from springs located immediately below the brow of the plateau to the north of the city; these areas are characterised by the presence of numerous toponyms incorporating the term "pınar" (= spring), demonstrating the abundance of drinking water, which is not present at the lower altitudes. The north-eastern and eastern aqueducts, instead, have been reconstructed up to the *Castellum Aquae*, located on a hill looking onto the city from the east;

the northern aqueduct arrived in the north area of the city, monumentalised between the late first and the third century AD. Some branches of this network served the farms closest to Hierapolis.

The main ancient highways leading out of the city were also reconstructed, particularly those connecting Hierapolis to Tripolis to the north-west and Laodikeia to the south-west. These roads were part of a single important route, mentioned both in the *Itinerarium Antonini* and in the *Tabula Peutingeriana*, which connected Pergamon and Sardis with Laodikeia, partly following a stretch of an ancient road of the Persian period. Long stretches of this route have been discovered, up to 8 m wide and paved with compacted gravel, bounded on both sides by two rows of travertine blocks; in this case too, an important contribution to the reconstruction of the road was offered by the satellite images, which made it possible to identify traces and anomalies resulting from buried stretches of the road. The route, about 19.5 km long, reached the valley of the *Lykos* just to the north of the village of Akköy, subsequently running north-west along the eastern edge of the valley, at the foot of the slope of the plateau north of Hierapolis. It partially coincides with an old street, today mostly rural road, which runs through the villages of Senekci, Çeşmebaşi, Gölemezli, Tepeköy and Mahmutlu, finishing at Yenicekent, on the eastern edge of which lie the remains of Tripolis; this road, documented also in the Corona photo of 1968, is today partially modified and paved in asphalt.

The road leading to Laodikeia, once it reached the valley of the *Lykos* just south of Hierapolis, proceeded in a south-westerly direction across the plain, probably along a route of about 10–11 km. The area in question was extensively transformed as a result of agricultural activity and modern road-building, so that it has not been possible to reconstruct with any accuracy the route of the ancient road. The analysis of the satellite images has however made it possible to discover that this road ran to the west of an ancient lake, today dried up, which has been identified at about 5 km to the south-west of Hierapolis, along the *Lykos*, 2 km to the north of Laodikeia. In the images, the area is characterised by a broad damp patch due to the residual moisture present in the terrain, linked to a pool of water that was still extant in the first half of the twentieth century. Subsequent drainage operations have wiped out any trace of this lake, which may be the one mentioned in an inscription discovered in Hierapolis itself. This was a copy in marble of a letter from the Emperor Hadrian (131 AD), who was seeking to intervene in a dispute between Laodikeia on one side, and Hierapolis and Tripolis on the other, concerning fishing rights in a lake lying within the territory of the former city. Thus the satellite images in this case also helped to reconstruct the paleo-geography of the area.

In the territory of Hierapolis various ancient villages have been identified which depended on the city; for some of these, thanks to discoveries of inscriptions, it has been possible to know their name, like for example Thiounta. By combining epigraphic and archaeological sources, an attempt is being made to delineate, at least for the Roman imperial period, the boundaries of the area controlled by the city, which are believed to have coincided with natural features. Perhaps to the north and north-west the territory was bounded by the river Meander, and to the west by the *Lykos*; to the south is believed to lie the above-mentioned lake controlled by Laodikeia, while to the east important geographical features are the Küçükçökelz massif and the rivers that descend southwards towards the *Lykos* (perhaps the above mentioned Küçük Dere) and northwards towards the Meander (Değirmen Dere and Koca Dere), where a deep river valley constitutes a clear natural boundary between the plateau of Uzunpınar, to the west, and the plateau of Çal, to the east. In all the villages identified, the strong rural character of the territory emerges, with remains of installations for the production of olive oil and wine. Only one has been identified in the valley of the *Lykos*, about 4 km to the south-east of Hierapolis, near the modern village of Küçükdereköy (see above), at the mouth of a narrow river valley crossed by a road that went up to the plateau of Uzunpınar, perhaps near the south-eastern boundary of the territory of the city. In contrast, numerous ancient villages have been identified on the broad plain to the north of Hierapolis, in the area of the modern towns of Kurtluca, Güzelpınar, Uzunpınar, Akçapınar, Çindere, Gözler and Dağmarmara. Before the start of the surveys, the area, which is believed to have constituted the rich agricultural hinterland of the city, had been almost entirely unexplored since the travels of W. M. Ramsay in the late nineteenth century. Among the villages that have been identified, mapped and documented, is the above-mentioned Thiounta, immediately to the north of the ruins of Eski Gözler, the modern town completely destroyed by an earthquake in the mid 1970s. The remains of the ancient village, with its white marble quarries recorded in various epigraphs of the necropolis of Hierapolis, were identified thanks to the discovery of some inscriptions; it occupies a series of terraces that descend towards the course of the Meander, about 22 km north of the city, at the northern edge of its territory.

On the plateau of Uzunpınar, among the most important discoveries, worthy of mention are two sacred areas. The first of these was related to *Karios/Kareios*, an indigenous divinity assimilated to Apollo by the Greeks who founded Hierapolis, who was worshipped in the main sanctuary of the city. The second of these is the sacred area of *Motaleis*, a population of the territory of Hierapolis, mentioned in some Greek inscriptions of Roman Imperial age, which lived in a district called "*Motalis*", that perhaps derived from the name of a village called "*Motala*" or similar.

The first is a small sanctuary identified about half way up the north-western slope of the Somakli Tepe, at the south-eastern edge of the plateau, just over a kilometre to the east of the village of Güzelpınar and about 11 km north-east of Hierapolis; here are the remains of a platform (c. 20 × 10 m), on which has shifted as the result of unauthorised excavations, were various fragmentary *stelae* in marble with dedications to the divinity *Karios/Kareios* and dating between first and third century AD. Remains referable to the sacred area of *Motaleis* are instead c. 20 km north-east of Hierapolis and occupy a small terrace located on the north-west of Yüksektepe, modest relief in the west part of the valley of Değirmen Dere, c. 2 km south of Dağmarmara; in summary, the archaeological findings visible on surface document the existence of two monumental marble buildings, close together and concentrated in an area of c. 4 ha, which, according to inscriptions found, must have been in use since at least from second to fourth century AD.

Fig. 6.11 The Höyük Tepe in a Corona KH-4A photo of 1968 and in a QuickBird-2 image of 2007

Regarding the most ancient phases of the occupation of the territory, worthy of mention are the discoveries of two important sites, that constitute the most ancient archaeological phenomena of the territory of Hierapolis. The first is in the plain of the *Lykos* River, about 2 km to the north-west of Akköy, 7.5 km from Hierapolis; it consist of a small tell (c. 0.5 ha), called Höyük Tepe, where an archaeological stratification has been identified which stretches from at least the Proto-historic age until the Roman imperial period. The hill is today cut by a modern country road and partially levelled because of agricultural works, instead the photographs taken by Corona KH-4A in 1968 show the tell completely preserved in its original morphology (Fig. 6.11). The second site is located on the Uzunpınar plateau, about 2 km south of the town, 10 km north-east of Hierapolis, and consists of a large settlement of almost 3 ha, dating to the Bronze Age. It is located on a promontory that dominates the confluence of two water courses and overlooks a spring (called Canpınar), still considered today the best of the whole plateau. On the surface numerous pottery shards dating to the Bronze Age are present, and many pebbles belonging to structures destroyed by recent agricultural activities.

Finally, the later phases of the settlement in the territory have also been documented, thanks to the identification of some abandoned villages of the Turkish period (see above), with their cemeteries and with stone houses that are poorly conserved but easily identifiable from the examination of the satellite images.

References

Altunel E (2000) L'attività sismica a Hierapolis e nelle zone limitrofe. In: D'Andria F, Silvestrelli F (eds) Ricerche archeologiche turche nella valle del Lykos. Congedo Editore, Galatina, pp 299–314

Ambraseys NN, Finkel CF (1995) The seismicity of Turkey and adjacent areas. A historical review, 1500–1800. Muhittin Salih Eren, Istanbul

Castrianni L, Di Giacomo G, Ditaranto I, Scardozzi G (2008a) Integrated technologies and methodologies for the reconstruction of the ancient topography of Hierapolis in Phrygia and its territory (Turkey). In: Proceedings of the 13th international congress "Cultural Heritage and New Technologies", workshop 13 – Archäologie und Computer, 03–05 Nov 2008, Vienna City Hall, pp 1–23

Castrianni L, Di Giacomo G, Ditaranto I, Scardozzi G (2008b) An online archaeological atlas: the webGIS for the monuments of Hierapolis in Phrygia. Archaeol Comput Newsl 69:1–8

Castrianni L, Di Giacomo G, Ditaranto I, Scardozzi G (2010a) High resolution satellite ortho-images for archaeological research: different methods and experiences in the Near and Middle East. In: Eppelbaum LV, Masini N, Soldovieri F (eds) Near surface geophysics for the study and the management of historical resources: past, present and future (EGU Session 2009). Adv Geosci 24:97–110

Castrianni L, Di Giacomo G, Ditaranto I, Scardozzi G (2010b) La cinta muraria di Hierapolis di Frigia: il geodatabase dei materiali di reimpiego come strumento di ricerca e conoscenza del monumento e della città. Archeologia e Calcolatori 21:93–126

D'Andria F, Scardozzi G, Spanò A (2008) Atlante di Hierapolis di Frigia. Ege Yayinlari, Istanbul

Di Giacomo G, Ditaranto I, Scardozzi G (2008a) Stereoscopic view, extracting DEM and cartography for archaeological purposes from Ikonos stereo pair: a case from Hierapolis of Phrygia territory (Turkey). In: Lasaponara R, Masini N (eds) Remote sensing for Archaeology and Cultural Heritage Management. Proceedings of the 1st international workshop (Rome, 30 Sept – 4 Oct 2008), Aracne, Rome, pp 41–44

Di Giacomo G, Di Giacomo GP, Scardozzi G (2008b) Integration between high resolution satellite images, GPS and Tablet PC with a new software for archaeological survey: the Ulixes system. In: Lasaponara R, Masini N (eds) Remote sensing for Archaeology and Cultural Heritage Management. Proceedings of the 1st international workshop (Rome, 30 Sept – 4 Oct 2008), Aracne, Rome, pp 339–342

Eisenbeiss H, Baltsavias E, Pateraki M, Zhang L (2004) Potential of Ikonos and Quickbird imagery for accurate 3D point positioning, orthoimage and DSM generation. In: XX ISPRS Congress (Istanbul, 12–23 July 2004). International Archives of Photogrammetry, Remote Sensing and Spatial Information Sciences, 35(B7):1250–1256

Gianinetto M, Scaioni M (2004) Estrazione di layer vettoriali per l'utilizzo cartografico di immagini satellitari ad alta risoluzione. In: VIII Conferenza Nazionale ASITA (Roma, 14–17 dicembre 2004), pp 1189–1194

Holland D, Marshall P (2003) Using high-resolution satellite imagery in a well-mapped country terrain. In: Proceedings of ISPRS-EARSeL Joint Work on "High Resolution Mapping from Space", Hannover

Jacobsen K (2006) Comparison of photogrammetric applications based on narrow angle line scanners with traditional photogrammetric methods. In: Everaerts J (ed) The future of remote sensing. Second international workshop (Antwerp, 17–18 Oct 2006)

Lasaponara R, Masini N (2007) Detection of archaeological crop marks by using satellite QuickBird multispectral imagery. J Archaeol Sci 34:214–221

Lasaponara R, Masini N, Scardozzi G (2008a) Satellite-based archaeological research in the ancient territory of Hierapolis (Turkey). In: Lasaponara R, Masini N (eds) Remote sensing for Archaeology and Cultural Heritage Management. Proceedings of the 1st international workshop (Rome, 30 Sept – 4 Oct 2008), Aracne, Rome, pp 11–15

Lasaponara R, Masini N, Scardozzi G (2008b) New perspectives for satellite-based archaeological research in the ancient territory of Hierapolis (Turkey). Adv Geosci 19:87–96

Lasaponara R, Masini N, Scardozzi G (2010) Elaborazione di immagini satellitari ad alta risoluzione e ricognizione archeologica per la conoscenza degli insediamenti rurali del territorio di Hierapolis di Frigia (Turchia). In: D'Andria F, Malfitana D, Masini N, Scardozzi G (eds) Il dialogo dei Saperi. Metodologie integrate per i Beni Culturali. Edizioni Scientifiche Italiane, Napoli, pp 479–494

Masini N, Lasaponara R (2007) Investigating the spectral capability of QuickBird data to detect archaeological remains buried under vegetated and not vegetated areas. J Cult Herit 8(1):53–60

Piccarreta F (2003) Aerofotogrammetria finalizzata all'archeologia. In: Guaitoli M (ed) Lo sguardo di Icaro. Le collezioni dell'Aerofototeca Nazionale per la conoscenza del territorio. Campisano Editore, Roma, pp 96–98

Piccarreta F, Ceraudo G (2000) Manuale di aerofotografia archeologica. Metodologia, tecniche e applicazioni. Edipuglia, Bari

Scardozzi G (2007a) Remote sensing and archaeological survey in the Hierapolis of Phrygia territory, Turkey. In: Remote sensing for environmental monitoring, GIS applications, and geology. Proceedings of SPIE Europe Remote Sensing, Conference 6749 "Remote Sensing and Archaeology" (Florence, 17–20 Sept 2007), vol 6749:04/1–12

Scardozzi G (2007b) L'urbanistica di Hierapolis di Frigia: ricerche topografiche, immagini satellitari e fotografie aeree. Archeol Aerea II:83–134

Scardozzi G (2007c) Hierapolis di Frigia. Applicazioni informatiche alle ricognizioni archeologiche e telerilevamento da satellite: l'esempio degli acquedotti della città. In: Scardozzi G (ed) Progetto FIRB 2001 "Il Mediterraneo antico e medievale come luogo di incontro tra Oriente e Occidente, Nord e Sud". Atti della Giornata studio sul tema "GIS e applicazioni informatiche alle ricerche archeologiche e storiche" (Roma, CNR, 5 luglio 2007). Archeologia e Calcolatori, 18:331–353

Scardozzi G (2008a) Hierapolis of Phrygia Project: integration of archaeological survey and remote sensing data. In: Lasaponara R, Masini N (eds) Remote sensing for Archaeology and Cultural Heritage Management. Proceedings of the 1st international workshop (Rome, 30 Sept – 4 Oct 2008), Aracne, Rome, pp 425–428

Scardozzi G (2008b) Old high resolution satellite images for landscape archaeology: case studies from Turkey and Iraq. In: Remote sensing for environmental monitoring, GIS applications, and geology. Proceedings of SPIE Europe Remote Sensing, Conference 7110 "Remote Sensing and Archaeology" (Cardiff, 15–18 Sept 2008), vol 7110/03:1–14

Scardozzi G (2009) The contribution of high resolution satellite images to the production of base-maps and cartographies for archaeological research in Turkey and Iraq. In: Remote sensing for environmental monitoring, GIS applications, and geology IX, (Berlin, 31 Aug – 3 Sept 2009) – Proceedings of SPIE Europe Remote Sensing, vol 7478:74780B1-12

Scardozzi G (2010a) Integrated geophysical methods for the knowledge of the Urban layout of Hierapolis in Phrygia (Turkey). In: Proceedings of the XIII international conference on ground penetrating radar, Lecce, Italy, 21–25 June 2010, pp 112–117

Scardozzi G (2010b) Oil and wine production in Hierapolis of Phrygia and its territory during Roman and Byzantine age: documentation from archaeological excavations and surveys. In: Olive oil and wine production in Anatolia during antiquity. Symposium proceedings (6–8 Nov 2008, Mersin, Turkey), Mersin, pp 277–302

Scardozzi G, Spanò A (2008) Remote sensing and GIS data managing for archaeological maps: the case of Hierapolis of Phrygia (Turkey). In: Lasaponara R, Masini N (eds) Remote sensing for Archaeology and Cultural Heritage Management. Proceedings of the 1st international workshop (Rome, 30 Sept – 4 Oct 2008), Aracne, Rome, pp 305–308

Toutin T (2004) Geometric processing of remote sensing images: models, algorithms and methods. Int J Remote Sens 25(10), 56:1893–1924

Vanhaverbeke H, Waelkens M (2002) The northwestern necropolis of Hierapolis (Phrygia). The chronological and topographical distribution of the travertine sarcophagi and their way of production. In: De Bernardi Ferrero D (ed) Saggi in onore di Paolo Verzone. Hierapolis. Scavi e Ricerche, IV. Giorgio Bretschneider, Roma, pp 119–145

Chapter 7
NASA Remote Sensing and Archaeology

Marco J. Giardino

Abstract Although not its primary mission, NASA's remote sensing missions have been providing archaeologists with useful and unique data since the launch of the very first earth observing missions. While imaging the Earth with a variety of multispectral and hyperspectral instrument mounted on both orbital and suborbital platforms, NASA scientists and collaborators from international universities have discovered, delineated and analyzed archaeological sites worldwide using remotely sensed digital imagery. Several of the ten NASA centers have collaborated with archaeologists to refine and validate the use of active and passive remote sensing for archeological use. The Stennis Space Center (SSC), located in Mississippi USA has been the NASA leader in archeological research. Together with colleagues from Goddard Space Flight Center (GSFC), Marshall Space Flight Center (MSFC), and the Jet Propulsion Laboratory (JPL), SSC scientists have provided the archaeological community with useful images and sophisticated processing that have pushed the technological frontiers of archaeological research and applications. Successful projects include identifying prehistoric roads in Chaco canyon, mapping the Lewis and Clark Corps of Discovery route, assessing prehistoric settlement patterns in southeast Louisiana, discovering the lost city of Ubar and mapping ancient irrigation channels at Angkor. The Scientific Data Purchase (SDP) managed from SSC solicited commercial remote sensing companies to collect archaeological data. Recently, NASA formally solicited "space archaeology" proposals through its Earth Science Directorate and continues to assist archaeologists and cultural resource managers in doing their work more efficiently and effectively. Hyperspectral data offers new opportunities for future archeological discoveries.

Keywords NASA • Stennis Space Center • Archaeology • Remote sensing • History • Hyperspectral

---

M.J. Giardino (✉)
National Aeronautics and Space Administration, Stennis Space Center, Mississippi, USA
e-mail: marco.j.giardino@nasa.gov

R. Lasaponara and N. Masini (eds.), *Satellite Remote Sensing: A New Tool for Archaeology*, Remote Sensing and Digital Image Processing 16,
DOI 10.1007/978-90-481-8801-7_7,

7.1 Introduction

NASA's contributions to archaeology began serendipitously as part of a more comprehensive earth observation program. NASA scientists and affiliated researchers from various universities realized that remotely sensed data acquired by satellites and research aircraft contained information that could further archaeological research world-wide. As data collection and processing techniques improved over the decades, archaeologists from numerous countries enhanced their research with remote sensing imagery.

On June 3, 1965, a Titan II rocket blasted off from Launch Complex 19 located at the Kennedy Space Center in Florida. Riding on top of the rocket was the Gemini 4 capsule carrying Commander James McDivitt and Pilot Edward White, II. White would conduct the first USA "spacewalk" or EVA (Extra Vehicular Activity) operation, lasting 36 min. More importantly for earth scientists, however, were the photographs of the Earth taken by the Gemini 4 astronauts. In fact, this Gemini mission became the first formal photographic experiment from space. The astronauts collected 207 color photographs of the Earth for geologic, geographic, and oceanographic studies using hand-held 70-mm Hasselblad 500-C camera with a Zeiss planar 80-mm f/2.8 lens and haze filter to reduce the intensity atmospheric scattering. Each image had nominal scales 1:2,400,000 and included about 140 km on a side. Gemini's earth images were part of the Synoptic Terrain Photography and the Synoptic Weather Photography experiments and were acquired between March 23, 1965 and November 15, 1966 (Gemini Missions III through XII).

Follow-on NASA manned missions continued and improved the imaging of our planet. The next important milestone in remote sensing from orbiting platforms occurred in 1968, when astronauts aboard the Apollo 9 capsule produced the first multispectral photography from space. Four Hasselblad cameras mounted in an instrument rack were engineered to trigger simultaneously over the same target. Each of the four cameras collected reflected light in specific wavelengths, including for the first time, the infrared portion of the electromagnetic spectrum (EMS).

Scientists, including archaeologists, did not miss the opportunity to investigate the potential of NASA's earth images, particularly those collected in the infrared. Over the next 50 years, archeologists used a variety of ever-more sophisticated techniques to identify, delineate and analyze archaeological sites and their surroundings. Throughout the age of archaeological exploration through remote sensing, NASA has contributed both directly and indirectly to furthering the use of satellite and suborbital platforms as tools for archaeological research.

7.2 First Archaeological Applications of NASA Remote Sensing

Among the first archaeological applications of NASA imagery was the identification of ancient Hohokom canal systems by Yehuda Kedar (UC Santa Barbara, Geography Department). Using black and white as well as infrared photography from the S065 multiband experiment flown on Apollo 11, Kedar was able to identify the linear features built by prehistoric people in the Phoenix Arizona area (Schaber and Gumerman 1969).

Two archaeological projects in Europe experimented with the application of similar technology around the same time as Kedar's work in Arizona. The University of Minnesota expedition to Messenia Greece in 1969 used a balloon to collect data using black and white color and infrared films with a Hasselblad El 500 developed for NASA's Apollo 11 Moon landing mission. The same camera and techniques were deployed by scientists from U.S. Army, the University of Missouri and the American Academy of Rome over the ancient harbor of Cosa, in Tuscany. The infrared portion of the spectrum especially provided new and exciting information from both these sites (MASCA Newsletter 1969:2–3). Writing for the journal *Science* in 1971, George Gumerman and Thomas Lyons declared: "the color infrared image, correctly interpreted, can guide the archeologist to previously undetected potential targets. . .the potential usefulness of infrared imagery to archeology consists of the fact that, like color infrared film, buried or obscure cultural features may absorb and radiate solar energy in amounts that differ from that of the surrounding soil matrix, thereby revealing the features on the imagery" (Gumerman and Lyons 1971:130).

In the United States, NASA was beginning to appreciate the potential benefit of studying ancient cultures through the imaging the earth from Earth's orbit. In 1970, Dr. Eugene Emme, NASA's first historian, encouraged a young NASA summer intern, hired by NASA HQ as a clerk-typist, to investigate the potential applications of remote sensing instruments for archaeological research. In pursuit of her assigned task, the intern, Mary Marguerite Scalera, contacted Thomas R. Lyons, who at the time was Assistant Director of the Technology Applications Center at the University of New Mexico; and Mr. Gary North, a Remote Sensing Specialist, Geographic Applications Program, U.S. Geological Survey.

Her brief and unpublished report summarized the use of NASA remote sensing instruments in six archaeological projects that were being conducted at the time. Scalera concluded her internal report with a highly optimistic forecast for the future of NASA "Space Age" technology to archaeology, a declaration that would prove prophetic (Scalera 1970).

Three years after Scalera's report, NASA initiated the Skylab missions (whose manned missions are denoted as Skylab I, II and III). Flying on Skylab was the Earth Resources Experiment Package (EREP) which included an array consisting of six 70 mm cameras each equipped with different film and filter combinations to record images in narrow band black-and-white (500–600 nm and 600–700 nm), broad band color (400–700 nm) and color and black-and-white near infrared. The nominal spatial resolution of the 190A camera system was 30 m for each square image which measured 163 km on the side. While the 190A system had been successful during the Apollo 9 mission, the other instruments located on Skylab were experimental and radiometric rather than photographic. Two of these, a spectrometer (S191) and a 10-band multispectral scanner (S192), operated in the infrared. The spectrometer recorded the wavelength and intensity of infrared radiation from selected small areas (0.45-km diameter) on the ground; the multispectral scanner simultaneously measured the intensity of infrared in ten wavelength ranges, scanning a swath 74 km wide centered on the spacecraft's ground track (Compton and Benson 1983; Belew and Stuhlinger 1973:148–157). By the time the last Skylab crew left the orbiting space station in February of 1974, crews had collected over 35,000 images of the Earth. These images, together with the vast majority of NASA imagery, are available from the United States Geological Survey (USGS) EROS data center. To date, Skylab imagery has not been widely used for archaeological research. Still, Skylab images were useful in identifying linear features, such as prehistoric irrigation canals in the southwestern United States (Ebert and Lyons 1980).

Just one year prior to Skylab, on July 23, 1972, NASA launched the first in a series of Earth Resources Technology Satellites (ERTS 1) which in many ways revolutionized how archaeologists today conduct their work. The ERTS Program and the satellites were later renamed Landsat to better represent the satellite program's prime emphasis on remote sensing of land resources. Seven Landsat satellites have been launched since 1972, with six of them achieving orbit and providing multispectral images from around the world. The USGS EROS Data Center (EDC) retained primary responsibility as the Government archive of Landsat data.

The sensors mounted on the various Landsat platforms (MSS or Multispectral Scanner; the TM or Thematic Mapper; the ETM or Enhanced Thematic Mapper and the ETM+ flown on Landsat 7) are well known and well documented (Short 1982). The MSS systems were the first global monitoring systems capable of producing multispectral data in a digital format. Since the early 1970s, the Landsat series of satellites has been widely used in archaeology as archaeologists gained experience and confidence in the use of multispectral data for a variety of archaeological applications (Miller 1974; McKee and Sever 1994:139–140).

The most extensive application of NASA satellite imagery to archaeological topics was carried out initially by the National Park Service (NPS). Under the leadership of Thomas Lyons and James Ebert, NPS archaeologists researched the feasibility of using satellite imagery to identify village sites in Alaska (Cook and Stringer 1975) and prehistoric sites and roadways in Chaco Canyon, New Mexico (Lyons and Hitchcock 1977). In 1977, Lyons and Avery published an influential guide to using remote sensing for archaeological research (Lyons and Avery 1977), followed by two additional and important publications on remote sensing and cultural resources (Lyons and Ebert 1978; Lyons and Mathien 1980). The amount of work on applying remote sensing to archaeology was already significant by the late 1970s so that a preliminary bibliography was feasible and necessary (Kruchman 1976).

In the 1980s, archaeology saw a notable increase in the use of NASA remote sensing imagery and some applications went beyond site identification toward the exploration of increasingly more complex issues, such as testing predictive models of site distribution. One such successful application was completed in Delaware (Wells et al. 1981; Custer et al. 1986). Here Custer and his colleagues subdivided a large portion of central Delaware into broad land cover and hydrological classes using Landsat MSS imagery. In their research, the archaeologists found that these modern land use classes were correlated with edaphic factors known to influence prehistoric settlement choices. Using the classified images as their base maps, Custer and his colleagues were able to statistically characterize the distribution of known sites in relation to these major land and water classes. The settlement model derived from this project proved to be over 90% accurate in predicting site locations in unsurveyed areas. In the following years, numerous archaeological projects using satellite imagery were undertaken throughout the world, each creating new analytical methods, identifying technological challenges and thereby providing NASA management with increased rationale for formalizing the Agency's role in archaeology (see for example, Johnson et al. 1988; Parcak 2009). Prominent among the pioneers of satellite remote sensing and archaeology in the United States are Scott Madry from University of North Carolina, Chapel Hill; Jay Johnson, University of Mississippi; Fred Limp, University of Arkansas; James Wiseman, Boston University and Tom Sever, NASA (Limp 1989; Wiseman 1984; Madry 1983; Johnson 2006).

7.3 NSTL Conference

In 1983, Dr. Thomas Sever of the NASA Stennis Space Center (then known as the National Space Testing Laboratory or NSTL) produced an internal NASA document whose objective was to assess the feasibility of "remote sensing techniques for delineating and inventorying archeological sites and features in order to expedite survey and excavation" (Sever 1983). Sever proposed imaging Chaco Canyon using two airborne multispectral scanners: the Thematic Mapper Simulator (TMS) and the Thermal Infrared Multispectral Scanner (TIMS) which are discussed in Sect. 7.4.1 (Sever and Wagner 1991).

NASA responded positively to Sever's proposal and in March of 1984, Sever joined James Wiseman in sponsoring a conference on remote sensing and archaeology to develop research projects that could explore the value of NASA imagery in archaeology (Sever and Wiseman 1985). The objective of the conference was "to bring together a group of archaeologists representing a broad spectrum of archaeological concerns to learn more about the recent advances in NASA's remote sensing technology and to discuss possible future applications of that technology in archaeology" (Sever and Wiseman 1985:1). The Conference was jointly funded by NASA, the National Science Foundation, and the National Geographic Society and was attended by 22 prominent professional archaeologists, including Patty Jo Watson (Washington University), John Yellen (Smithsonian), the late Glyn Isaac (Harvard University), the late Richard MacNeish (Boston University) and Robert McAdams, then Provost at the University of Chicago.

The participants concluded the conference convinced that NASA remote sensing technologies would benefit archaeological research and that NASA and the archaeological community should become involved in joint research projects. During their meetings, the conferees had sought consensus on which areas of archaeological research, and what kinds of archaeological problems, should be the focus of the first remote sensing applications. They generally agreed that these included: site location and point resource, the range of paleo-environmental resources, ecology, mapping, hypothesis formulation (predictive modeling) and testing, cultural resource management, and general ethnological investigations. The group concluded that the most immediate applications of remote sensing technology and image processing, should be in the analysis of the environments of human societies and their interrelationships thereby generating and testing models both of human behavior and environmental change. The Conference participants encouraged those types of projects that would bring these applications to bear on endangered resources, worldwide. Finally, the conferences discussed "ground truth" and agreed that this phase of remote sensing projects should include not only direct observation aimed at confirming or rejecting a proposed spectral identification but also "some amount of ground search in areas where the computer images do not indicate by spectral signature the kind of site or natural resource that is an object of the research" (Sever and Wiseman 1985:71–72).

7.4 Sub-orbital Remote Sensing

Starting in the early 1980s, and continuing to the present, Landsat imagery has been the preferred platform for archaeological research in the United States. SPOT data offered better spatial resolution and was utilized in a few archaeological projects in the United States. But mostly, archaeologists interested in satellite data gravitated toward Landsat, particularly bands 4-3-2 (IR, red, green) (Parcak 2009; McKee and Sever 1994).

Initially, the use of sub-orbital or aerial platforms to collect multispectral and hyperspectral remote sensing data was not broadly considered or available to the archaeological community even though aerial photography was widely used as an archaeological tool. Archaeologists generally required data with high spatial resolutions, and so were interested in collecting imagery from lower altitudes. Archaeologists were intrigued also by the potential offered by spectral resolution, including the thermal wavelengths which were part of the Landsat sensor arrays. The advantage of collecting data using airborne platforms included the ability to vary the ground sampling distance (GSD) by varying platform altitude and the ability to fly on demand. It was in this arena, that NASA's Stennis Space Center (SSC) pioneered several efforts that advanced the use of remotely sensed information in archaeological projects (Parrington 1983).

7.4.1 *Stennis Space Center Sub-orbital Program*

In the early 1980s, two SSC airborne sensors became the workhorses of archaeological remote sensing: the Thematic Mapping Simulator (TMS) and the Thermal Infrared Multispectral Scanner (TIMS). TMS simulated the Landsat Thematic Mapping (TM) instrument, with a slightly higher spatial resolution. The sensor collected 8-bit data in 12 bands from visible to thermal. In order to produce imagery with 30 m spatial resolution, the TMS was flown at 12,000 m above mean terrain elevation. The TMS scanner system was developed by Daedalus (known as the AADS-1268 scanner) and flew on ER-2 aircraft. Color IR photography was acquired simultaneously with 60% forward overlap for stereoscopic viewing.

These instruments were used on several projects with mostly positive results. For example, archaeologists through the study of remotely sensed imagery were able to detect archeological features in Western Kentucky (Mid-America Remote Sensing Center 1983); Anasazi roads in Chaco Canyon (Sever 1983, 1990); earthwork features at the Poverty Point in Louisiana (Gibson 1984, 1987, 1989; Sever and Wiseman 1985); and protohistoric sites in northern Mississippi (Johnson et al. 1988). At Chaco Canyon, the TMS was also flown at an altitude of 4,267 m resulting in 10-m spatial resolution which was deemed at the time necessary to identify prehistoric roadways. Processing of the TMS data over Chaco Canyon using ELAS software (also developed at Stennis) revealed that image enhancement techniques were successful in detecting the existence of prehistoric roadways particularly in the thermal band (10.43–12.33 μm) (Sever 1983:12).

The other important airborne sensor developed by NASA and used in archaeology during the 1980s was the Thermal Infrared Multispectral Scanner (TIMS). This instrument was a six-channel multispectral spectrometer which measured radiation only in the thermal infrared region of the EMS (9,400–12,200 nm) at intervals between bands varying from 400 nm between 8.2 and 9.4 μm and in 800 and 1,000 nm between 9.4 and 12.2 (Kahle and Abbott 1985; Bennett et al. 1986). TIMS measured thermal radiation in degrees centigrade better than a tenth of a degree, which, after atmospheric correction, resulted operationally in a spectral resolution of ½°C. The TIMS was especially important in "thermal" archaeology because it was multispectral in the thermal region, recording emitted energy in multiple bands.

Developed by NASA's Jet Propulsion Laboratory (JPL) and flown on the NSTL LearJet, the TIMS was funded by the NASA geology program in 1981 and built by Deadalus Enterprises, Inc. of Ann Arbor, Michigan. Like all airborne sensors, the TIMS' GSD was controlled by the aircraft's altitude so that flying at 2,000 m the nominal pixel resolution was 5 m, a vast improvement over the resolution of Landsat sensors, and more within the range required by archaeologists. The higher spatial resolution came at a cost, however, since numerous scan lines needed to be mosaicked to approximate the image size of Landsat data and thus provide the synoptic view, also required by many archaeological research designs. Consequently, archaeologists began exploring the process of sub-sampling Landsat imagery with higher resolution aerial imagery, a technique which has been refined over the decades and, especially with the advent of modern image processing techniques, has allowed the fusion of multi-scale data to reap the unique benefits of the various image collecting techniques.

During the early 1980s, NASA flew the TIMS in two archaeological projects, one over Chaco Canyon looking for prehistoric roads, the other at Poverty Point Site. This latter site is an impressive 400-acre site consisting of a complex assortment of mounds, concentric ridges and "avenues" dating to 1500 B.C. E. The TIMS data, particularly Band 3 (9.0–9.4 μm), when processed with high-pass filters, dramatically highlighted the aisles to the northwest and southwest (Gibson 1984; Rickman and Kalcic 1982). In addition TIMS data demonstrated that soil conditions at Poverty Point are characterized by phosphate concentration that are possibly indicative of human habitation. These phosphate "signatures" are sufficiently strong to be regularly detected and correctly resolved through spectral analysis.

In 1987, NSTL designed and built the Calibrated Airborne Multispectral Scanner (CAMS), a nine channels broad-band multispectral instrument measuring energy between 0.45 and 12.5 μm. Two years later, NSTL developed the Airborne Terrestrial Applications Sensor or ATLAS which combined the spectral characteristics of the TIMS and CAMS in a single instrument. ATLAS, like its predecessors, was an airborne multispectral instrument, but unlike the CAMS and TIMS, the ATLAS gathered reflected and emitted energy in 14 bands from 0.45 to 12.2 mm all in one instrument. The ATLAS (IFOV 2.0 mrads) collected data with ground resolutions ranging from 2.5 to 25 m depending on the altitude of the NASA Stennis Learjet 23 which served as its platform. Two on-board blackbodies served as calibration instruments for the thermal bands. Color infrared photography was collected concurrently with the multispectral data, using a Zeiss RMK A 15/23 Camera loaded with Kodak 1443 Aerochrome III CIR. The ATLAS collected 8-bit data on 5 GB analog tapes, requiring numerous landings for data downloads. The data, once collected, needed to be processed through an analog-digital converter. With the development of the ATLAS, NSTL (now renamed the Stennis Space Center) improved and expanded on its pioneering work in archaeological remote sensing. Archaeologists from numerous institutions both in the US and aboard eagerly requested project specific ATLAS data for their research.

The data collected from all four airborne sensors (TMS, TIMS, CAMS and ATLAS) were reformatted into the Earth Resources Laboratory Applications Software (ELAS) format. The ELAS software package, developed by Stennis Space Center researchers, consisted of a Fortran-based operating system which provided advanced processing capability for remotely sensed digital data (Junkin et al. 1981). This image processing software was used extensively to develop archaeological information products until the advent of commercial software packages like ERDAS Imagine and ENVI in the early 1990s. Before the advent of faster, more capable computers and more interactive and user-friendly image processing software, scientists depended on a team of IT specialists and on time-consuming (and therefore, quite expensive) image processing techniques to retrieve information from the remotely sensed data using ERDAS and the early processors. As computing technology rapidly developed in the 1980s and early 1990s, image processing became faster, cheaper and easier, thanks in large part to the development of the coprocessor. These types of microprocessors, once only found in main frame computers, began to be included in many desktops during this period. The Intel 80386 introduced in 1985, provided a very significant advantage for remote sensing researchers, especially when combined with the optional "math" coprocessor, since scientists could now run floating point functions with much greater speed, resulting in lower costs for processing large digital data sets, such as those collected by multispectral scanners.

Still, many challenges remained at the end of the 1980s before remote sensing could become an efficient and effective tool in the archaeologist's kit. For example, in a 1993 review of remote sensing in southeastern US archaeology, Fred Limp noted that one limitation of remote sensing applications in archaeology was that the imagery available was poorly suited for detecting small targets. Spatial resolution could be improved by flying a sensor "low and slow" but the data collected in this fashion required extensive correction especially when dealing with the random attitudinal movements of the aircraft as it completed each individual flight line. In addition, individual flight lines of data needed to be stitched together or mosaicked before most archaeological analysis could be initiated. In part to address these issues of interest to archeologists, Stennis Space Center personnel increased their search for improving spatial resolution while reducing processing time. A new program managed at Stennis in the late 1990s improved significantly the availability of high spatial resolution data which could be used in archaeology with a minimum of cumbersome pre-processing.

7.5 NASA Scientific Data Purchase (SDP) Program

NASA's Scientific Data Purchase (SDP) Program implemented by the Stennis Space Center Earth Science Application Directorate in fiscal years 1998–2002 resulted in several advances in the use of remote sensing in archaeology. SDP was one in a series of congressionally funded demonstration projects aimed at evaluating the utility of commercial remote sensing data for scientific purposes (NASA/SSC 2003). A number of commercial vendors were selected to provide scientific data for a variety of research and application projects. Most relevant to archaeological applications were: Positive Systems of Whitefish Montana who collected airborne high spatial resolution multispectral data with its ADAR 5500 camera (Airborne Data Acquisition and Registration); Earth Watch, Inc. (formerly Digital Globe) of Longmont Colorado who partnered with Intermap Technology to provide radar and elevation data from STAR 3i X , an airborne Interferometric Synthetic Aperture RADAR (InSAR) system which collected radar imagery at 2.5 m. and Space Imaging, LLC (now GeoEye) of Thornton, Colorado that provided 1 m pan and 4 m multispectral 11-bit data in the visible and the near-IR wavelengths data from its orbiting IKONOS satellite.

As part of the SDP program, NASA partnered with several institutions to deliver high quality commercial data products for archaeological research. In this way, Stennis provided a variety of data products for Boston University's projects in Aksum, Ethiopia and Holmul, Guatemala; for NASA's Marshall Space Flight Center archaeology work in Central America, and for UCLA who in partnership with the Centro Interdipartimentale di Servizi di Archeologia Istituto Universitario Orientale, Naples Italy conducted archaeological research of early Greek and Roman settlements in the Italian peninsula and finally for Rutgers University in support of paleoanthropological research in Koobi Fora, Kenya (Dibble 2006). Also, SSC and Environmental Protection Agency's Gulf of Mexico Program (collocated at the Stennis Space Center in Mississippi) collaborated with state agencies in Veracruz to further the analysis of several important Olmec sites including San Lorenzo, Laguna de los Cerros, Tres Zapotes, and El Zapotal through the use of commercial data.

One project supported through the SDP and later continued through funding by the Stennis Space Center coincided with the 200th anniversary of the Lewis and Clark Corps of Discovery Expedition to explore the Western United States.

7.5.1 Lewis and Clark Bicentennial Project

The NASA Lewis and Clark project coincided with the Corps of Discovery Expedition bicentennial (2003–2006) celebration. The objective of the project was to demonstrate the potential of integrated remote sensing and GIS for site-specific studies of Lewis and Clark encampments. NASA SSC partnered with Montana University and the Columbia Gorge Discovery Center to map and identify Lewis and Clark sites using both IKONOS and ATLAS airborne data. The project resulted in map products that reconstructed 360° views of the landscape traversed by Corps of Discovery Expedition in the early nineteenth century, and compared their journal entries and historical documents to modern landscapes using remote sensing imagery to provide accurate historic views of the landscape explored by the Expedition.

For several years, Ken Karsmizki of the Columbia Gorge Discovery Center had led the search for Lewis and Clark sites along the Expedition's route. He provided NASA with copies of the original Clark maps together with a prioritized list of archaeological research objectives. Remote sensing scientists at Stennis Space Center enhanced satellite imagery (initially LANDSAT and AVHRR) with elevation models to project the historic Lewis and Clark maps onto the current topography. William Clark who produced the vast majority of the maps collected during the Corps of Discovery Expedition (1804–1806) often denoted the location of bluffs on his charts which unlike the river courses remained relatively stable over the centuries. The elevation models developed by SSC included the location of

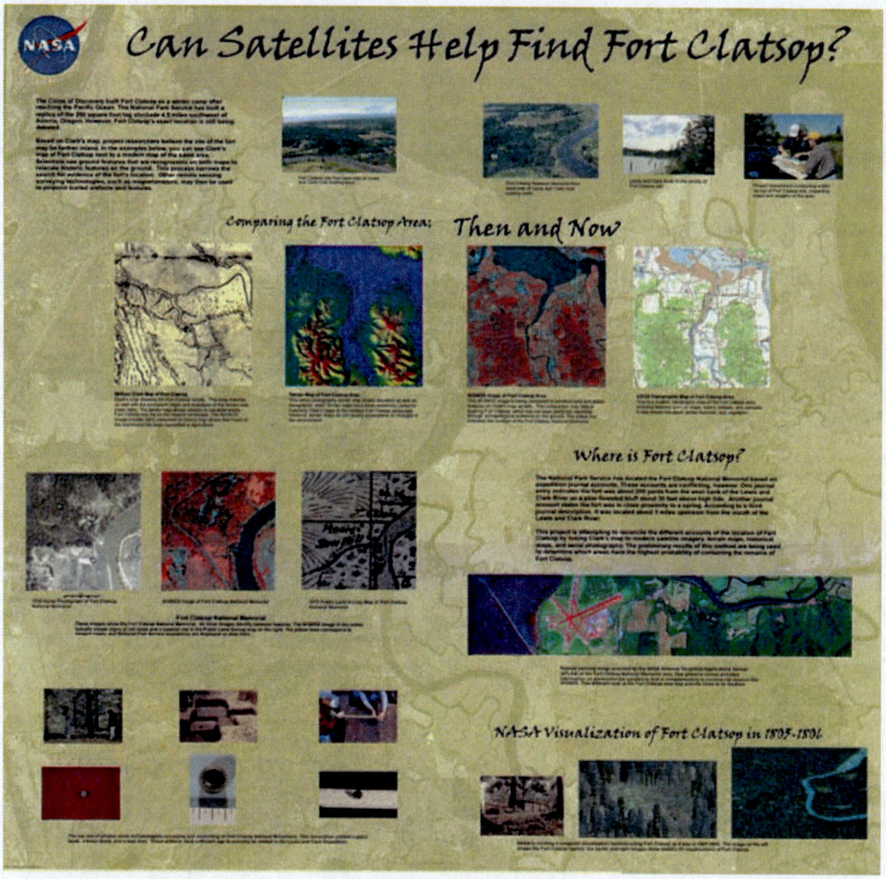

Fig. 7.1 Remote sensing products developed by NASA/SSC for the Lewis and Clark Expedition project. Collaboration between archaeologists and NASA remote sensing scientists resulted in more accurate maps of the possible locations of the Corps of Discovery camp sites and travel routes. *Lower right panels* illustrate the virtual rendering of Fort Clatsop derived from historic records and satellite imagery. This poster and the one that follows as Fig. 7.2 below were developed as part of a 200 Anniversary Exhibit on the Lewis and Clark Expedition held in New Orleans in 2003

these bluffs and so served to refine the predicted locations for several important sites such as Fort Clatsop (Karsmizki et al. 2003a), the location of the Iron Boat (Karsmizki et al. 2003b) Rock Fort Camp Site, the Dalles, Oregon (Karsmizki et al. 2004) (see Fig. 7.1).

The co-registration of modern digital images with Clark's historic maps completed during the Lewis and Clark project, provided a very valuable product for focusing

Fig. 7.2 Remote sensing products developed for the Lewis and Clark Expedition project. Collaboration between archaeologists and NASA remote sensing scientists resulted in more accurate maps of the possible locations of the Corps of Discovery camp sites and travel routes

the ground search for specific sites. This approach was implemented subsequently by NASA in other projects, including that conducted at Gainesville, Mississippi (Sect. 7.6). Another product developed during the Lewis and Clark project was the virtual rendering of several historic Corps of Discovery sites, especially Fort Clatsop. These virtual products enhanced scientific research and served to educate and inform the general public about the project and its results (see Fig. 7.2).

By the conclusion of the Lewis and Clark project, a total of 30 different NASA space borne remote sensing data sets were effectively processed. These included MODIS 8 day average reflectance; SRTM hill shaded DEM data at 750 m spatial resolution; orthorectified mosaics of Landsat TM and ETM data, 11 scenes of ASTER imagery plus 4 datasets from the International Space Station and 9 from the Space Shuttle platforms. In addition, pan-sharpened bi-products of 19 commercial high-resolution satellite image data sets (7 from IKONOS and 12 from QuickBird) were processed and ingested into the ArcSDE. Finally, three scenes of geo-referenced ATLAS multispectral bands 12,6,2 (TIR, NIR, Green) at spatial resolutions ranging from 2.5 to 4 m were provided to the archaeologists for Fort Clatsop in Oregon, the Dalles in Oregon, and over the Upper Portage Camp in Montana. When co-registered to the IKONOS data, the ATLAS data sets provided a method to view spectral response in the thermal wavelengths, not available from IKONOS and QuickBird.

7.6 Gainesville, Mississippi and the Stennis Space Center Buffer Zone Archaeology

Lessons learned during the Lewis and Clark Project were applied to the study and preservation of several historic localities situated within the Stennis Space Center. The Gainesville site is one of five historic towns that were evacuated when the area became part of the Stennis Space Center Buffer Zone and Fee area in 1962 with the advent of the Saturn rocket program (see Fig. 7.3). Gainesville had been the county seat of Hancock County, Mississippi, during the mid-nineteenth century and even though no buildings were left standing, many subsurface features were still in situ.

Armed with high-spatial resolution multispectral data from orbiting platforms, NASA researchers co-registered historic plats and maps to the digital data (IKONOS, Quick Bird and airborne ATLAS) to identify historic boundaries and structures (see Fig. 7.4). Digital Elevation Models (DEMs) were developed with the use of LIDAR and RADAR imagery which, when utilized in combination with X and Y coordinates derived from high spatial resolution imagery, produced accurate 3-D renderings of sites and the surrounding environment.

This approach provided more precise locations of these historic sites and features, produced very accurate site maps and allowed development and testing of predictive site location models. NASA is developing the Gainesville site as an archaeological validation and verification site for the application of a variety of remote sensing data collection and interpretation methods including ground penetrating radar, magnetometry and resistivity; and hyperspectral data from an airborne platform.

Fig. 7.3 Location of the NASA Stennis Space Center in Southwest Mississippi, USA. The *red outline* depicts the Acoustic Buffer Zone; the *yellow outline* illustrates the Fee Area

Fig. 7.4 Co-registration of 1837 Plat of Gainesville to modern satellite imagery (IKONOS). This method greatly facilitates planning and implementing archaeological research designs by identifying the location of historic places

7.7 Coast 2050 Cultural Resources Survey

One final notable archaeological project undertaken by archaeologists at the Stennis Space Center was the remote sensing survey of 50,000 acres of Coastal Louisiana for the detection of archaeological sites (Giardino 2009). The survey, conducted in collaboration with the U.S. Army, Corps of Engineers in New Orleans was a part of the "Coast 2050" program which was funded to understand and to mitigate the massive losses of wetlands (and therefore archaeological resources) occurring in the Southeast Louisiana. NASA sought to advance the use of remote sensing as an efficient and effective tool for identifying cultural resources located in the coastal wetlands where traditional surveying techniques are complicated by the difficult terrain.

As in previous similar surveying efforts, NASA utilized both LANDSAT and high-resolution hyperspectral and multispectral data from an airborne platform to develop predictive models of site location within the project area. Site files located at the State Archaeologists' Office in Baton Rouge were reviewed and those with relatively accurate location information were entered into a GIS. This enabled the location of sites on LANDSAT imagery. The study area was then flown with the ATLAS scanning radiometer at a spatial resolution of 4 m to refine site locations. The data was analyzed to extract spectral response curves for a variety of sites including shell middens and earthen mounds and the results were used as the basis for supervised classifications whose purpose was to identify high probability localities for undiscovered sites. Although this approach was not entirely new, it is was augmented by the use of hyperspectral data which provided new information on the physical properties of land features, particularly vegetation and soils. One hundred and thirteen high potential localities of which more than 73% were validated through field work.

7.8 Space Archaeology Solicitation

In 2008, for the first time in its history, NASA released a solicitation specifically addressing the use of remote sensing in archaeology. NASA archaeological research became formalized, rather than serendipitous. Seven projects were selected for funding including one by Douglas Comer of the Cultural Site Research and Management, Inc. to develop software that will specifically identify archaeological site signatures from images acquired by means of NASA sensing platforms, in particular ASTER and Hyperion. Another by William Middleton of the Rochester Institute of Technology will analyze hyper- and multi-spectral satellite imagery from the Hyperion and ALI for the analysis of archaeological landscapes in Oaxaca and John Weishampel of the University of Central Florida will using hyperspatial satellite imagery and airborne LiDAR data to study ancient Maya land use at Caracol, Belize.

The common theme among several of these NASA funded projects is hyperspectral remote sensing which is increasingly being investigated by archaeologists as the next potentially promising technology for research. A new NASA solicitation was released in 2009 and its resulting awards are pending, and hypespectral and LiDAR projects will probably be prominent once again.

This paper has not addressed the valuable contributions of NASA active remote sensing projects in enhancing archaeological discovery and research. These have been documented elsewhere (see for example Freeman et al. 1999; Moore et al. 2007). Clearly, active sensors like imaging RADAR provide valuable data in most conditions, and terrains.

7.9 The Future

Remote sensing technology and image analysis are currently undergoing a profound shift in emphasis from broad classification to detection, identification and condition of specific materials, both organic and inorganic. In the last few years, remote sensing platforms have grown increasingly capable and sophisticated. Sensors currently in use, or nearing deployment, offer significantly finer spatial and spectral resolutions than were previously available. Paired with new techniques of image analysis, this technology may make the direct detection of archaeological sites a realistic goal (Jones and Giardino 1997; Wiseman and El-Baz 2007).

As of 2009 twenty-one NASA Earth Observing satellites are operating with five in development and six under study. One in particular, ASTER is increasingly being utilized by archaeologists due to its high spatial resolution and broad spectral capabilities (Altaweel 2005). ASTER is one of the five state-of-the-art instrument sensor systems on-board the Terra satellite launched in December 1999 as a cooperative effort between NASA and Japan's Ministry of Economy Trade and Industry (METI). The ASTER instrument provides improved remote sensing imaging capabilities compared with the older Landsat Thematic Mapper. ASTER's 14 multispectral bands monitor land temperature, land use, and vegetation patterns at a spatial resolution of 90 to 15 m. ASTER is capable of producing stereoscopic images and detailed terrain height models. ASTER imagery has been collected already over several important archaeological sites including L'Anse aux Meadows in Newfoundland, where the remains of a Viking village were discovered in 1960; the ruins of Machu Picchu, Mount Ararat, Teotihuacan, the Nazca Lines in Peru and Olduvai Gorge.

Even more exciting is the potential for adopting hyperspectral data in archaeological research. NASA's Hyperion sensor currently on an extended mission on the EO-1 platform provides relatively high resolution (30 m) hyperspectral images across 220 spectral bands (from 0.4 to 2.5 μm). Airborne hyperspectral sensors like NASA's Jet Propulsion Laboratory Airborne Visible InfraRed Imaging Spectrometer (AVIRIS) provide similar hyperspectral coverage at variable spatial resolutions (nominally at 17 m GSD). HYDICE (HYperspectral Digital Imagery Collection Experiment) collects data across 210 bands over the range similar to Hyperion and AVARIS (0.4–2.5 μm) but with greatly improved spatial resolution (1–4 m depending on the aircraft altitude and ground speed).

These new technologies offer archaeologists even more opportunities for research and analysis. One possible area where hyperspectral data, can be very valuable is phytoarchaeology which is defined as the analysis of relationship between vegetation and archaeology (Brooks and Johannes 1990:9). Three specific areas of phytoarchaeology can greatly benefit from hypespectral imagery: (1) the identification of specific plant species that are associated with archaeological sites; (2) comparative plant physiology or the determination of plant stress or vigor and (3) creation of a vegetation variability index.

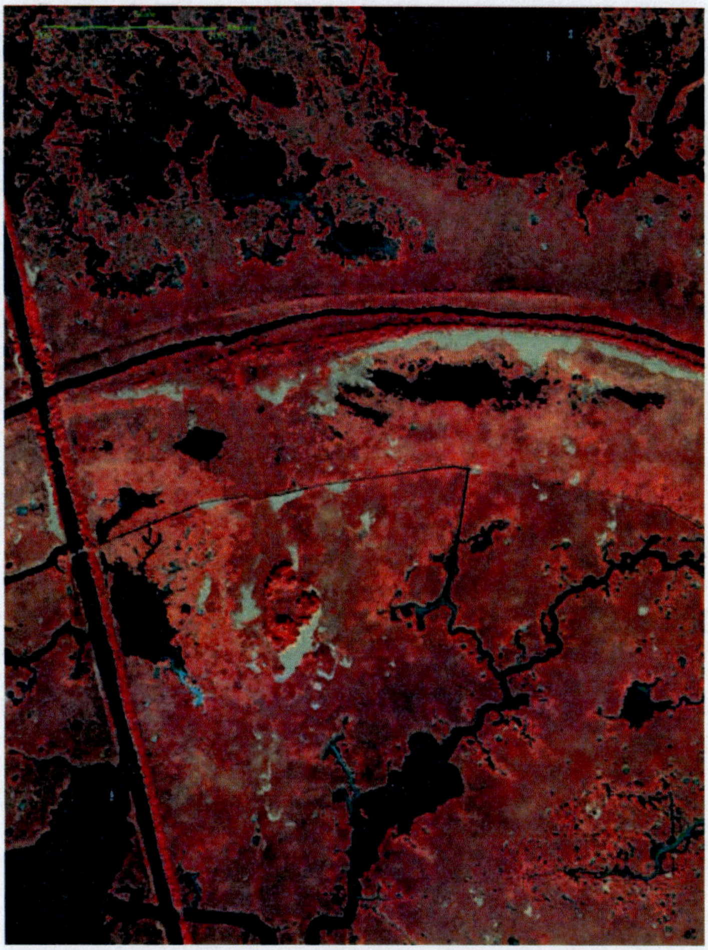

Fig. 7.5 ATLAS image (RGB) of the Grand Chenier Site in Plaquemines Parish, Louisiana USA. Clearly visible are the large oak trees on the 11 prehistoric mounds which are arranged around a plaza. The *faint red line* running NW from the mound site toward the bayou is a prehistoric causeway whose existence was undiscovered until remote sensing images were analyzed. At its northern terminus, the causeway intersects another prehistoric site located along the ancient bank of the bayou. The archaeological site is a shell midden which is sufficiently high to provide the proper environment for the growth of live oaks which are visible in the imagery

For example, in Louisiana's coastal marsh landscapes, "upland" live oaks and associated plants like hackberries are often the first and most reliable evidence of a buried archaeological site (Brown 1936; see Fig. 7.5). These trees are readily identifiable through their spectral response curves in remote sensed images, and can be further classified to specie with hyperspectral data.

Secondly, the presence of plant stress (or vigor) indicators in remotely sensed data (generally in area of the "red edge" [0.68–0.75 μm]) may serve as a proxy for the presence of archaeological deposits. Hyperspectral data can be useful in identifying stress in vegetation, particularly when coupled with laboratory research. For example, in one experiment, the ratio of R_{695} to R_{760} or R_{800} was the most consistent stress indicator among several ratios tested (Carter 1994; Giardino and Haley 2006:51). Hyperspectral data can provide data in these narrow regions of the EMS.

Finally, hyperspectral data, in conjunction with algorithms for unmixing the spectral constituents of pixel, will be useful toward identifying sites through a third technical approach: development and use of a "vegetation variability index (VVI)".

The premise underpinning the vegetation variability index is that vegetation on archaeological sites will exhibit higher specie variability resulting from anthropogenic factors such as plant collecting, horticulture, waste production and disposal than the off-site communities (Traviglia 2008; Dunn 1983; Eleuterius and Otvos 1979). Measuring variability in plant species over a given region becomes more feasible when using hyperspectral data and concomitant processing techniques.

7.10 Conclusion

In her report 1970 report, NASA summer intern Mary Scalera concluded (1970:26): "The space age has broadened the archaeologist's horizons. In fact, fantastic as it may sound, many have contemplated the expansion of archaeology to extraterrestrial discoveries. Archaeologists have been confronted with such questions as, 'What would you do if a discovery of past civilization was made on Mars?'" I have no doubt that should that event unfold, NASA will be ready to apply all its remote sensing tools and techniques to a new chapter of archaeological exploration.

References

Altaweel M (2005) The use of ASTER satellite imagery in archaeological contexts. Archaeol Prospect 12(3):151–166

Belew LF, Stuhlinger E (1973) Skylab: a guidebook EP-107. U.S. Government Printing Office, Washington, DC

Bennett J, Smith L, Laustrup M (1986) The red river valley archaeological project. TIMS user workshop JPL, California Institute of Technology

Brooks RR, Johannes D (1990) Phytoarchaeology. Dioscorides Press, Portland

Brown CA (1936) The vegetation of the Indian mounds, middens, and marshes in Plaquemines and St. Bernard Parishes. Department of Conservation-Louisiana Geological Survey 8:423–440

Carter GA (1994) Ratios of leaf reflectances in narrow wavebands as indicators of plant stress. Int J Remote Sens 15:697–703

Compton DW, Benson CD (1983) Living and working in space: a history of Skylab. The NASA History Office Series Scientific and Technical Information Branch National Aeronautics and Space Administration, Washington, DC

Cook JP, Stringer WJ (1975) Feasibility study for locating archaeological village sites by satellite remote sensing. Final report for contract NASS 21833, ERTS project 110-N. Greenbelt, Maryland: Goddard Space Flight Center

Custer JF, Eveleigh T, Klemas V, Wells I (1986) Application of LANDSAT data and synoptic remote sensing to predictive models for prehistoric archaeological sites: an example from the Delaware coastal plain. Am Antiquity 51:572–588

Dibble LJ (2006) Holocene landscapes at Koobi for a, Kenya: application of geographic information science to archaeological questions of landuse and subsistence. PhD Dissertation, Rutgers University, May 2006

Dunn ME (1983) Coquille Flora (Louisiana): an ethnobotanical reconstruction. Econ Bot 37:349–359

Ebert JI, Lyons TR (1980) Prehistoric irrigation canals identified from Skylab III and landsat imagery in Phoenix, Arizona. In: Lyons TR, Mathien FJ (eds) Cultural resources remote sensing. Southwest Cultural Resources Center, National Park Service, Albuquerque, pp 209–228

Eleuterius LN, Otvos EG (1979) Floristic and geologic aspects of Indian middens in salt marshes of Hancock County, Mississippi. Sida Contrib Bot 8:102–112

Freeman A, Hensley S, Moore E (1999) Analysis of radar images of Angkor, Cambodia. IGARSS Proc 5:2572–2574

Giardino MJ (2009) NASA, remote sensing and archaeology: an example from Southeast Louisiana. EARSel Newsl 80:24–28

Giardino MJ, Haley BS (2006) Airborne remote sensing and geospatial analysis. In: Johnson J (ed) Remote sensing in archaeology: an explicitly North American perspective. University of Alabama Press, Tuscaloosa, pp 47–77

Gibson JL (1984) The earthen face of civilization: mapping and testing at poverty point. Office of the State Archaeologist, Baton Rouge

Gibson JL (1987) The ground truth about poverty point: the second season, 1985. Center for archaeological studies report 7. University of Southwestern Louisiana, Lafayette

Gibson JL (1989) Digging on the dock of the bay(ou): the 1988 excavations at poverty point. Center for archaeological studies report 8. University of Southwestern Louisiana, Lafayette

Gumerman GJ, Lyons TR (1971) Archaeological methodology and remote sensing. Science 172:126–132

Johnson JK (2006) Remote sensing in archaeology: an explicitly North American perspective. University of Alabama Press, Tuscaloosa

Johnson J, Madry S, Sever T (1988) Remote sensing and GIS analysis in large scale survey design in North Mississippi. Southeastern Archaeol 7(2):124–131

Jones P, Giardino MJ (1997) Recent advances in remote sensing technology. Paper presented at the 54th annual meeting of the southeastern archaeological conference, Birmingham, 5–8 Nov 1997

Junkin BG, Pearson RW, Seyfarth BR, Kalcic MT, Graham MH (1981) Earth resources laboratory applications software (ELAS). Report no. 183. NASA, National Space Technology Laboratories, Earth Resources, NSTL, Mississippi, USA

Kahle AB, Abbott E (eds) (1985) The TIMS data user's workshop 18–19 June 1985. NASA Jet Propulsion Laboratory, California Institute of Technology, Pasadena

Karsmizki KW, Spruce J, Giardino MJ (2003a) Archaeological remote sensing: searching for Fort Clatsop from space. In: Abstract in proceedings for society of historical archaeology 2003 conference. Providence, 14–19 Jan 2003

Karsmizki KW, Spruce J, Weymouth JW, Giardino M, English RW (2003b) Searching for Lewis and Clark's Iron Boat: remote sensing on the ground and from space. In: Abstract in proceedings for 2003 society of American archaeology conference, Milwaukee, 9–14 April 2003

Karsmizki KW, Spruce J, Giardino MJ (2004) Use of remote sensing and GIS in search for Lewis and Clark's rock fort camp site of The Dalles, Oregon. Presentation given at 2004 ESRI conference, San Diego, 9–13 Aug 2004

Kruchman L (1976) Remote sensing and archeology: a preliminary bibliography. RSEMS, Newsletter of the remote sensing committee of the Association of American Geographers 3(2)

Limp FW (1989) The use of multispectral digital imagery in archaeological investigations. Arkansas Archaeological Survey Research Series 34

Lyons TR, Avery TE (1977) Remote sensing: a handbook for archeologists and cultural resource managers. National Park Service, (U.S. Government Printing Office), Washington, DC

Lyons TR, Ebert JE (eds) (1978) Remote sensing and non-destructive archeology. Cultural Resources Management Division, National Park Service, Washington, DC
Lyons TR, Hitchcock RK (eds) (1977) Aerial remote sensing techniques in archeology. Reports of the Chaco Center, no. 2, National Park Service and University of New Mexico, Albuquerque
Lyons TR, Mathien FJ (eds) (1980) Cultural resources remote sensing. Cultural Resources Management Division, National Park Service, Washington, DC
Madry S (1983) Remote sensing in archaeology. Archaeology, pp 18–19
McKee BR, Sever TL (1994) Remote sensing in the arenal region. In: Sheets PD, McKee BK (eds) Archaeology volcanism and remote sensing in the arenal region, Costa Rica. University of Texas Press, Austin
Mid-America Remote Sensing Center (1983) Determination of the utility of remote sensing data for land use/cover analyses in the Lower Appalachia Region-assessing the utility of remote sensing data for archaeological site reconnaissance. NASA Contract NAS13-200, Murray State University
Miller WF (1974) Applications of remote sensing in archeological site identification. Institute of Environmental Studies, Mississippi State University, Starkville, Mississippi
Moore E, Freeman T, Hensley S (2007) Spaceborne and airborne radar at Angkor: introducing new technology to the ancient site. In: Wiseman J, El-Baz F (eds) Remote sensing in archaeology interdisciplinary contributions in archaeology. Springer, New York, pp 185–216
MASCA Newsletter (1969) Balloon and airfoil photography. Applied Science Center for Archaeology, University of Pennsylvania, Nov 1969, 5(2):2–3
NASA/SSC (2003) The NASA scientific data purchase final report. Contract no. 13–650, National Aeronautics and Space Administration, John C. Stennis Space Center, Mississippi, Sept 2003
Parcak SH (2009) Satellite remote sensing for archaeology. Routledge, New York
Parrington M (1983) Remote sensing. Ann Rev Anthropol 12:105–124
Rickman D, Kalcic M (1982) Noise removal by Principal Component Analysis. Report no. 203, NASA National Space Technology Laboratories, Earth Resources Laboratory, NSTL, MS
Scalera MM (1970) Aerial archaeology in the space age. NASA Historical Division, Office of Policy, NASA-TM-X-68502, Aug 1970
Schaber GG, Gumerman GJ (1969) Infrared scanning images: archaeological application. Science 164(3880):712–713
Sever TL (1983) Feasibility study to determine the utility of advanced remote sensing technology in archeological investigations. Report no. 227, (NASA-TM-88650), NSTL, ERL, Dec 1983
Sever TL (1990) Remote sensing applications in archeological research: tracing prehistoric human impact upon the environment. Unpublished Ph.D. dissertation, University of Colorado, Boulder. University Microfilms, Ann Arbor, MI
Sever TL, Wagner DW (1991) Analysis of prehistoric roadways in Chaco Canyon using remotely sensed digital data. In: Trombold C (ed) Ancient road networks and settlement hierarchies in the New World. University Press, Cambridge, pp 42–52
Sever TL, Wiseman J (1985) Conference on remote sensing: potential for the future. NASA Earth Resources Laboratory, NSTL, MS, Jan 1985
Short NM (1982) The landsat tutorial handbook: basics of satellite remote sensing. NASA Reference Publication 1078, Scientific and Technical Branch, National Aeronautics and Space Administration, Washington, DC
Traviglia A (2008) A view of greater Angkor: a multi-scalar approach for investigating the Khmer forests. In: Lasaponara R, Masini N (eds) Advances in remote sensing for archaeology and cultural heritage management. Aracne Press, Rome, pp 23–26
Wells I, Custer J, Klemas V (1981) Locating prehistoric archaeological sites using landsat. In: 15th International symposium on remote sensing of environment, Ann Arbor, 11–15 May 1981, Proceedings. vol 2. Environmental Research Institute of Michigan, Ann Arbor, pp 771–780
Wiseman JR (1984) Archaeology in the space age. Context 4(1–2):1–3, Boston University
Wiseman J, El-Baz F (2007) Remote sensing in archaeology. Springer, New York, NY

Chapter 8
Satellite-Based Monitoring of Archaeological Looting in Peru

Rosa Lasaponara, Maria Danese, and Nicola Masini

Abstract Illegal excavations represent one of the main risk factors which affect the archaeological heritage all over the world, in particular in those countries, from Southern America to Middle East, where the surveillance on site is little effective and time consuming and the aerial surveillance is non practicable due to military or political restrictions. In such contexts satellite remote sensing offers a suitable chance to monitor this phenomenon. The chapter deals with the results obtained on some areas of Cahuachi (Peru) by using a time series of very high resolution satellite images. The rate of success in detecting changes related to archaeological looting has been fruitfully improved by adopting a semiautomatic approach based on spatial autocorrelation.

Keywords Archaeological looting • Spatial autocorrelation • Cahuachi • Peru

8.1 Introduction

Archaeology contributes much more than other disciplines to the history and knowledge of ancient civilizations. It stimulates high interest in ancient artefacts and objects, and thus, unfortunately, strongly boosting illegal searching for treasure which causes irreversible damage to archaeological sites.

R. Lasaponara
Institute of Methodologies for Environmental Analysis, CNR-IMAA,
C. da S. Loya, 85100 Tito Scalo, PZ, Italy
e-mail: lasaponara@imaa.cnr.it

M. Danese • N. Masini (✉)
Institute of Archaeological and Architectural Heritage, CNR-IBAM,
C. da S. Loya, 85050 Tito Scalo, PZ, Italy
e-mail: m.danese@ibam.cnr.it; n.masini@ibam.cnr.it

R. Lasaponara and N. Masini (eds.), *Satellite Remote Sensing: A New Tool for Archaeology*, Remote Sensing and Digital Image Processing 16,
DOI 10.1007/978-90-481-8801-7_8,

In many countries of Southern and Central America, Asia and Middle East looting and clandestine excavations affect archaeological heritage more than other man-made and natural risks, denying this cultural heritage to new generation (Atwood 2006).

Clandestine excavation activity is mainly linked to illicit trade of antiquities in Europe and North America (Brodie and Renfrew 2005; Luke 2006). To contrast this phenomenon since 1956 the General Conference of the United Nations Educational, Scientific and Cultural Organization recommended all the Member States to take "all necessary measures to prevent clandestine excavations and damage to monuments and also to prevent the export of objects thus obtained" (UNESCO 1956).

The 1956 UNESCO Recommendation increased the international cooperation in adopting repressive measures and in returning objects derived from clandestine excavations or theft to their countries of origin. In particular, since the 1960s most of the UNESCO Member States took measures which obliged museums to ascertain that acquired archaeological objects were not procured by clandestine excavations, theft or any other methods regarded as illicit by the competent authorities of the country of origin.

To contrast site looting and illegal trade of archaeological objects we also cite the role of excavation services, museums and their representative associations which lend assistance in order to ensure or facilitate the recovery of objects derived from clandestine excavations or exported in infringement of the legislation of the country of origin. This was much more strictly after the adoption in 1970 by the UNESCO of the "Convention on the Means of Prohibiting and Preventing the Illicit Import, Export and Transfer of Ownership of Cultural Property" (UNESCO 1970).

The looting phenomenon is much more dramatic during wars or armed conflicts, as occurred in Iraq during the two Gulf Wars. At such regard, Elizabeth C. Stone comments that "total area looted was many times greater than all the archaeological investigations ever conducted in southern Iraq" (Stone 2008). Media reports described the massive looting in broad daylight and destruction of the Iraqi museums and other cultural institutions. Between 2003 and 2004, several buried ancient cities have been completely eaten away by crater-like holes (http://www.savingantiquities.org/feature_page.php?featureID=7), and many other archaeological sites would be pillaged without the valuable activity of the Italian Carabinieri, responsible for guarding archaeological sites in the region of Nassyriah (Fig. 8.1).

In spite of a new ethical environment against the acquisition of unprovenanced antiquities followed to the UNESCO conventions (Brodie and Renfrew 2005) much more efforts need to be addressed to contrast the looting and smuggling of antiquities which increasingly affect archaeological monuments and sites all over the world.

An effective approach used in Europe is based on the cooperation of public departments for the management and protection of cultural heritage, superintendences, heritage protection services and non-profit archaeological groups, which compose a synergic network for direct surveillance on site.

However such approach is time consuming, expensive and not suitable for remote archaeological sites, characterized by difficult accessibility.

Fig. 8.1 Aerial surveillance of archaeological sites in Iraq by Italian Carabinieri, responsible for guarding cultural heritage in the region of Nassyriah (Courtesy of Italian Carabinieri)

To cope with such difficult conditions, in many countries the protection activity is carried out by using remote sensing. In particular, aerial images allow the monitoring of the most valuable archaeological heritage to contrast the clandestine excavations, to limit the impact of illegal urbanization and anthropic activities.

Since the 1930s of the twentieth century aerial surveillance has been a common manner to monitor the phenomenon of archaeological looting and damage. Aerial images had been long appreciated by archaeologists and conservators as it permits not only the discovery of unknown sites but also the monitoring of know sites and the estimation of risk factors, such as environmental (landslides, flooding) and man-made (i.e. the impact of urban and industrial infrastructures) issues.

The effectiveness of aerial prospection as manner to contrast and monitor the archaeological damage depends on the possibility to carry out the surveillance on a regular time basis with a systematic scheduling. To optimize performance, costs and time the approach used is based on : (i) the integration of vertical and oblique photographs (the latter emphasize microrelief or shadow-marks caused by looters), (ii) and the gaining of aerial views with different vectors such as helicopter, airplane, hot air balloons, scaffolds or cameras attached to kites.

Aerial observation is used with effectiveness in the US and in most of European countries for monitoring archaeological sites and territories of cultural interest.

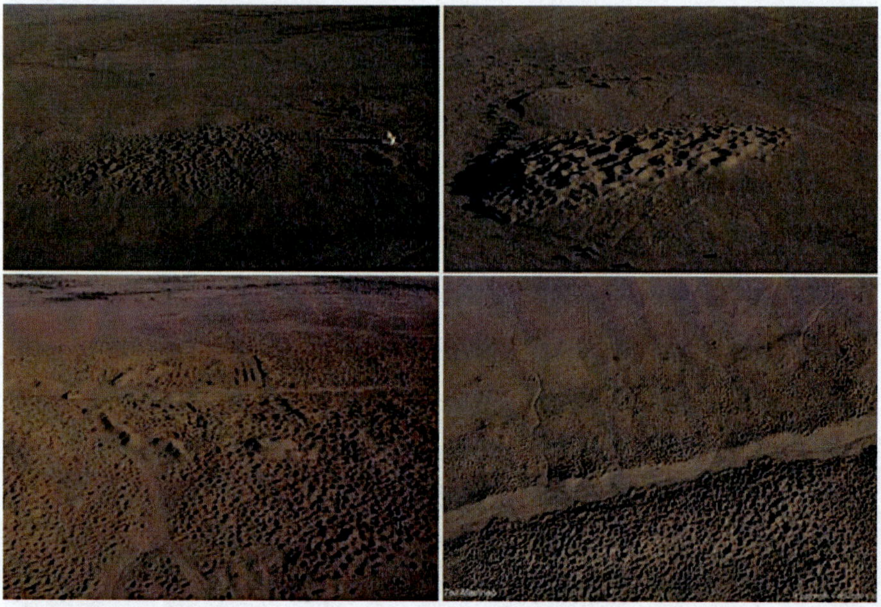

Fig. 8.2 Some archaeological sites looted in Iraq during the second Gulf War: Abbas al Kurdi (*upper left*), Sifr (*upper right*), Jokha (*lower left*), and Tell Medinah (*lower right*) (Courtesy of Italian Carabinieri)

Unfortunately, it is non practicable in several countries due to military or political restrictions, unless aerial surveillance is carried out within of peace-keeping missions as in the case of Nassyriah in Iraq, where the Italian Carabinieri Protection Nucleus, between May 2003 and January 2004, surveyed several sites from late Uruk to early Islamic targeted just before and after the 2003 invasion. Aerial survey put in evidence the significant damages of numerous archaeological sites, such as Tell-Medinah, Abbas al-Kurdi, Jokha, Sifr, Tell Schmid, Umm al-Aqarib, Zabalam (Fig. 8.2). The Italian Carabinieri's activities opened a way to reduce or even prevent further damages to the archaeological heritage in the following years.

Today the problem is to assure a continued surveillance on site and by means of aerial observation. Both of them are actually little practicable for security reasons and because they are very expensive and time consuming. In particular, the georeferencing is time consuming not only for large archaeological areas but also small archaeological sites, when frequent revisitation times are required. Finally, aerial surveillance is little effective for huge areas and for difficult environmental settings (desert, rain forest).

In such contexts very high resolution (VHR) satellite imagery such as Ikonos, QuickBird2 (QB2), WorlView1 (WV1), GeoEye and WorldView (WV2) offer a suitable chance thanks to their global coverage and a frequent re-visitation time.

Recently, QB2 satellite data have been used in Middle East to survey and quantify the looting damage on the archaeological heritage of Southern Iraq during the last armed conflicts (Stone 2008). The available VHR satellite time series, from 2002 January to 2005 November, allowed to address many issues such as (i) analysis of the chronology of looting of several sites, (ii) estimation of the percentage of damaged area for each site, (iii) evaluation of size and the number of holes and (iv) distribution of damage in relation with size and historical periods of archaeological sites (Stone 2008).

Declassified satellite imagery acquired in the 1960s and early 1970s by Corona represent a further data source not only for detecting archaeological features (Fowler 1997), but also for monitoring cultural resources from the 1960s of the twentieth century to the present. To such regard, in the regions where declassified satellite scene are available, multitemporal observations in the last 40 years put in evidence the dramatic loss of archaeological sites caused by illegal urban expansion, infrastructures, mechanized agriculture and looting.

For example, in some areas in Egypt investigated by Sarah Parcak (2007), maps available from the nineteenth century and high resolution satellite data time series, put in evidence that at least around the 88% of the overall sites were lost over the past 200 years, with the majority of destructions occurring within the last 40 years (Parcak 2007). By using SPOT4, with four bands at 20-m ground sample distance (GSD), Landsat 7 (with seven bands at 30-m GSD) and QB2 together with declassified Corona KH-4B scenes acquired in 1972 (1.8-m GSD), Parcak (2007) proposes a model for documenting the impact of plunder and demographic changes on ancient tells in the East Delta of Northern Egypt and in Middle Egypt. To obtain this model the author compared visual analysis with band combinations, principal component analyses, supervised and unsupervised classification classifications.

A significant improvement regards the use of very high resolution satellite for studying clandestine excavations has been provided by Lasaponara and Masini (2010). The authors devised a semiautomatic data processing approach mainly based on the use of geospatial analysis. The integration of remote and GIS technologies was applied to the archaeological heritage in the desert coastal areas of Southern Peru. The high satisfactory results put in evidence the great potentialities of satellite earth observation for monitoring of archaeological looting. The spatial resolution generally well fits with the dimensions of illegal excavations and the geospatial analysis enables us to easily identify traces and patterns linked to archaeological looting. This encouraged the authors of this chapter to further improve the semi automatic object extraction procedures. An application on the Ceremonial Centre of Cahuachi is herein shown.

8.2 Archaeological Looting in Cahuachi

Since 2007, two institutes of the Italian CNR have been using VHR satellite data to support archaeological investigations in some sites of southern and northern desert coastal areas of Peru (for additional information on the scientific investigations of CNR in Peru, the reader is referred to Chap. 14; Masini et al. 2009a, b; Lasaponara et al. 2011), where the clandestine excavations are a plague, very difficult to face by only on-site and aerial surveillance. The measures set up by Peruvian Government, in terms of aids awareness and contrast of illicit art trade, reduced this problem in some regions. However much efforts should be still addressed to guarantee an effective monitoring of cultural sites and areas, which can be effectively approached by using the novel technologies of Earth Observation, as it has been done in Cahuachi since 2008.

Cahuachi is the largest adobe Ceremonial Centre in the World, built in the southern desert of Peru by the Nasca Civilizations. The archaeological area is characterized by around 40 semi-artificial mounds, spread out on the south bank of the Nasca river and facing the Pampa de San Jose, where the majority of the famous Nasca Lines were etched (see Fig. 14.1 in Chap. 14). The archaeological investigations in the last 25 years (Orefici 1992, 1993; Orefici and Drusini 2003; Silverman 1988, 1993) allowed the understanding of the functional and cultural evolution of the site between 400 B.C. and 400 A.D. It was at the beginning a shrine (*Huaca*), then a ceremonial centre and later the Theocratic Capital of the Nasca State. The difficult environmental setting of the Nasca territory favoured an intense ceremonial activity with rituals, precious offerings and sacrifices to propitiate the gods, to have rich harvests and prevent natural disasters such as earthquakes and flash floods (Orefici 2009). The enormous quantity of precious offerings and rich tombs has been very tempting target for looters since the nineteenth century. Thus, a vast amount of polychrome Nasca ceramic was unearthed and sold to antique collectors in Lima. However, the looting in Cahuachi increased dramatically in the early twentieth century, after the first excavation campaign carried out by Max Uhle. The looting increased dramatically in the next 25 years during which the looters (named in Peru *huajeros*) dug about 30,000 tombs (Silverman 1993).

If in the early of the twentieth century the *huajeros* worked mainly individually, in the subsequent decade they started to work in teams for their own gain or for second parties. The looting was considered by people as a work to take pride in so much so distinguished between "looters of finesse" and the "brutal profaners" (Silverman 1993).

Illegal excavations do not only damage the cultural heritage, they also cause a loss of context which makes interpretation of remains unearthed by archaeologists very difficult.

Looters' holes are usually recognizable by circular pits (Fig. 8.3), somewhat filled with sand, and by scattered remains such as human and animal bones, sherds of the pottery sells broken by *huajeros* or left by Nasca as part of their mortuary rituals. A tomb can be recognized by *huajeros* by a surfacing big *huarango* post stood upright linked to an underneath wood roof which covered a square or rectangular adobe-walled chamber.

After the WW2 mounds and flat areas follow to be eaten away by crater-like holes, as proved by comparing aerial photographs of 1947, 1952 and 1970. From the 1990s the looting starts to decrease. However it is still a big problem to face by direct surveillance. Moreover, the difficulty to recognize points of reference makes also the aerial surveillance little effective and time consuming, especially for extensive areas, as in the case of Cahuachi.

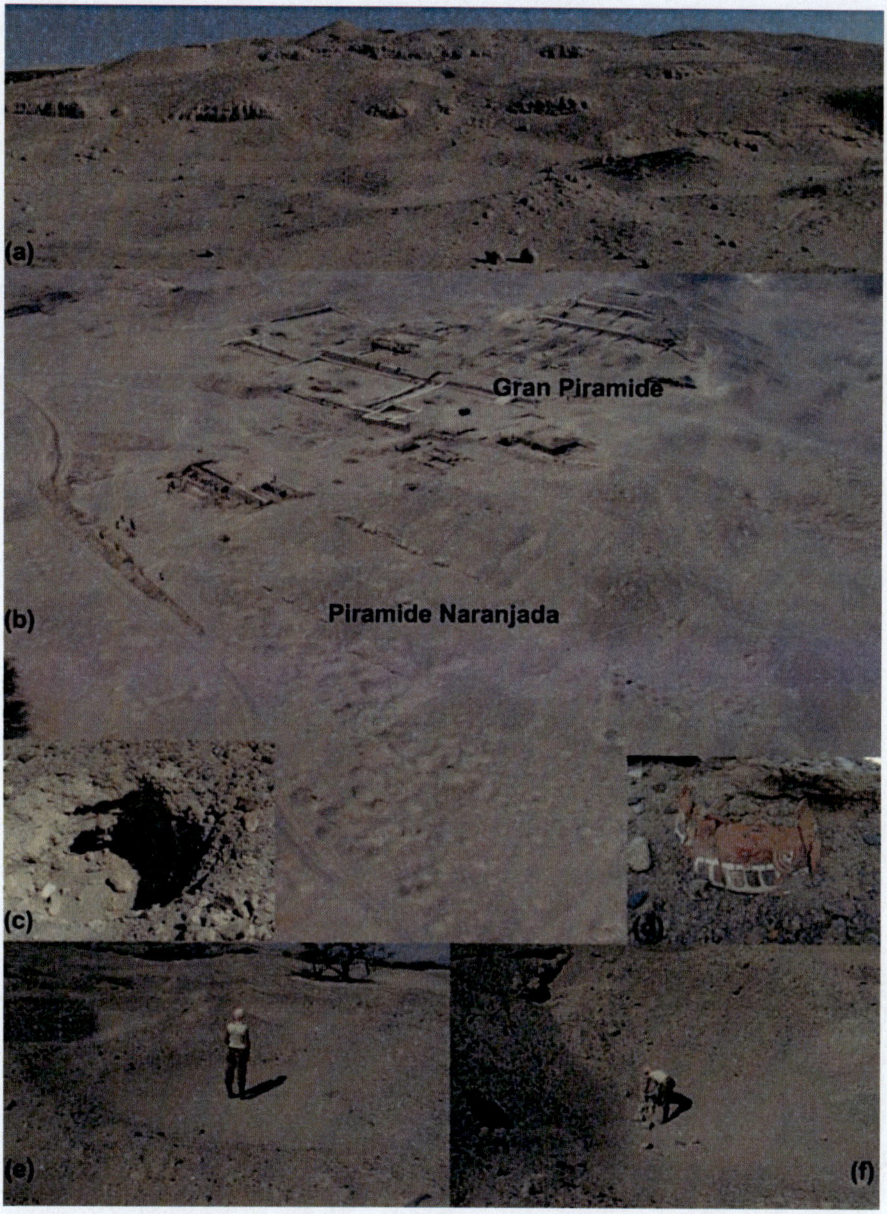

Fig. 8.3 Archaeological looting in Cahuachi. The looted areas are characterized by circular holes (**a**), with diameters ranging from 1 m (**c**) to 6–10 m (**e–f**) which affect large areas (**a**) of Cahuachi, included the pyramids (**b**). Clandestine excavations are mainly linked to illicit trade of Nasca ceramics (**d**) in Europe and North America

Fig. 8.4 Cahuachi. The black rectangular box indicates the investigated test site characterized by illegal diggings, located between the core of the ceremonial Centre and the *Rio Nasca*

Such considerations suggested to develop a monitoring method based on the use of very high satellite imagery, which is tested near the archaeological area (see Fig. 8.4).

8.3 Methodological Approach

The reliability of the satellite imagery in detecting changes linked to looting has been previously experienced and assessed by using a time series of panchromatic and multispectral QB2 and panchromatic WV1 images. The QB2 data used for this study were acquired on the 16th September 2002 with Ground sample distance (GSD) of 61.90 cm and the 25th March 2005 with GSD = 63.40 cm. WW1 data were acquired on the 31st July 2008 with GSD = 58.10 cm.

The multitemporal observation put in evidence a differential and selective occurrence of looting. Some areas are intensively plundered, whereas other came through *huajeros*. The rate of success in detecting looted areas has been assessed by field surveys carried out on some test sites. The evaluation has shown a rate of success was very high in some areas and unsatisfactory for other areas. This suggested follow an approach based on the integration of image processing and geostatistics, which achieved remarkable results for a wide spectrum of applications (Curran and Atkinson 1998; Hyppänen 1996; Atkinson and Lewis 2000; López-Granados et al. 2005 are just some examples). This chapter shows the results obtained by means of an approach based on the spatial autocorrelation, already experienced for archaeological site discovery (Ciminale et al. 2009).

Spatial autocorrelation measures the degree of dependency among events, considering at the same time their similarity and their distance relationships. For event we mean the number of spatial occurrences in the considered variable s_i (more information on spatial autocorrelation approach adopted in this case study are included in Chap. 2). In the context of image processing, the spatial event is the pixel, so spatial autocorrelation statistics are calculated considering geographical coordinates of its centroid.

In case of a non-autocorrelated distribution of a given variable s_i the probability $P(s_i)$ to have a s_i event in the point with (x, y) coordinates is computed. For an autocorrelated distribution, fixed a neighbourhood for each event, spatial autocorrelation expresses how $P(s_i)$ is modified from the presence of other elements belonging to same variable n inside that neighbourhood. In particular, two effects cause the presence of autocorrelation in a spatial distribution: (1) first order effects which measure how the expected value (mean of events number) varies in the space, (2) second order effects which they concern local interactions between events and are measured by covariance variations (the reader is referred to formulas 2.38 and 2.39 in Chap. 2, respectively; see also Gatrell et al. 1996).

Even if the definition of these two effects is clear, to separate their study in the practice is not possible. Under their effect, a distribution can be: (i) clustered, (ii) uniform and (iii) random, for which the autocorrelation will result positive, negative and null, respectively. In other words, we can have respectively: (i) attraction between events (when they are near and similar); (ii) repulsion between events (when, even if they are near, they are not similar); (iii) and no spatial effects, neither about the position of events, neither their properties (for additional information, see Fig. 2.14 in Chap. 2).

According to what said since now, the study of spatial autocorrelation requests to know: (i) the quantitative and (ii) the geometric nature of dataset.

In the context of image processing, the quantitative aspect is given by the intensity equal to the value of each spectral band. So, it is necessary to measure the degree of dependency among spectral features. Whereas, the conceptualization of geometric relationships is the lag distance that is the distance between events which can be equal to or a multiple of pixel size.

For what concern the method, Euclidean distance (see formula 2.40 in Chap. 2) is the most used. For what concern direction, three are the methods used. They consider contiguous events and include them in the calculation only if they are in the admitted direction, take their name from the game of chess. They are called tower, bishop and queen contiguity, thus taking their name from the game of chess.

As a whole, the output is a new image which contains a measure of autocorrelation for each pixel. Such measure could be done by computing global and local indicators.

Global indicators of autocorrelation measure, with one summarizing value, if and how much the dataset is autocorrelated.

One global indicator of autocorrelation is the Moran's I (Moran 1948). If I $\in$ $(-1; 0)$ there's negative autocorrelation; if I $\in$ $(0; 1)$ there's positive autocorrelation; if I converges to 0 there's null autocorrelation (see formula 2.47 in Chap. 2).

Another global indicator is the Geary's C (Geary 1954).

C $\in$ $(0; 2)$; if C $\in$ $(0; 1)$ there's positive autocorrelation; if C $\in$ $(0; 2)$ there's negative autocorrelation; if C converges to 1 there's null autocorrelation (see formula 2.49 in Chap. 2).

Local indicators of autocorrelation allow us to understand where clustered pixels are, by measuring how much are homogeneous features inside the fixed neighbourhood.

In this study we used three indicators: the Local *Moran's I* (Anselin 1995), the Local *Geary's C* (Cliff and Ord 1981), and the Local *Getis-Ord Gi* (Getis and Ord 1992; Illian et al. 2008), defined according to formulas 2.50, 2.51, and 2.52 in Chap. 2, respectively.

These indicators show a different concept of spatial association:

1. In the local *Moran's I* high value of the index means positive correlation both for high values both for low values of intensity;
2. With the Local *Geary's C* it is possible to detect areas of dissimilarity between events;
3. In the *Getis and Ord's Gi* high value of the index means positive correlation for high values of intensity, while low value of the index means positive correlation for low values of intensity.

Geostatistical analysis tools are available in several commercial softwares, such as GIS and image processing ones. We used ENVI packages for the current study.

8.4 Data Processing and Results

The time series of panchromatic and multispectral satellite images described in Sect. 8.3 has been used to map looting in Cahuachi from 2002 to 2008. The investigated area (see Fig. 8.5), strongly damaged by illegal diggings, is located between sector B and the zone where the excavation are in progress (Gran Piramide and Piramide Naranjada). Looting pits are characterized by circular shape (0.7–10 m diameter), somewhat filled with sand, and by scattered remains

Fig. 8.5 Cahuachi. Panchromatic data time series: (**a**) 2002 QB2; (**b**) 2005 QB2; (**c**) 2008 WV1. The black box denotes the area where the method based on spatial autocorrelation has been tested. It is located between sector B and the area where the excavation are in progress (*Gran Piramide* and *Piramide Naranjada*)

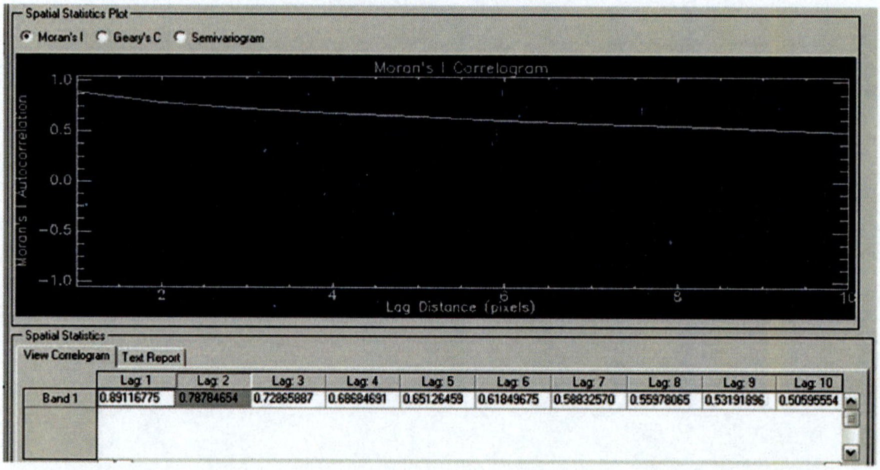

Fig. 8.6 Results obtained with global Moran's I and lag distance between 1 and 10 computed on 2002 panchromatic image

(human and animal bone, pottery fragments) (see Fig. 8.3). Some parts of them are illuminated, others are in shade.

Consequently all these characteristics, pixels with holes show very different values of reflectance, so we supposed to find a break in autocorrelated zones (soil without holes). The comparative visual inspection of the available satellite dataset put in evidence that the panchromatic images are more suitable than pansharpened spectral bands to emphasize the pitting holes. This is due to the fact that for the study area there are no significant spectral variations in the four bands of QB imagery.

On the basis of these results, we focused only on satellite panchromatic scenes. The reliability of the detection was evaluated by field survey carried out in November 2008 on some test sites selected on mounds and flat areas. The evaluation has shown that the rate of success was very high for flat areas (higher than 90%) but unsatisfactory for mounds (40–70%), due to the effect of wind erosion and geomorphological features.

To overcome this drawback, an approach, based on local spatial autocorrelation statistics applied to panchromatic imagery, such as *Moran's I*, *Geary's C*, and *Getis-Ord Gi* index, has been employed.

The first step to use the local indicators is to choose parameters to introduce in the calculation: the lag distance and the rule of contiguity.

About the rule of contiguity the queen's contiguity was chosen, because the analysis should be done in all the directions also for the curve configuration of holes.

To find the optimal lag distance global *Moran's I* and *Geary's C* were used. These indexes were calculated for different values of lag distances.

The best value is the lag that maximizes *Moran's I* (Fig. 8.6) and minimizes *Geary's C* (Fig. 8.7), allowing to capture in the best way the autocorrelation of the image. The lag chosen for all the 3 years is 2.

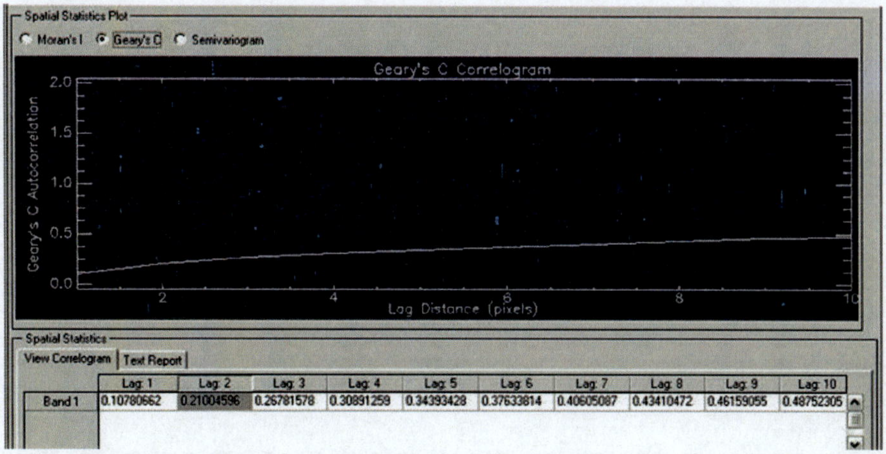

Fig. 8.7 Results obtained with global Geary's C and lag distance between 1 and 10 computed on 2002 panchromatic image

At this point the local indicators of spatial association were calculated, using the queen's contiguity.

Results obtained (Fig. 8.8) enable us to recognize and quantitatively characterize patterns of spatial dependence at multiple scales, thus making them useful in detecting archaeological features.

In particular the local *Geary's C* index (Fig. 8.8d) allowed us to better represent the rough surface, while *Getis and Ord Gi* (Fig. 8.8e) needed a classification, before to be interpreted and used.

The classification of *Getis and Ord GI* (see Fig. 8.9a, b) shows that the clusters with the best results are those characterized by low values of reflectance and, at the same time, of *Gi* and those characterized by great values of Gi. These clusters were then converted to polygons with the aim to obtain the map of the looting phenomenon (Fig. 8.9c).

As a whole, the detection of looting pits on mounds has been significantly improved (75–90%) by applying local spatial autocorrelation statistics. Such improvement is still more evident if we compare the panchromatic satellite time series with the correspondent time series processed by local spatial autocorrelation statistics (see Figs. 8.9 and 8.10, respectively).

In detail:

Figures 8.10a–c show the panchromatic scenes (2002, 2005 and 2008) related to four mounds characterized by the typical circular pits dug by grave looters. Such traces of looting are visible thanks to the micro-relief, but the spatial resolution is not enough to appreciate significant variation between 2002 and 2008.

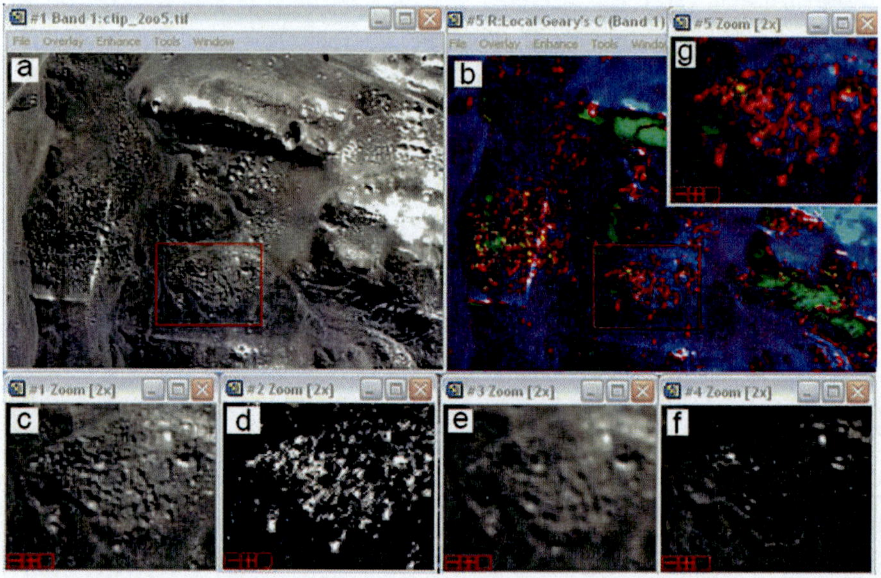

Fig. 8.8 Results obtained for 2005. (**a**) Panchromatic image, (**b**) RGB composition of the Geary's C, the Getis and Ord's Gi, and the Moran's I results. The figure shows also a zoom of the panchromatic image with holes (**c**), the Geary's C (**d**), the Getis and Ord's Gi (**e**), the Moran's I (**f**), and finally the RGB (see **a**)

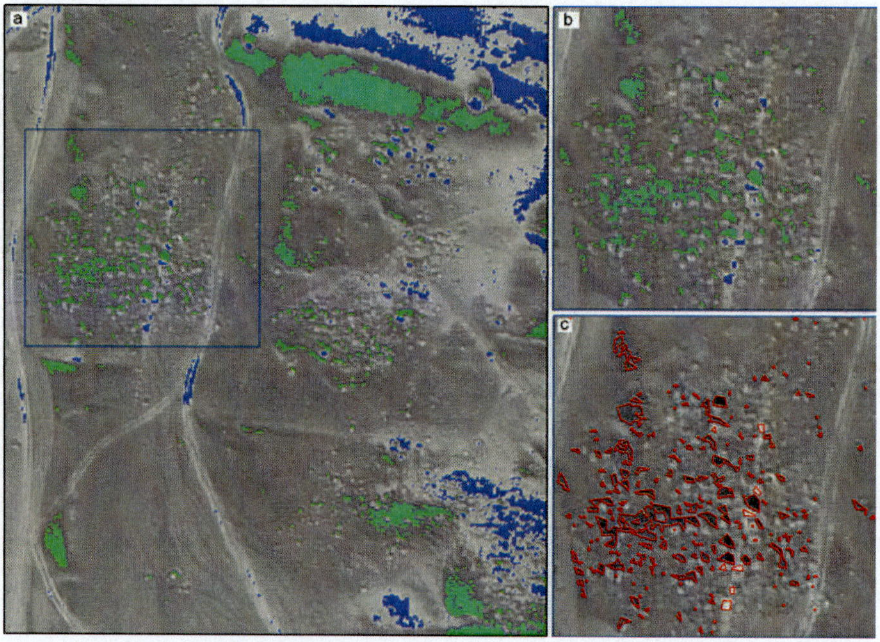

Fig. 8.9 Overlay between panchromatic image and clusters obtained with Getis and Ord's (**a**). In the zoom area (**b**) it is possible to see how clusters of low values and clusters of high values fit the holes distribution. (**c**) Conversion to vector of clusters found

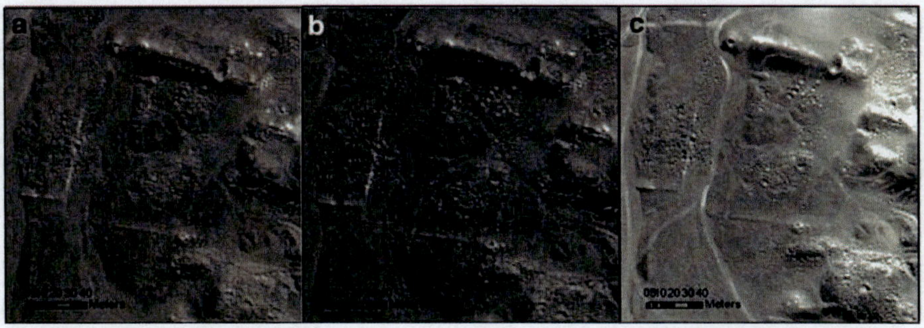

Fig. 8.10 VHR satellite panchromatic image provided in (**a**) 2002 (QB2); (**b**) 2005 (QB2); (**c**) WW1

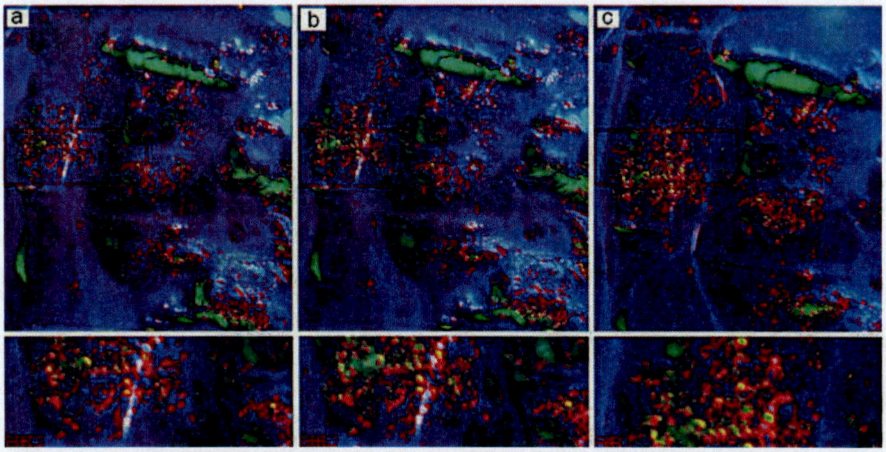

Fig. 8.11 RGB composition of Moran, Getis and Geary indices (R: Geary; G: Moran; B: Getis) applied to panchromatic images of 2002 QB (**a**), 2005 QB (**b**) and 2008 WW1 (**c**). RGB composition of Moran, Getis, and Geary indices enhances the edges of pits (*circled with magenta*). The multitemporal comparison of the three RGB images clearly shows an increasing number of pits from 2002 to 2008 and, therefore, the intensification of the looting phenomenon over the years

Figures 8.11a–c show the RGB composition of Moran; Getis; and Geary indices which does emphasize these pits enhancing their edges (yellow coloured). The multitemporal comparison of the three RGB images clearly shows an increasing number of pits from 2002 to 2008 and, therefore, the intensification of looting phenomenon over the years.

8.5 Final Remarks

"It is ironic that the fascination with the past, which motivates all possible public behaviour toward archaeological resources, also causes so much damage and destruction" (McAllister 1991). Archaeological looting is recognized as one of the most a serious threats to cultural resources throughout the world in both recorded and unrecorded sites. Many problems are associated with illegal excavations among them: (i) damaging of archaeological sites, (ii) loss of artefacts, (iii) destruction of the context of artefacts and therefore irreplaceable loss of valuable information, (iv) and denying this cultural heritage to new generation (Atwood 2006).

To contrast and limit this phenomenon a systematic monitoring is required. Up to now, the protection of archaeological heritage from illegal diggings is generally based on a direct or aerial surveillance, which are time consuming, expensive and not suitable for extensive areas. VHR satellite images offer a suitable chance thanks to their global coverage and frequent re-visitation times.

Unfortunately the spatial resolution is still a limit in detecting all the changes related to clandestine excavations. To overcome such limit a semiautomatic data processing approach based on spatial autocorrelation has been devised.

The chapter showed the results of a test performed in the Nasca Ceremonial Centre of Cahuachi in Peru. By using this approach, the detection of looting pits by VHR satellite images has been significantly improved.

The high satisfactory results encourage to continue in jointly using of VHR satellite data and geostatistics to identify traces and patterns linked to archaeological looting.

References

Anselin L (1995) Local indicators of spatial association LISA. Geogr Anal 27:93–115

Atkinson PM, Lewis P (2000) Geostatistical classification for remote sensing: an introduction. Comput Geosci 26:362–371

Atwood R (2006) Stealing history: tomb raiders, smugglers, and the looting of the ancient world. Reprint, Griffin

Brodie N, Renfrew C (2005) Looting and the world's archaeological heritage: the inadequate response. Annu Rev Anthropol 34:343–361

Ciminale M, Gallo D, Lasaponara R, Masini N (2009) A multiscale approach for reconstructing archaeological landscapes: applications in Northern Apulia (Italy). Archaeol Prospect 16:143–153

Cliff AD, Ord JK (1981) Spatial processes, models, and applications. Pion, London

Curran P, Atkinson PM (1998) Geostatistics and remote sensing. Progr Phys Geog 22(1):61–78

Fowler MJF (1997) A ColdWar spy satellite image of Bury Hill near Andover, Hampshire. Hamps Field Club Archaeol Soc Newslett 27:5–7

Gatrell AC, Bailey TC, Diggle PJ, Rowlingson BS (1996) Spatial point pattern analysis and its application in geographical epidemiology. T I Brit Geogr 21:256–271

Geary RC (1954) The contiguity ratio and statistical mapping. Incorp Statist 5:115–145

Getis A, Ord JK (1992) The analysis of spatial association by use of distance statistics. Geogr Anal 24:189–206

http://www.savingantiquities.org/feature_page.php?featureID=7

Hyppänen H (1996) Spatial autocorrelation and optimal spatial resolution of optical remote sensing data in Boreal forest environment. Int J Remote Sens 17(17):3441–3452

Illian J, Penttinen A, Stoyan H, Stoyan D (2008) Statistical analysis and modelling of spatial point patterns. Wiley, West Sussex, 534 p

Lasaponara R, Masini N (2010) Facing the archaeological looting in Peru by local spatial autocorrelation statistics of Very high resolution satellite imagery. In: Taniar D et al (eds) Proceedings of ICSSA, The 2010 international conference on computational science and its application, Fukuoka-Japan, 23–26 Mar 2010, Springer, Berlin, pp 261–269

Lasaponara R, Masini N, Rizzo E, Coluzzi R, Orefici G (2011) New discoveries in the Piramide Naranjada in Cahuachi (Peru) using satellite, ground probing radar and magnetic investigations. J Archaeol Sci 38:2031–2039. doi:10.1016/j.jas.2010.12.010

López-Granados F, Jurado-Expósito M, Peña-Barragán JM, García-Torres L (2005) Using geostatistical and remote sensing approaches for mapping soil properties. Eur J Agron 23:279–289

Luke C (2006) Diplomats, Banana Cowboys, and Archaeologists in Western Honduras: a history of the trade in Pre-Columbian materials. Int J Cult Property 13(1):25–57

Masini N, Lasaponara N, Orefici G (2009a) Addressing the challenge of detecting archaeological adobe structures in Southern Peru using QuickBird imagery. J Cult Heri 10S:e3–e9. doi:10.1016/j.culher.2009.10.005

Masini N, Rizzo E, Lasaponara R (2009b) Teledeteccion y Investigaciones geofísicas en Cahuachi: primeros resultados. In: Nasca. El Desierto de los Dioses de Cahuachi. Graph Ediciones, Lima, pp 250–277

McAllister M (1991) Looting and vandalism of archaeological resources on federal and Indian lands in the United States. In: Smith GS, Ehrenhard JE (eds) Protecting the past. CRC Press, Boca Raton, pp 93–99

Moran P (1948) The interpretation of statistical maps. J R Stat Soc A Stat 10:243–251

Orefici G (1992) Nasca. Archeologia per una ricostruzione storica. Jaca Book, Milano

Orefici G (1993) Nasca: arte e società del popolo dei geoglifi. Jaca Book, Milano

Orefici G (2009) Cahuachi, the largest adobe ceremonial centre in the world. In: Nasca. El desierto del los Dioses de Cahuachi. Graph, Lima, pp 36–59

Orefici G, Drusini A (2003) Nasca: hipótesis y evidencias de su desarrollo cultural. Centro Italiano Studi e Ricerche Archeologiche Precolombiane, Brescia

Parcak S (2007) Satellite remote sensing methods for monitoring archaeological tells in the Middle East. J Field Archaeol 32(1):65–81

Silverman H (1988) Cahuachi: non-urban cultural complexity on the South Coast of Peru. J Field Archaeol 15:403–430

Silverman H (1993) Cahuachi in the ancient Nasca world. University of Iowa Press, Iowa City

Stone EC (2008) Patterns of looting in southern Iraq. Antiquity 82:125–138

UNESCO (1956) Recommendation on international principles applicable to archaeological excavations. New Delhi

UNESCO (1970) Convention on the means of prohibiting and preventing the illicit import, export and transfer of ownership of cultural property. Paris

Part III
Palaeoenvironment and Archaeology: The Contribution of Satellite Observation

Chapter 9
Uncovering Angkor: Integrated Remote Sensing Applications in the Archaeology of Early Cambodia

Damian Evans and Arianna Traviglia

Abstract For more than a decade the multinational (Australian, French, Cambodian) Greater Angkor Project has been investigating the rise and fall of medieval urbanism at Angkor, in Cambodia, using a diverse range of techniques, including extensive use of remotely sensed imagery to find, map and analyse elements of urban form. The research activities have focussed on the role of Angkor's elaborate water management system in the demise of the urban complex, and has recently been expanded to include nearby 'secondary' settlement complexes such as provincial centres and ephemeral capitals. In such a research agenda, it is crucial to gain a full understanding of the original hydrological layout of the Angkor basin, in order to provide essential insights into human modifications to the natural hydrology and topography. To this end, a number of multispectral satellite images (including QuickBird and ASTER) were processed and analysed to identify palaeo-environmental traces and anthropogenic features relevant to the identification of remnants of the original fluvial system. Vegetation indices (VI), Vegetation suppression and Principal Component Analysis (PCA) were adopted as the primary procedures in order to detect relevant traces over differing environments such as perennially forested zones, scrubland and barren terrain. The outcome of this work has been to add significant chronological resolution to the current map of Greater Angkor.

Keywords Angkor • Cambodia • Multispectral data • QuickBird • ASTER • Vegetation Indices • Vegetation suppression • Principal Component Analysis (PCA)

D. Evans (✉) • A. Traviglia
Department of Archaeology, School of Philosophical and Historical Inquiry, Faculty of Arts, University of Sydney, A18 Brennan-MacCallum Building, Sydney, NSW 2006, Australia
e-mail: damian.evans@sydney.edu.au; arianna.traviglia@sydney.edu.au

R. Lasaponara and N. Masini (eds.), *Satellite Remote Sensing: A New Tool for Archaeology*, Remote Sensing and Digital Image Processing 16,
DOI 10.1007/978-90-481-8801-7_9,

9.1 Introduction

Throughout the colonial period, projects of archaeological exploration and mapping in Indochina were critically important components of the French colonial enterprise (Evans 2007). Archaeological maps were used to validate the spatial extents of French hegemony, and also played a key role in shaping and defining the modern nation-states that emerged from that era. The rise of new methods and theories in archaeology from the 1920s to 1950s created enormous potential to broaden the focus beyond the great monuments, and to investigate the intriguing fragments of evidence suggesting that cities that may have surrounded them. However, various historical and political currents conspired to create a situation where, until very recently, sites like Angkor were very poorly understood as places where people actually lived (Fletcher 2000–2001).

In the 1950s the French archaeologist B.-P Groslier began to articulate the controversial hypothesis that Angkor (Fig. 9.1) was a ‘hydraulic city’, a constellation of suburbs given coherence by a large-scale infrastructural network designed principally for rice irrigation (Groslier 1952). In his view, the eventual failure of that system precipitated the collapse of Angkorian civilisation (Groslier 1979). Groslier understood that a long-term program of settlement archaeology was needed to broaden the perspective beyond the great monuments and provide a firm basis for assessing his theory (Groslier 1960).

The ascendancy of the Khmer Rouge in the early 1970s and the ensuing decades of civil conflict meant that Groslier’s ambitious cross-disciplinary research program was never achieved. In spite of the empirical void that this created, and the impossibility of conducting field research at Angkor, the ‘hydraulic city’ hypothesis was subject to intense scrutiny and criticism by a number of scholars (e.g. Acker 1998; Kiernan 2001; Moore 1995; Stott 1992; Van Liere 1980).

Critics argued that Groslier’s hypothesis was the legacy of a colonial research agenda which undervalued or ignored indigenous conceptions of water management, and that there was no evidence to support his claims of a vast urban network or a hydraulic system capable of feeding up to a million people. Careful consideration of these criticisms, however, reveals that they are largely unfounded and can be logically and empirically refuted (Evans 2007). This, and the re-opening of Cambodia to archaeologists, in turn paves the way for a renewal of research on the role of Angkorian hydraulics in the collapse of the classical Khmer civilisation.

In a major recent contribution to Groslier’s agenda, Pottier’s (1999) survey of the southern and central areas of Angkor in the 1990s documented a system of occupation mounds, local temples and household ponds stretching between, and far beyond, the major temples. As a result of this, the Greater Angkor Project (GAP), a collaborative multi-disciplinary research program interested in the rise and fall of Angkorian urbanism, was formed (Fletcher and Pottier 2002). This study collates all existing map data and extends the mapping coverage to the entire water catchment area for the settlement of Angkor, based on field surveys and the analysis of a diverse range of remote sensing datasets. The results of this research

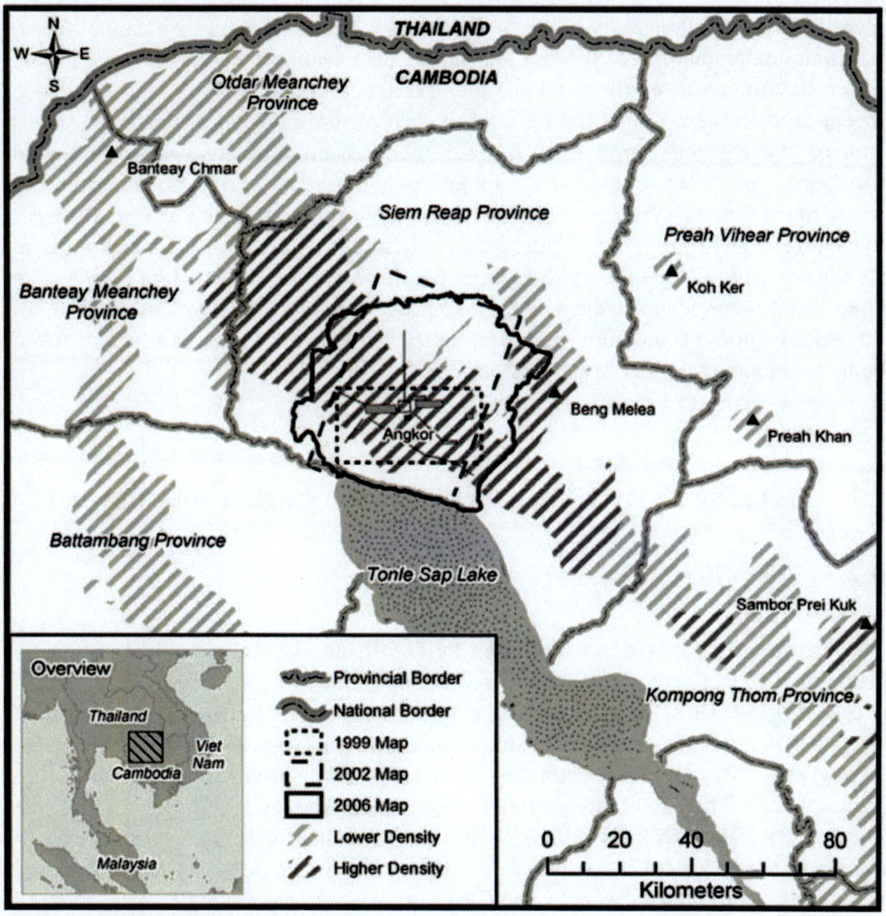

Fig. 9.1 Early Khmer temple complexes and cities mentioned in this paper, with approximate density of minor site distribution

(Evans et al. 2007) confirm several key aspects of Groslier's thesis, including the existence of an extended network of 'hydraulic suburbs' and a vast and intricate water management system capable of sustaining a significant number of additional people (Pottier 2000). It shows extensive anthropogenic changes to the landscape, including intensive land-use across nearly 2,000 km^2, with clear potential to impact substantially on the regional ecology.

In spite of these considerable advances, much work remains to be done. The three principal areas of research of the Greater Angkor Project are: remote sensing and mapping of cultural features; palaeobotany; and detailed, finer-grained study—principally through coring and excavation—of key components of the water management system such as the major canals and the inlets and outlets of Angkor's great reservoirs or baray. Throughout the course of this work, a need has been identified for supplementary projects to address related issues that are crucial for an adequate understanding of the rise and fall of urbanism in the Angkor area. In this paper, having reviewed the evidence thus far accumulated for Angkor, we present recent outcomes from ongoing research in two of these related projects: an extension of the methods applied at Angkor to other major early urban centres in Cambodia, in order to gain a comparative perspective of the development of water management systems within the Khmer world; and a technical analysis of the nature of anthropogenic modifications to natural hydrology on the Angkor plain. Remote sensing is the core methodology for these two studies, both of which, like GAP itself, demonstrate the need to employ aerial archaeology as part of an integrated suite of methods and within the framework of a carefully crafted deductive archaeological investigation.

9.2 Mapping Early Khmer Settlements: Angkor and Beyond

9.2.1 The Site of Angkor

The greater Angkor region, as defined by GAP (based on a watershed model by Kummu 2003), covers an area of approximately 3,000 km^2, with the majority of the archaeological sites arrayed across a plain that stretches from the sandstone outcrops of the Kulen hills to the shores of Cambodia's great lake, the Tonle Sap. Topographically, the plain is punctuated by only four small hills, none of which are higher than 200 m (Japan International Cooperation Agency 1999). It contains three major rivers, all with their primary source in the Kulen area: the Puok, the Siem Reap and the Roluos. The modern flows of these rivers are often artificial, a legacy of Angkorian hydraulic works. The Siem Reap river, for example, was diverted several times during the Angkorian period and its present-day course is almost entirely canalised south of its junction with the Puok (Engelhardt 1995, 20). The Angkor area has a sharply seasonal monsoon climate. Rainfall is plentiful but highly variable from year to year, and fully 90% of it falls between 1 May and 1 December (Acker 1998, 9). In October or November, when the Mekong's flow subsides, the Tonle Sap river reverses its current and drains the great lake until it is nearly a quarter of its peak size (Acker 1998, 7), and the distance between the maximum and minimum waterlines in the Angkor area is roughly 10 km. The soils range from sandy and relatively infertile near the Kulen to soils rich with alluvium in the area inundated annually by the lake, often under laid with a thick lateritic clay (Acker 1998, 9; Bronson 1978, 266).

Retention and exploitation of the receding floodwaters of the Tonle Sap may have played a role in the intensification of urbanism in the area: although prehistoric sites are scattered throughout the greater Angkor area (Evans 2007; Moore 2000), with at least one dating to the Bronze Age (2870 BP +/− 60, see Pottier et al. 2005, 62), the earliest well-dated pre-Angkorian sites lie within 15 km of the lake's high water mark, and the majority of these within 5 km (Pottier et al. 2004). By the time of the height of classical Angkor, in the eleventh to thirteenth centuries AD, more than a 1,000 temples dotted the landscape between the Kulen and the Tonle Sap (Evans et al. 2007; Ministère de la culture et des beaux-arts and École française d'Extrême-Orient 2007), with the cultivation of rice in small, walled fields as the cornerstone of the economy (Baty 2005; Pottier 2000). The temples were an important component of the settlement pattern and of the economy, acting as nodes within a redistributive network that had the massive and well-known sandstone temples at the centre of Angkor as its apex (Hall 1992). With the exception of a brief hiatus in the tenth century, the site of Angkor, broadly defined, was the capital of the Khmer Empire from the early ninth century AD to the middle of the fifteenth century AD. At various stages within that period the Empire encompassed much of mainland Southeast Asia, with a network of elevated earthen roadways connecting Angkor to thousands of small, temple-based settlements as well as major provincial centres such as Banteay Chmar, Koh Ker, Wat Phu, Preah Vihear and Preah Khan of Kompong Svay (Hendrickson 2007; Jacques and Lafond 2007). For reasons that remain partially obscure, Angkor entered into a period of decline in the fourteenth to fifteenth centuries AD, its empire disintegrating in the process. The capital of the Khmer kingdom moved to region of what is now Phnom Penh (Chandler 2000), and for centuries vegetation continued to reclaim much of the landscape of Angkor until twentieth century land clearance exposed the majority of the landscape to the view of surveyors and archaeologists (Evans 2007; Gaughan et al. 2008).

9.2.2 Groslier's Hydraulic City Hypothesis: Methods and Theory

The pioneering inventories of Aymonier (1900–04) and Lunet de Lajonquière (1902–11) were landmark achievements in the history of Khmer archaeology, each documenting hundreds of major archaeological sites—mainly temples with obvious architectural remains—that were previously unknown to the scholarly community. Lunet de Lajonquière's work, in particular, generated the first scientific and properly surveyed map of the Angkor area; notwithstanding incremental changes from the 1930s to 1960s, it was to remain definitive template for representations of the archaeological landscape for nearly a century (Evans 2007). This and subsequent maps documented little more than the largest of the temples, reservoirs and canals. The first compelling evidence for settlement around the great monuments, which emerged in the 1930s in the form of observations about small ponds and occupation mounds, remained largely ignored and unpublished until the 1950s, in spite of the golden opportunity provided by the large number of aerial archaeological missions undertaken at Angkor in the 1930s. Because of the relentless focus on the temples and their immediate precincts, the belief that Khmer cities were formally bounded and spatially well-defined (Goloubew 1936), and the lack of scholarly interest in Trouvé's (1933) pioneering study of water management, those flights were largely opportunistic and exploratory in nature. Importantly, they lacked any kind of deductive framework within which systematic surveys and the documentation or analysis of extra-mural settlement features might be considered useful.

It was only in the 1950s that the French archaeologist Bernard-Philippe Groslier, of the EFEO, began to trace the outlines of such a theory. Building on his experiences conducting aerial surveys in the Mekong Delta under the direction of Louis Malleret, he proposed the idea that the canals that had been partially mapped at Angkor were part of a vast and intricate hydraulic system permeating an extended 'suburban' network of settlement space revolving around the little-noticed mounds and ponds. In his view, the water management system of the Angkor area was responsible both for the great success of Angkor due to surplus rice production and also, when the system ultimately failed, for the city's rapid decline (Groslier 1952, 1979). Recognising that the aerial surveys of the 1930s amounted to little more than "experiments" (Groslier 1960, 20), he also proposed and began to implement a broad-based, comprehensive archaeological programme to test his hypothesis which included not only remote sensing and mapping but also palynology, architectural studies and excavations.

Because of the rise of the Khmer Rouge and the ensuing decades of conflict, this programme was never completed, and the data that might prove or disprove his hypothesis were never collected. In spite of this, a number of English-language publications from 1980 to 1990s roundly rejected Groslier's claims. W.J. Van Liere, for instance, argued that the hydraulic system was not used for subsistence and pointed to the lack of outlets or distributors from the great reservoirs, as well as the overall lack of sophistication and efficacy within the system by the standards of modern engineering. He also makes a point later pursued in more detail by Moore (1989) and Acker (1998): that the irrigation system, such as it was, would not have fed a substantial number of additional people. Alternatively, these critics proposed, we should consider Groslier's theory as a legacy of colonialism and begin to address the water management system from the indigenous point of view, as a component of the ritual landscape of Angkor and not as a functional agricultural system. Although these re-evaluations of Groslier's theory were based on scant evidence, one could argue that the cumulative effect was to lay the foundations of a null hypothesis for the 'hydraulic city' concept: if it were not true, then it should be expected that a systematic and well crafted archaeological study would reveal no reservoir outlets, no distributor channels to Angkorian fields, an unsophisticated system incapable of delivering reliable water supplies to those fields, a system that would clearly have been designed for an exclusively ritual purpose, and a system that could not have impacted substantially on subsistence or demography.

The argument of Stott (1992) and Kiernan (2001) that Groslier's theory was a product of the colonial imagination is perhaps superficially true, as one can see from a close reading of his works on the issue (Evans 2007); however, it should be noted that this is fundamentally irrelevant to the question of its veracity. Fortunately,

Stott's (1992, 47) assertion that the 'hydraulic city' hypothesis is "untestable" is quite untrue: Once additional testable elements are added to the foundations that Van Liere, Stott, Moore, Acker have provided—for example, evidence of the existence a very large population at Angkor, of an extensive and integrated urban infrastructure, of a practical dependency on hydraulics, and of failure within the hydraulic system coinciding with the decline of classical Angkor—all that remains is to undertake the required program of data collection.

9.2.3 The 1990s: A Period of Renewal in Angkorian Archaeology

The Paris Peace Agreements of 1991 once again opened the doors for archaeological research in Cambodia, for the first time in nearly 20 years. The EFEO's Siem Reap office re-opened almost immediately, and one of the first objectives of its newly appointed director, Christophe Pottier, was to undertake a systematic inventory of the monuments of the Siem Reap area in support of both EFEO research objectives and the UNESCO-funded Zoning and Environmental Management Project, or ZEMP (Pottier 1993; Royal Angkor Foundation 1995). By using a series of 1:25,000 scale aerial photographs of Cambodia taken in the early 1990s (FINNMAP 1992–1993), Pottier was able to identify and catalogue a very significant number of additional sites in the vicinity of Siem Reap—sites which did not exist in any previous inventory, but which could be clearly recognised as temples due to the characteristic spatial configuration of their remnant moat, mound and pond features on the surface. Throughout the 1990s, Pottier continued to refine the mapping work for the central and southern areas of Angkor using a combination of stereoscopic analysis of the aerial photographs and systematic ground survey, and culminating in the publication of a (1999) doctoral thesis titled *Carte Archéologique de la Région d'Angkor: Zone Sud*. After canvassing all of the previous studies and archival references (many from as early as the 1930s) that mention the distributed network of ponds and mounds around the ceremonial centre, the work also presented a definitive methodology for cataloguing and verifying sites such as 'village temples' through analysis of subtle variations in surface topography and by documenting occupation debris such as ceramic scatters. Pottier's work also had several important implications for the 'hydraulic city' thesis (Pottier 2000). It documented a series of outlet and distribution mechanisms from the great reservoirs in the central zone, which permeated a network of Angkorian rice fields down slope. It also revealed an almost continuous field of small-scale occupation around, but also far beyond, the major monuments at the core of Angkor. Pottier (2000) has further argued that the amount of irrigable land down slope of the reservoirs was substantial, and may have provided for up to 150,000 additional people on an annual basis.

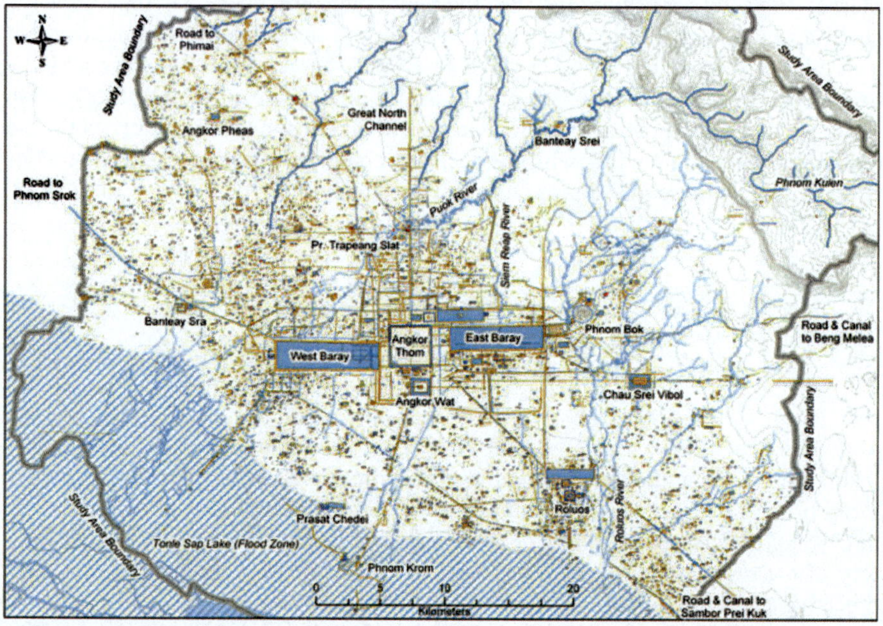

Fig. 9.2 An overview of the comprehensive map of Angkor generated from remote sensing and survey (After Pottier 1999 and Evans 2007)

In 1999 the University of Sydney began a longstanding collaboration with Pottier and the EFEO with a view to expanding the mapping to the northern areas of Angkor and complementing the remote sensing and survey methods with a program of paleobotanical research and excavation (Fletcher and Pottier 2002). This cross-disciplinary and multinational (French, Australian and Cambodian) program forms the basis of the Greater Angkor Project. One of the first steps completed was to digitise Pottier's (1999) paper maps and the entire range of available historical mapping and inventory sources into a GIS database. In 2002, a preliminary northward extension of Pottier's mapping was completed using 5 m resolution airborne synthetic aperture radar (AIRSAR) data (Evans 2002). Although necessarily coarse given the spatial resolution of the data, this study indicated that a similar settlement pattern of road and canal infrastructure, along with a network of ponds and local temples, also existed in the area between the monumental core and the Kulen Hills to the north. From 2002 to 2007, a comprehensive and more detailed map of that extended area was created using AIRSAR, 2.5 m SPOT 5 data and aerial photography (Fig. 9.2); Pottier's maps of the central and southern zones were also extended to the east and west to include the entire watershed area of the river resources exploited by Angkorian engineers (Evans 2007; Kummu 2003). The study area was divided into 3,000 one km^2 cells, each of which was analysed using all of the available data sources and the results plotted into the GIS using heads-up digitising over the 2.5 m resolution

SPOT 5 base, at a nominal scale of 1:10,000, but otherwise following the same methodological and classificatory standards developed by Pottier.

Again, many of the findings were relevant to the ongoing assessment of Groslier's thesis. It was confirmed that the low-density settlement pattern continued to the north of Angkor, and that a highly complex network of canals and dykes had been put in place by Angkorian engineers in order to channel water down to the storage devices at the centre of the settlement complex. A continuous field of occupation extended in all directions beyond the area previously mapped by Pottier, although the infrastructural network that defines Angkor as an operational entity was largely limited to a 1,000 km^2 region beyond the ceremonial centre of Angkor Thom. Annual excavations by the GAP team have focussed on finer-grained analyses of key junctions and structures within that network (Fletcher et al. 2003, 2008), while pollen evidence has revealed elements of both chronology and land-use in some areas of the urban zone (Penny 2006; Penny et al. 2007). Inlets and/or outlets have now been identified for all of the major reservoirs (Fletcher et al. 2008, Roland Fletcher, personal communication, 2011). Map data, pollen data and excavations have revealed numerous instances of failure and adaptation within an extremely complex and sophisticated water management system, including during the classical Angkorian period (Evans 2007; Fletcher et al. 2008; Penny 2006). Importantly, the distribution of occupation features and agricultural systems indicates that land-use within the Angkor period was characterised by highly intensive and extensive re-engineering of the natural environment, suggesting that the vast scale of urbanism at Angkor would have impacted profoundly on the local environment, including in ways detrimental to the proper functioning of the water management system (Evans 2002, 2007). Field verification of the hundreds of additional temple sites identified during this mapping process continues (Evans et al. 2007), with an expected completion date of 2012.

9.2.4 Banteay Chmar

The temple complex of Banteay Chmar is located approximately 100 km to the northwest of Angkor, in Banteay Meanchey Province. Although it features one of the most impressive sandstone monumental complexes in all of the Khmer Empire, its relative remoteness and the traditional focus on the site Angkor has contributed to a situation where little is known about the archaeological context of the monuments (Jacques and Lafond 2007, 246–54). Numerous accounts of the area remark on the relative poverty of the landscape and the relative scarcity of water (Richards 2007), although the temple complex includes a large reservoir measuring 1,700 × 800 m, similar in morphology and dimensions to the great baray of Angkor.

Previous studies had noted the existence of two linear features, probably canals, converging on this temple complex (Pottier 2004), and a number of potentially Angkorian linear features to the north of the main temple (Hendrickson 2007), but no detailed analysis or mapping of surface features had, until recently, been completed.

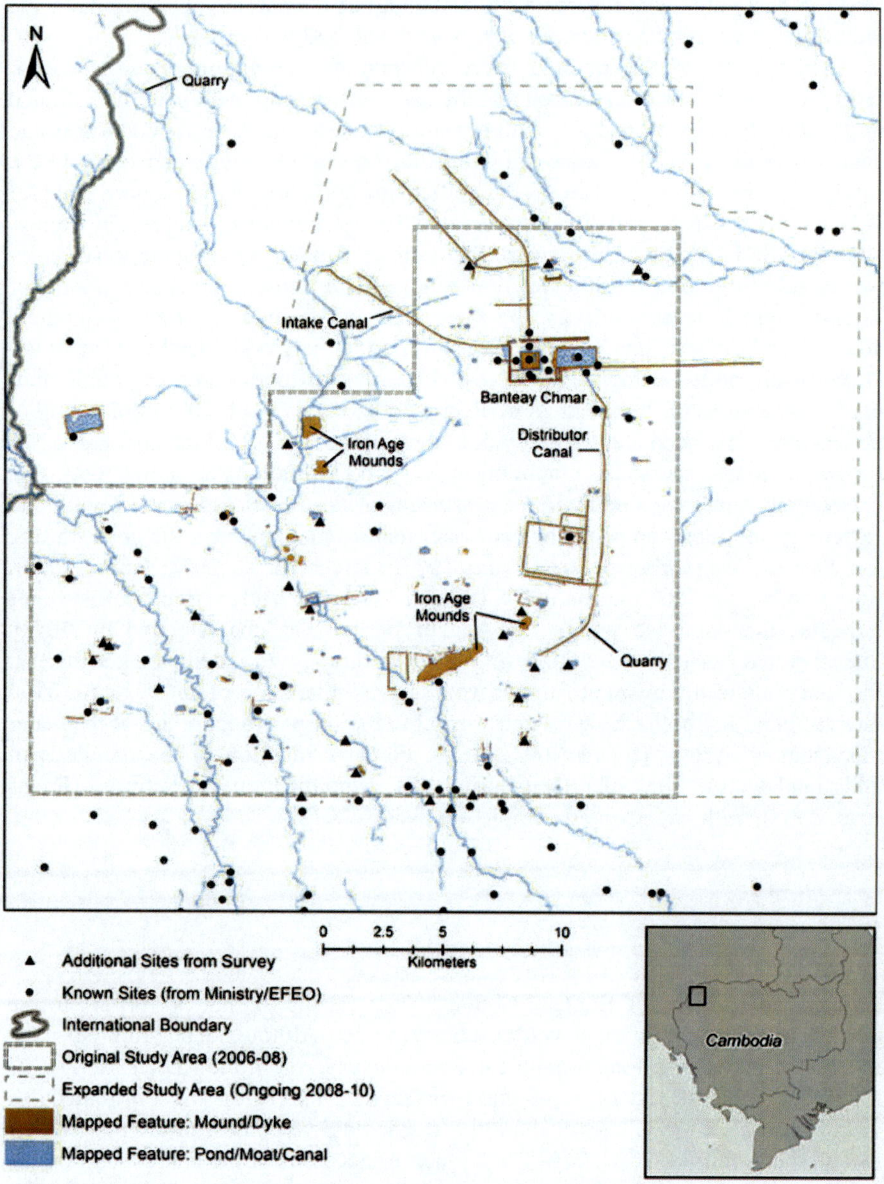

Fig. 9.3 Preliminary archaeological map of the Banteay Chmar region (After Richards 2007)

Richards' preliminary (2007) study was the first to systematically address this issue, applying the same methods used at Angkor by Evans (2007) to a study of the structure of urbanism and hydraulics at the site (Fig. 9.3). The datasets used in this study were the 1:25,000 scale black and white aerial photo coverage

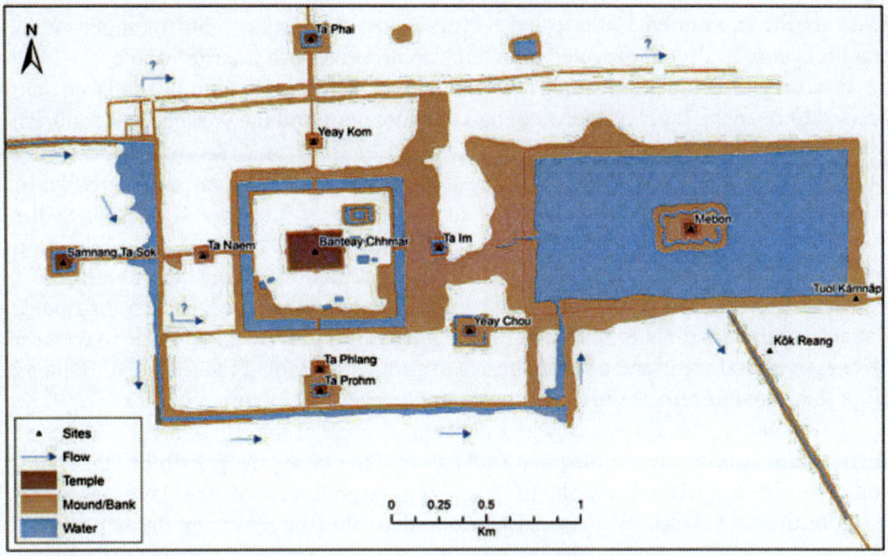

Fig. 9.4 Settlement patterns in the vicinity of the Banteay Chmar temple complex

(FINNMAP 1992–1993) used previously at Angkor, and 0.61 m resolution pansharpened multispectral data from the QuickBird satellite.

The survey, which did not involve extensive ground verification, tentatively identified a number of additional sites beyond those already documented as points in the most recent inventory of temples (Ministère de la culture et des beaux-arts and École française d'Extrême-Orient 2007). Numerous small-scale water management features were associated with these 'local shrines' which, as at Angkor, stretched far beyond the well-known monuments and did not specifically cluster near the central monumental zone. Unlike Angkor, however, this system of low density settlements was not permeated or well-connected by an extensive network of infrastructure with the exception of the areas immediately to the north of the temple.

Given the lack of field verification in Richards' study and the inability to distinguish many linear features from the (extensive, but undocumented) Khmer Rouge era hydraulic works of the 1970s, further field seasons were undertaken in 2008 and 2009. These new surveys have made use of 0.5 m resolution GeoEye-1 pansharpened multispectral data acquired in 2009, and of declassified Corona spy satellite imagery from the 1960s in order to identify and analyse the structure of the pre-Khmer Rouge hydraulics. This work is ongoing, but some preliminary outcomes can be presented here (Fig. 9.4). Dozens of additional temple sites have been verified in the study area beyond those already documented in the official inventory (Ministère de la culture et des beaux-arts and École française d'Extrême-Orient 2007). The linear feature approaching Banteay Chmar from the northwest is a canal designed to capture the flow of several small streams and direct them towards the monument. This canal traverses uneven terrain, and for much of its length is essentially an elevated aqueduct, reinforced with laterite rubble.

Where it joins the outer wall of the Banteay Chmar temple complex, a large dam is used to raise the level of the water and channel it around the western and southern sides of the temple, from whence it flows into the reservoir. A series of masonry culverts permit the passage of the water underneath roadways and enclosure walls. Similarly, a system of embankments to the north of Banteay Chmar have the function of harnessing seasonal streams; undoubtedly, they served a dual function as roads, and connect numerous small temple installations to the central temple.

The major linear feature emerging southwards from the reservoir is a distributor canal, also lined with laterite rubble, which proceeds regularly down slope to connect with agricultural areas and a second series of dams surrounding the temple of Banteay Top; this canal terminates in a previously undiscovered Angkorian quarry.

In its totality, the system of water capture, storage at the urban centre, and then distribution strongly resembles that of Angkor. This raises the possibility that here, too, existed a system capable of irrigating large areas of the landscape and contributing substantially to subsistence and to the longevity of the settlement. Further research is concentrating on the chronology of the development of the urban landscape using analysis of ceramic material found at temple sites.

9.2.5 Koh Ker

The site of Koh Ker presents something of an anomaly in Khmer archaeology: whereas other major centres are known to have been redeveloped at various stages, Koh Ker presents, by and large, an impression of uniformity in its inscriptions, in the structure and layout of its inhabited space, and in its architectural and art historical styles (Jacques and Lafond 2007, 107–33; Parmentier 1939). Parmentier's landmark (1939) study included a map of Koh Ker (Fig. 9.5) which has remained fundamentally unchanged and unchallenged until the present day (Jacques and Lafond 2007, 108; Ministère de la culture et des beaux-arts and École française d'Extrême-Orient 2007). His study documents around 60 temples in an extended area around Koh Ker, with most of them situated near the large reservoir or Rahal at its centre. Parmentier's map also included a number of linear features around the north of the main temple zone which he interpreted as the remains of a large enclosure delimiting the bounds of the inhabited space and principal rice growing areas.

The University of Sydney's involvement in research at Koh Ker began with a brief field survey and low-level aerial reconnaissance using an ultra light aircraft in 2003. More recently, it has been the focus a detailed and comprehensive survey of the landscape, on the ground and by ultra light, assisted by 1 m resolution Ikonos data acquired in 2001 and 0.5 m resolution TerraSAR-X radar imagery acquired in 2009. As a result of this work, a new map of Koh Ker and its region has recently been completed (Fig. 9.6). Again, although analysis and interpretation of the map is ongoing, several

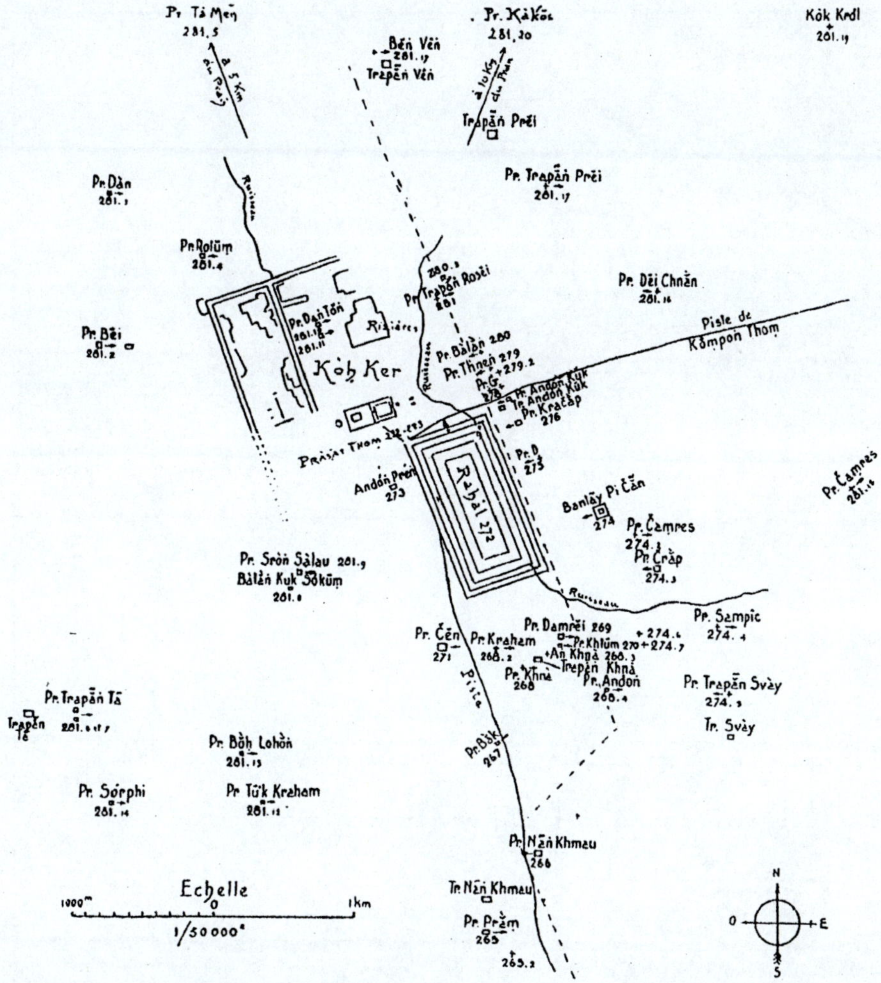

Fig. 9.5 Parmentier's (1939) map of Koh Ker

preliminary results from Koh Ker affirm the value of remote sensing applications when used in concert with ground survey and analysis of surface material such as ceramic debris. The most important outcome is the identification of the 'enclosure wall' as a water storage device, rather than a structure designed to delimit urban space. Parmentier's initial interpretation of the linear structures as an enclosure was based on the assumption that, like all other formal enclosure walls of the Angkor period, they were near-perfectly straight; although two sides of the proposed rectilinear enclosure were missing, Parmentier argued, their absence could be partly explained by the existence of the Rahal, whose northern bank functioned also the southern wall of the enclosure (no explanation was provided for the missing eastern wall).

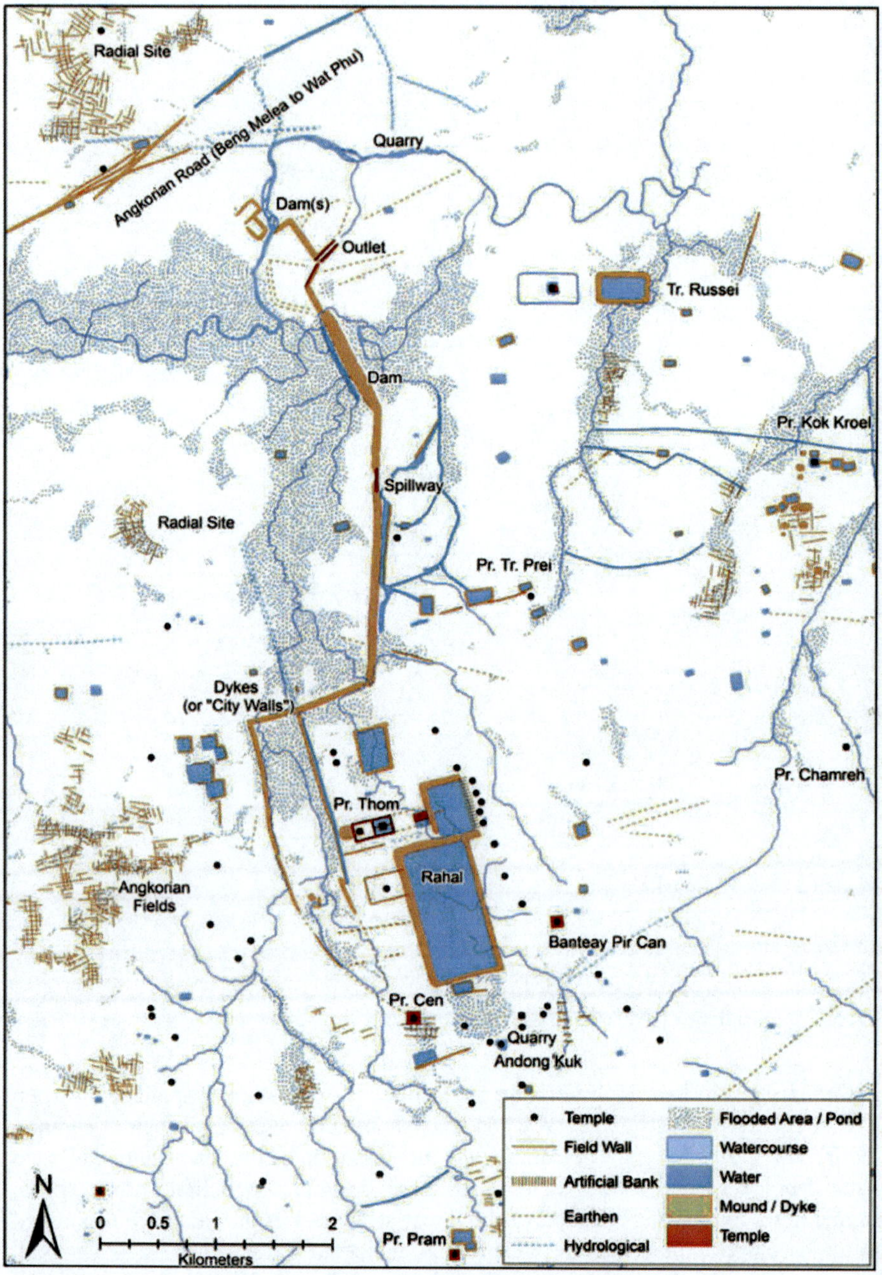

Fig. 9.6 A new map of the archaeological landscape of Koh Ker

The new map of Koh Ker, however, reveals that Parmentier's western enclosure walls are in fact curvilinear; moreover, they continue south far beyond the north bank of the Rahal, following the natural topography of the landscape and terminating at a point where several small streams enter the temple zone.

Although these embankments are now fragmentary and have been heavily eroded in places by these streams, intact walls would have served to channel water northward and effectively dam the natural flow against the northern (east–west) dyke of Parmentier's 'city wall'.

This configuration strongly suggests that these embankments are in fact water storage and capture devices. Furthermore, the evidence for large-scale damming of water is unambiguous immediately to the north of this misidentified 'enclosure'. Here, a massive artificial ridge, reinforced with laterite rubble, stretches some 4.5 km north from the central monuments, stopping the flow of several natural watercourses along its length. At approximately the halfway mark a large, sloping masonry structure 190 m in length is clearly a spillway; at the northernmost extension of the ridge there is a reinforced masonry outlet. Both the spillway and the outlet have many commonalities in size and structure to similar (tentatively tenth century) features identified by GAP during excavations at Angkor (Fletcher et al. 2008). These results suggest that, rather than being a well-defined urban space centred around a single storage reservoir, Koh Ker was a settlement whose existence depended heavily on numerous, large-scale water control devices. Relative to the size of the temples and the presumed population of the area, an enormous amount of engineering effort was invested in the management of water at this site. Significantly, the dam structures do not conform in any way to conventional notions of sacred geography within the Hindu-Buddhist urban landscape, and provide an unequivocal example of the pragmatic, functional, non-ritual nature of hydraulic engineering during the classical Angkorian period. However, no remnant field systems are down slope of the outlet to the large northern dam, suggesting that a large population may have relied on the water for basic subsistence agriculture rather than the production of surplus yields through irrigation and multi-cropping. Analysis of surface finds of ceramics are ongoing; contrary to conventional wisdom about Koh Ker, preliminary results (including imported wares dating from the fourteenth to nineteenth centuries) indicate a long and complex history of occupation.

9.3 Case Study: Mapping Pre-Angkorian Hydrology

9.3.1 Issues and Research Strategy

In order to augment the chronological resolution of the current map of Angkor and to disentangle the layers within the complex palimpsest of features within it, a clear understanding of the original hydrological layout of the Angkor basin was regarded as a crucial aspect to be addressed. An analysis of the historical palaeo-hydrological

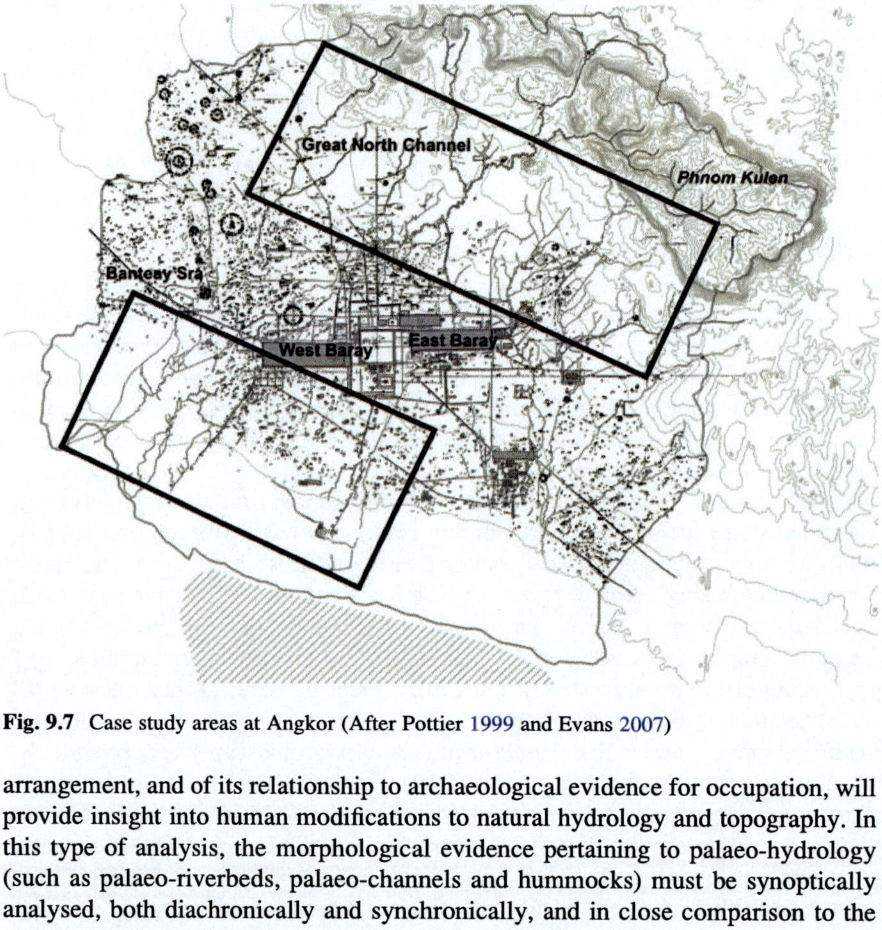

Fig. 9.7 Case study areas at Angkor (After Pottier 1999 and Evans 2007)

arrangement, and of its relationship to archaeological evidence for occupation, will provide insight into human modifications to natural hydrology and topography. In this type of analysis, the morphological evidence pertaining to palaeo-hydrology (such as palaeo-riverbeds, palaeo-channels and hummocks) must be synoptically analysed, both diachronically and synchronically, and in close comparison to the nature and distribution of archaeological features (Fig. 9.7).

In the effort to identify palaeo-environmental traces and anthropogenic features relevant for the identification of remnants of the original fluvial system, a number of multispectral satellite images were acquired. The difficulties inherent in the detection of the traces of the primal fluvial arrangement in this complex environment are predictable: concealed under the dense canopy or obliterated on bare soil surfaces by heavy monsoon precipitation, the features that indicate ancient rivers are often difficult to discern.

Identification of the relevant traces in this ecological context by means of remote sensing is possible due to a range of factors, including: the shadows projected by subtle topographic features; variation in the growth or type of vegetation; broad damp marks on bare soils; and the disposition of topographic elements or spatial logic of archaeological features. Wide surfaces occupied by dense forest form a camouflage that inhibits the identification of features that might be normally apparent. Remote sensing-supported detection of features in the open floodplain can also present several difficulties, particularly when the topographic variations that define dried riverbeds are absent or too subtle to be identified, when they have been washed out by floods and precipitation, or when contemporary water flow has obscured surface wetness related to ancient water flow.

The identification of the large primeval fluvial systems is relatively easy, as some of the systems are still preserved by minor seasonal rivers and by geological formations resulting from the substantial discharge on the surface. The recognition of minor streams, on the other hand, presents several issues, principally the difficulty in distinguishing between palaeo-riverbeds and dried modern seasonal riverbeds on the bare soil floodplain during the dry season. This season lasts from November to April, and is generally the optimal timeframe for acquiring multi-spectral images. Seasonal rivers and streams are very common in this area, and when flow ceases during the dry season the traces they leave on the surface are very similar—and consequently very difficult to distinguish from—the remnants of ancient rivers. Digital topographic datasets for the area (Japan International Cooperation Agency 1999; Royal Angkor Foundation 1995) tend to have poor accuracy and limited coverage of seasonal streams, and cannot therefore be considered reliable for validation of analyses. It is therefore necessary to apply a multi-temporal and multi-dataset approach to generate a range of data products to enable the monitoring of seasonal flows over time.

Consequently, several image processing techniques were applied to a number of different satellite images: QuickBird, ASTER, Landsat TM, Ikonos. Due to their intrinsic spectral characteristics and their high spatial resolution, QuickBird and ASTER (Fig. 9.8) data provided the most significant results. Further analysis and processing is still in progress and, while a complete map of the hypothetical ancient hydrology of the area has not been completed yet, preliminary outcomes have nonetheless contributed substantially to an understanding of the genesis and development of the Angkorian water network.

9.3.2 *Methodological Approach and Image Processing*

The methodological approach adopted for the identification of features was based on a clear delineation of specific outcomes to be achieved through remote sensing, and on the classification of the landscape into uniform ecological contexts within which to apply various strategies. Among the differing environments in the palaeo-hydrological basins of the region, the most frequently occurring are: the perennially forested zones (in different degrees of density); vast swathes of wet-season rice fields; and abundant patches of unproductive cleared land. From the outset, an effort was made to clearly define the technical and methodological limitations presented by each type of environment, and to attempt to devise specific solutions that would permit rigorous analysis of remotely sensed data in all areas.

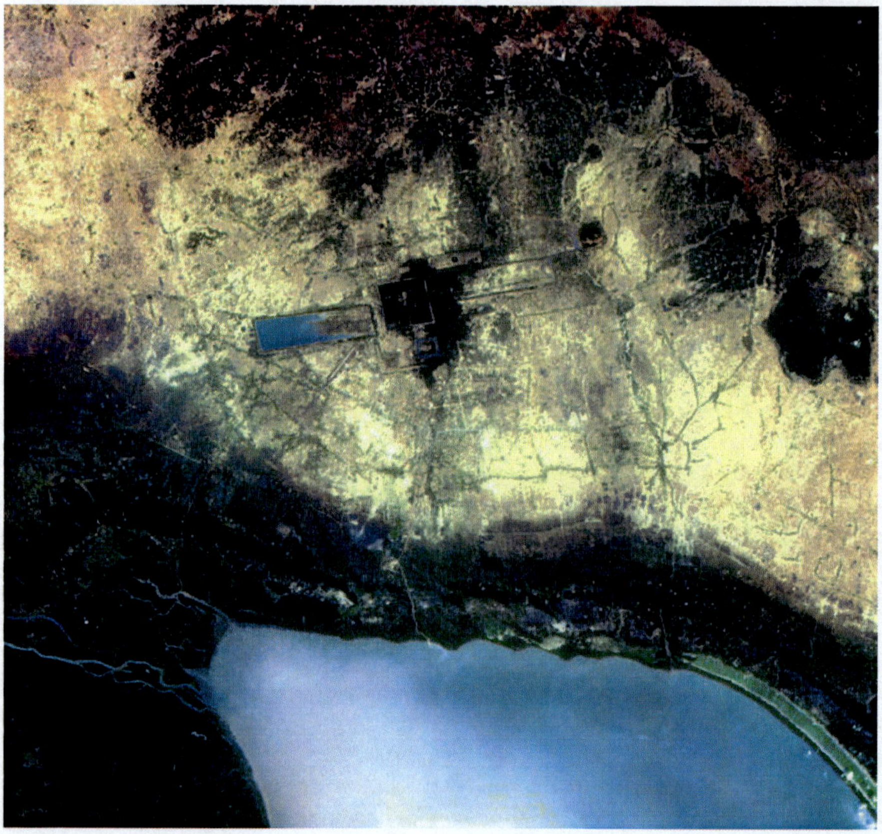

Fig. 9.8 ASTER scene in synthetic RGB true colour display

9.3.3 Technical Specifications on Satellite Data

The principal dataset for this project was Digital Globe QuickBird imagery. The QuickBird sensor concurrently collects panchromatic and multispectral imagery with spatial resolutions, respectively, of 61/72 cm and 2.44/2.88 m, depending upon the off-nadir viewing angle (0–25°). Information in the panchromatic sensor is collected at the visible and near infrared wavelengths and has a bandwidth of 0.450–0.900 μm. The multispectral sensor acquires data in four spectral bands ranging from Blue to Near-Infrared (NIR). The scenes were acquired towards the end of the Cambodian dry season: two in January and April 2004, and one more recently in January 2009 (Table 9.1).

Imagery from ASTER (Advanced Spaceborne Thermal Emission and Reflection Radiometer), one of the sensors on board NASA's Terra satellite, was also used. The sensor supplies various resolution images in 14 bands of the electromagnetic spectrum, ranging from visible to thermal infrared (TIR). Reflected radiation is

Table 9.1 QuickBird metadata

QuickBird	Group 1	Group 2
Site	Central Greater Angkor	East Greater Angkor
Product name:	Standard imagery	Standard imagery
Product option:	Bundle (multispectral and panchromatic) imagery	Bundle (multispectral and panchromatic) imagery
Collection range (Y-M-D)	Start 2004-01-06; stop 2004-04-23	Start 2009-01-11; stop 2009-01-11
Acquisition date (Y-M-D)	2004-01-06	2009-01-11
Cloud cover (%)	Not assigned	0
Terrestrial coordinates	Nwlat = 13.69600000 Nwlong = 103.72200000 Selat = 13.17700000 Selong = 104.01100000	Nwlat = 13.69587965; Nwlong = 104.00525887 Selat = 13.17698318 Selong = 104.05946677
Quantisation	16 bits	16 bits
Spectral bandwidth	Panchromatic: 450–900 nm Blue: 450–520 nm Green: 520–600 nm Red: 630–690 nm Nir: 760–900 nm	Panchromatic: 450–900 nm Blue: 450–520 nm Green: 520–600 nm Red: 630–690 nm Nir: 760–900 nm
Swath width	16.5 km	16.5 km
Mul resolution (m)	2.44 at nadir; 2.88 m 25° off-nadir	2.44 at nadir; 2.88 m 25° off-nadir
Pan resolution (m)	0.61 at nadir; 0.72 at 25° off-nadir nadir	0.61 at nadir; 0.72 at 25° off-nadir nadir

measured in 3 bands between 0.52 and 0.86 μm (VNIR), in 6 bands from 1.6 to 2.43 μm (SWIR), and emitted radiation in 5 bands in the 8.125–11.65 μm wavelength region (TIR). The spatial resolution of VNIR, SWIR, and TIR is 15, 30 and 90 m respectively. The numerous tiles of ASTER imagery used for this study (L1A reconstructed unprocessed instrument data v003) were acquired at various stages in 2002 and 2003 (Hendrickson 2007) (Table 9.2).

9.3.4 Mapping Under the Canopy: Vegetation Quality Assessment Indices (VI – Vegetation Indices)

One of the major limitations in undertaking the process of detection of palaeo-hydrological features or archaeological remains is forest canopy, which covers large tracts of the study area in the north-east of Angkor. A large area is also covered by lighter vegetation such as scrubland and sparse forest, which presents similar difficulties for recognition of palaeo-hydrology. Within this environmental context, however, hydrographical and artificial features are potentially identifiable by slight differences in the biomass and in other properties of the vegetation cover,

Table 9.2 ASTER metadata

ASTER	
Site	Greater Angkor
Product name:	L1A
Product option:	Uncorrected data, correction coefficients included with the granule
Collection range (Y-M-D)	2004-02-17
Acquisition date (Y-M-D)	2004-02-17
Cloud cover (%)	Not assigned
Quantisation	VNIR 8 bits; SWIR 8 bits; TIR 12 bits
Spectral bandwidth	Green: 520–600 nm Red: 630–690 nm Nir: 760–860 nm Swir: 1,600–2,430 nm Tir: 8,125–1,1650 nm
Swath width	60 km
Mul resolution (m)	VNIR 15; SWIR 30; TIR 9

which can be identified in the satellite imagery through spectral processing operations. Vegetation indices (VI) derived from remote sensing data were consequently adopted as prime procedure in this research. VI provide measurements of variability in the amount, growth and vigour of vegetation, and have thus proven valuable for detecting natural and archaeological deposits that augment or limit the growth of the vegetation (Lasaponara and Masini 2006, 2007; Traviglia 2006). The presence of extraneous elements in the texture of the subsoil can have a strong impact on the development of the vegetation, determining the manifestation of "marks" over the vegetation canopy. The vegetation becomes, in this way, a mediating element of subsurface heterogeneity.

A wide range of vegetation indices, calculated typically from arithmetical operations operated on the spectral reflectance values of two or more bands, have been developed to characterize vegetation canopies in the last 30 years (see e.g. Jackson 1983; Jensen 2000; Logan and Strahler 1983; see also Perry and Lautenschlager 1984, however, for an argument that most indices are essentially equivalent). Efforts have been made to compensate for recurrent external factors like illumination, atmosphere and soil backgrounds that affect several of the indices. When adopting VI for feature identification, the availability of a range of techniques for their computation gives the opportunity to test multiple indices in order to identify the most profitable in terms of feature recognition. A flexible approach to using VI is advisable when analysing vast portions of territory that have different types of vegetation coverage. The computation of different kinds of vegetation indices can in fact lead to distinctly different outcomes, depending on the vegetation coverage type and the density of the canopy; both are key aspects to take into account when selecting an appropriate process for evaluation of the biomass of a certain area. Environmental variables such as these can strongly affect the data quality. Normally, vegetation indices provide a monochromatic image that emphasises variations in the strength of the vegetation, where pixels related to more vigorous vegetation show high values due to the high reflectance of the NIR and pixels related to less vigorous vegetation have lower values. Differences in vegetation strengths are displayed as brightness within an image, in which healthier vegetation is brighter than non-healthy vegetation.

In order to detect remnants of the natural palaeo-hydrological system under the varied vegetation cover of north eastern area Angkor, a number of vegetation indices were calculated and compared. Each were devised on the basis of environmental and morphological characteristics of the macro-sections into which the area had been subdivided (dense forest, light forest, and scrubland). Tests using at least three different vegetation indices were performed for each macro-area; each time, the results obtained were contrasted with each other to establish the best process for vegetation health evaluation in that specific macro-area, and to verify the degree of improvement in the visibility of traces they offered. VI trials (Fig. 9.9) were executed both on QuickBird and ASTER data and produced images that occasionally revealed contradictory results, which can be attributed to many different factors including the dissimilar intrinsic characteristics of the sensors and variations in the acquisition date of the imagery. Of the VI tested, four are referred to specifically in this article: Difference Vegetation Index (DVI), Normalized Difference Vegetation Index (NDVI), Ratio Vegetation Index (RVI) and 2nd Modified Soil Adjusted Vegetation Index (MSAVI2). DVI and NDVI were chosen as relevant VI tests for their well known capability to reflect the range of vegetation quality, while VI and MSAVI2 were selected for their capacity to address signal distortions, caused by soil background, that are superimposed on or interact with information related to vegetation.

In case of ASTER images NIR and Red are the bands B3 and B2, respectively; for QuickBird images NIR and Red are the bands B4 and B3, respectively. The DVI (Difference Vegetation Index, occasionally referred to as Vegetation Index, VI) (Richardson and Wiegand 1977; Tucker 1979) consists of a subtraction operation involving Red and NIR pixel values, and is calculated through the formula:

$$DVI = NIR - Red \tag{9.1}$$

in order to evaluate the degree of variation between the values of the used bands. Following computation of this index, valuable results were achieved in assessing different types of vegetation cover, in particular the condition of open canopies (and patchy plant cover), but offered less valuable outcomes in close canopy situations. On the basis of a generic suitability for diverse contexts, DVI was largely applied and used as main reference image in the evaluation of the other processes results.

In areas of heterogeneous but generally dense canopy the NDVI (Normalized Difference Vegetation Index) (Nogi et al. 1993; Rouse 1974; Tucker et al. 1985) has a well-known capacity for detecting changes in vegetation cover, and also for compensating for changing illumination conditions and other extraneous factors. It generally proved to have better sensitivity than DVI in this specific environmental context.

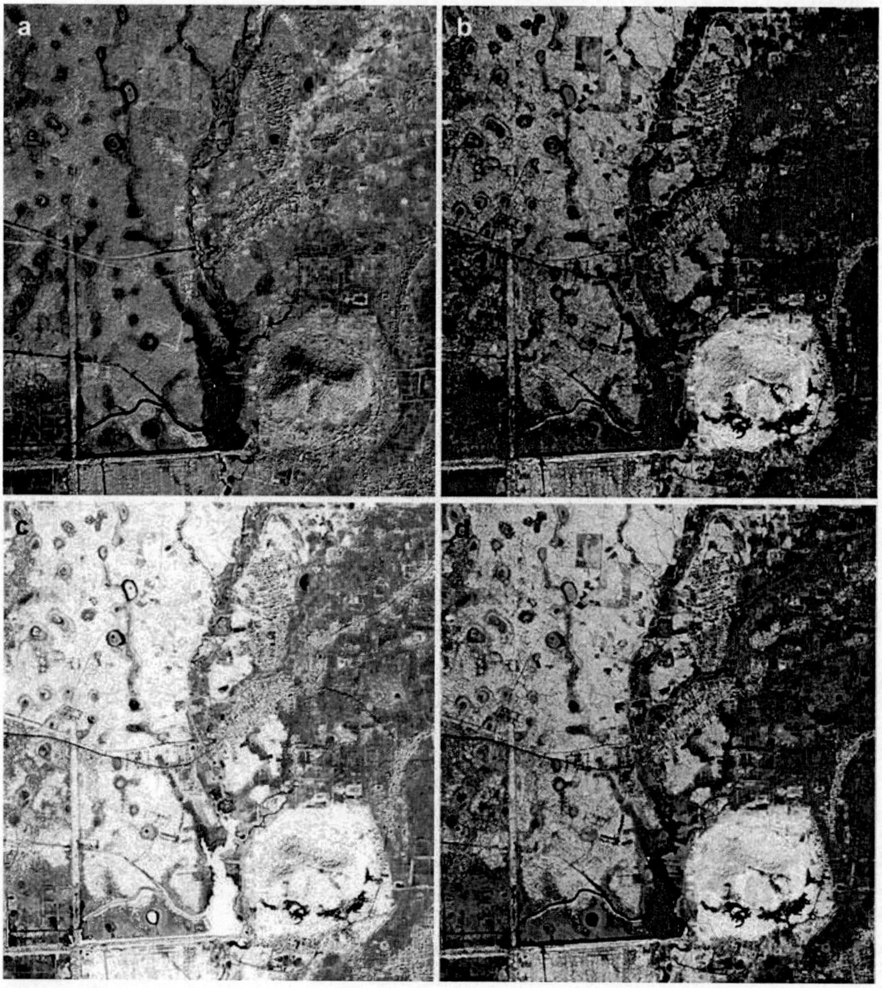

Fig. 9.9 Comparison among vegetation indexes: (**a**) Red band; (**b**) DVI; (**c**) RVI; (**d**) NDVI

This index involves the sum of, and also the difference between, Red and NIR bands according to the formula:

$$NDVI = \frac{(NIR - Red)}{(NIR + Red)} \tag{9.2}$$

The strength of NDVI lies in its "ratio" concept, which reduces the impact of some of the types of disturbance, for example variations in illumination and topography, in the final calculation. The principal disadvantage was its sensitivity to canopy background variations in cases of low or sparse vegetation (<30%).

In general, on test cases over areas of sparse vegetation, soil background conditions exerted considerable influence on partial canopy spectra: for some of those circumstances, simple ratio (SR or Ratio Vegetation Index, RVI) (Jordan 1969; Rouse 1974) of Red and NIR bands, according to the formula

$$SR = NIR/Red \tag{9.3}$$

appeared effective in normalising the effect of soil background reflectance variation.

Equally in those cases, the calculation of MSAVI2 (2nd Modified Soil Adjusted Vegetation Index) (Qi et al. 1994), an iterated version of MSAVI2, proved to be an effective strategy. This Vegetation Index is computed through the formula:

$$MSAVI2 = \frac{\left[2 \times (NIR + 1) - \sqrt{(2 \times NIR + 1)^2 - 8 \times (NIR - RED)}\right]}{2} \tag{9.4}$$

The MSAVI2 was created to reduce the reflectance effects of the soil background, which can introduce substantial alterations in the spectral response of the vegetation. In the test areas, these effects were manifested particularly in zones where the coverage of the vegetation was lean and/or scattered (i.e. <30/40%). Many of the areas targeted for study were characterised by this vegetation coverage, and consequently the index showed promising results. The sensitivity to variations in the atmosphere, which alters the light seen by the instruments, may however explain some unsatisfactory results obtained with a minority of the images. Such atmospheric distortion can in fact produce variations in the calculated values of the vegetation indices, and no preliminary atmospheric correction was undertaken on the target images.

9.3.5 Unveiling Features Under the Canopy: Vegetation Suppression

In addition to the computation of VIs as a recognition method, a test aiming to reveal features partially masked by vegetation was undertaken using Vegetation Suppression, a module included in the ITT ENVI 4.5 software package. The process performs a so-called "forced invariance" (Crippen and Blom 2001) to remove the spectral signature of vegetation from spectral data using information from Red and Near-Infrared bands, and without requiring any specific *a priori* knowledge of the scene. The method, by suppressing the expression of the vegetation component, facilitates an improved analysis of the underlying geological and archaeological features.

The process clearly demonstrated its potential in the area located about 1 km NE of the NE corner of Angkor Wat moat, within the Angkor World Heritage

Site, where a number of features presumably related to the Angkorian water management network were identified through visual analysis on the processed images. The area is covered by an open canopy forest, although not particularly dense, as well as a modest understorey. It is an ideal situation for this type of process: after the vegetation data has been suppressed, the remaining pixel values in the imagery of this area present both lithological and vegetative information. Ground surveys performed on the area demonstrated the presence of a number of artificial mounds occasionally topped with fragments of stone blocks. This particular area will become a focus for GAP field work activities in the future, including clearing transects through the forest to assess the correspondence between remotely sensed data and the physical evidence, and to assess the nature of the detected features.

At the moment it appears that a possible road-embankment, oriented EW and in apparent spatial connection with the Phnom Bakheng temple, joins with a V shaped connection the SE road-embankment heading to the Indratataka *baray* in the Roluos Group. The feature so formed is of interest when looked at as part of the larger water management system, as it shares a very similar shape with the embankment which forms the outlet of the water network grid located just south of the SW corner of the West *baray* (Fig. 9.10). The implication may be that in the ninth and tenth century configuration of the area SW of the East *baray* there was a similar system of which the detected feature could represent one outlet.

The ground survey has shown that the SE road-embankment heading to the Indratataka baray does approach the Phnom Bakheng road-embankment and that both are associated with substantial EW channels both on the northern and the southern sides of the Phnom Bakheng embankment. A study of the area east of the West *baray* also revealed the previously unknown length of a major hydraulic feature. Remote sensing data and excavation have shown that there was a buried exit channel running eastwards from the middle of the east bank of the West *baray*.

Vegetation suppression has shown that this linear feature runs on eastwards into the space of Angkor Thom. As it could not have functioned as a channel once the moat of Angkor Thom was dug, this feature presumably predates Angkor Thom i.e. is pre late 12th C. and presumably is the extent of the exit channel of the West *baray*.

Interestingly it runs for about 600 m, almost as far as the NS road that leads to the Baphuon. This suggests that in its 11th C. configuration that *baray*, the West Mebon at its centre, the axial eastern exit and the architectural ensemble of the Baphuon formed one vast piece of engineered landscape.

9.3.6 *Accentuating Surface Changes: Principal Component Analysis (PCA)*

Elements of the palaeo-hydrological system within barren terrain, and zones of very limited vegetation cover, are best accessed through an analysis of tonal variation within the examined surfaces. Evidence of past hydrology and occupation can

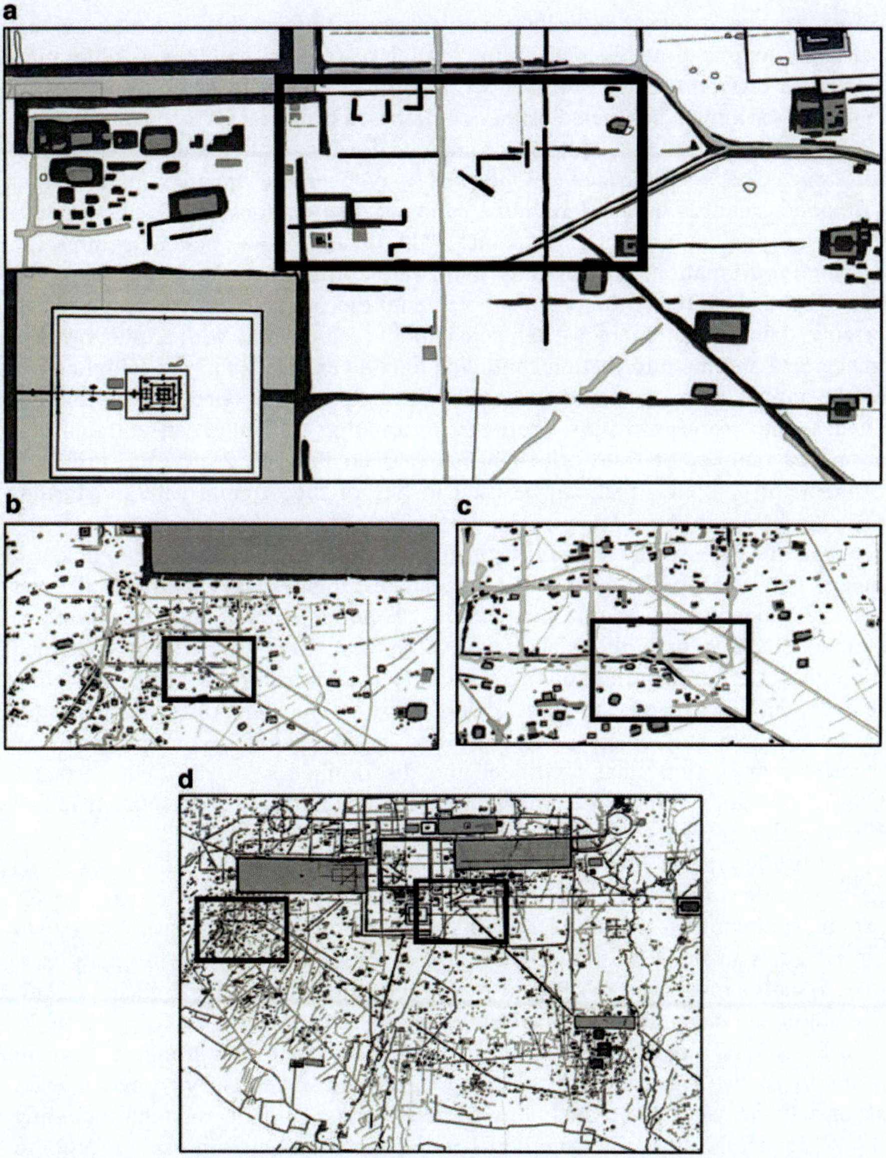

Fig. 9.10 Features identified using vegetation suppression: (**a**) V shaped joint between the EW road-embankment and the SE road-embankment to the Indratataka baray; (**b**) V-shaped outlet of the water network grid located SW of the West baray; (**c**) a detail; (**d**) the S Angkor water management system and the symmetric outlets connected to the SE oriented canals (Base map after Pottier 1999 and Evans 2007)

sometimes be detected by subtle variations in topography, but also in this particular application by identifying spatial patterning within soil coloration related to different filling soils, or within soil moisture. In order to more efficiently discriminate between dissimilar surfaces, a Principal Component Analysis (PCA) (Lillesand et al. 2004) was computed for the available data, as spectral differences between surface components may be more apparent in Principal Component images than in individual bands or their composites. Transformation of the original multispectral data into PCA data products better captures the essential information, and by improving the differentiation of landform and geomorphic features it therefore supplements the information contained in the original data. In fact, there is often considerable redundancy within multispectral image data and the information contained in one variable is largely a duplication of the information contained in another. Principal Component transformation creates a new series of bands where the information content is concentrated into a smaller number of transformed bands (named PC1, PC2 etc., i.e. Principal Component 1, 2 etc.) that can be used in lieu of the original data and further processed (Fig. 9.11).

Over the 97% of the variability in pixel values of the available QuickBird images (Fig. 9.11) and almost 96% of the ASTER images was captured in the first two Components and only respectively a 1% and a 2% was found in the third Component. For both the image types, the first three Components obtained through the PC transformation were visually analysed both as black and white images and as combined to form a RGB false colour composite. In this latter case very colourful images were obtained, where colours, combining freely, created combinations that facilitated the discrimination of different surfaces. QuickBird data's Principal Components were used also in a subsequent image classification process.

A specific type of PC transform, the Selective Principal Component Analysis (SPCA) (Siljestrom 1997), was performed on the ASTER scenes to make up for generic shortcomings within the process, specifically from information loss due to unused components. It was found that some of the details visible in the analysis of the original, individual ASTER bands were somehow obscured and made inaccessible through conventional PC analysis.

A Selective PCA was therefore executed for subgroups of correlated bands in order to reduce the information loss (Fig. 9.12). After carrying out a PCA separately on one group including visible bands, and on another including SWIR bands, the first PCs from both groups were combined to band 3 (NIR) to create a false colour composite (PC1 of bands 1 and 2 was assigned to Blue, PC1 of SWIR to Red and NIR to Green). The resulting image product was then subjected to visual interpretation, and noteworthy information related to geomorphologic aspects of the landscape was recorded and used to map a significant number of features that may be related to the original hydrological system.

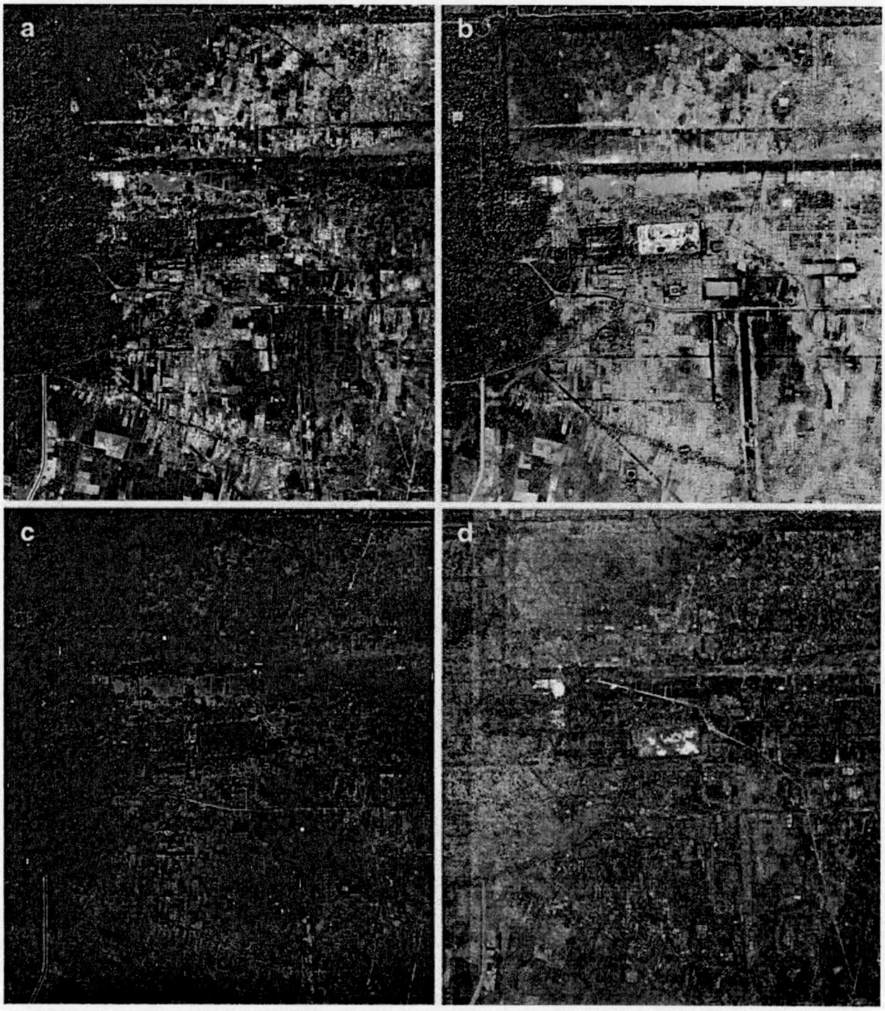

Fig. 9.11 Details of PC1 (**a**), PC2 (**b**), PC3 (**c**), PC4 (**d**) of a QuickBird scene: the amount of information decreases in the higher orders components

9.3.7 Preliminary Results

The evaluation of QuickBird and ASTER scenes and their transformed derivatives was performed in a GIS environment and followed by mapping. The anthropogenic features and the topographical anomalies detected during on-screen analyses were mapped at a nominal scale of 1:2,000 and given a series of attributes to encapsulate relevant information. Metadata retained in this process included: the image process(es) that facilitated identification; possible interpretation of the

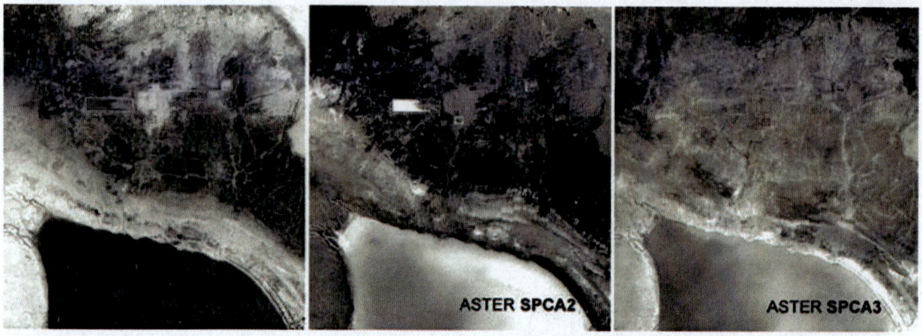

Fig. 9.12 First 3 components of a SPCA performed on VNIR bands of ASTER data

feature, degree of visibility; factors in the recognition of the trace; and archaeological reliability, i.e. an evaluation of the quality of the observed features. The new features and their metadata will allow further analysis and interpretation by way of comparison with pre-existing topographical and cultural datasets (Evans 2002, 2007; Japan International Cooperation Agency 1999; Pottier 1999; Royal Angkor Foundation 1995), and will provide the basis for a prioritised strategy of ground verification.

In order to move towards an objective framework for assessing the interpretations of palaeo-hydrographic traces, a SRTM3 DEM (90 m nominal resolution) was used as a reference dataset and systematically compared to the remotely sensed data and the interpretive data products. This process provided important information that contributed to an understanding of the feature patterns. Among the principal palaeo-hydrologic structures identifiable at this preliminary stage, it is worth noting in particular the trace of a palaeo watercourse flowing with an orientation of E-NE/W-SW about 3 km N of an intermittent side-stream of the present Puok River, oriented in similar way and almost parallel to the detected palaeo-feature. The persistence of this ancient watercourse during the Angkorian period appears to be partly validated by an analysis of the distribution of reservoirs and other human occupation features in the surrounding area. Their positioning, in fact, implies the presence of a large, curved entity, like the flow of a river, which they border precisely edge (Fig. 9.13).

9.4 Outcomes and Conclusions

The overview of remote sensing applications above reveals the crucial importance of the method for an understanding of the nature of early cities in this region. Elements of urban structure such as temples, occupation mounds and ponds are frequently difficult to discern on the ground, but in the Cambodian environment are readily identifiable from the air. As we have shown at sites such as Angkor and Banteay Chmar, incorporating aerial archaeology into research on site

Fig. 9.13 Trace of a palaeo-riverbed E of the Puok river

distributions and urban form can lead to very substantial increases in the number of features identified and mapped, when compared to methods which rely primarily or exclusively on ground-based investigations. Of course, the field verification of features identified using remote sensing is still absolutely essential for a full understanding of the landscape, but the studies at Angkor undertaken by the Greater Angkor Project have clearly proven the value of prefacing traditional survey and excavation methods with a comprehensive and systematic evaluation of all of the available remotely sensed imagery. Angkor is a vast urban complex that will, in all likelihood, never see a majority of its sites excavated; once most of the surface features have been mapped from the air, however, areas of critical importance within the network of features can be identified and effectively prioritised for ongoing investigation on the ground.

Moreover, as the case study at Angkor has shown, remote sensing can continue to contribute substantially to ongoing research objectives even when the focus of a project has shifted to ground-based techniques. Although the initial phases of GAP revolved around the qualitative, manual interpretation of imagery, sophisticated image processing techniques continue to play an important role in refining the maps of Angkor and adding detail to areas of critical importance. Whereas previously remote sensing applications at Angkor have concentrated on the identification of the large-scale and medium-scale structure of the settlement, ongoing work is concentrating on the identification of extremely subtle features while also addressing weaknesses of the purely qualitative approach, for example by applying complex processing algorithms to uncover features hidden beneath dense canopy.

One of the key lessons to be learned from an overview of the history of aerial archaeology in Cambodia is the importance of applying the method in an appropriate theoretical context. For many decades in the mid-twentieth century, archaeologists had at their disposal all of the tools necessary to undertake research that may have placed Cambodia at the forefront of developments in world archaeology, for example the emerging settlement archaeology of the 1940s and 1950s (Billman 1999). Instead, this opportunity was lost, as questions well-suited for the application of landscape-scale remote sensing techniques—for example, the spatial layout of extra-mural habitation, or the structure and function of the canal system—were largely ignored even by scholar-aviators in favour of the traditional focus on walled cities, sacred geographies and the immediate surrounds of the great sandstone monuments (Evans 2007).

Paradoxically, nonetheless, most researchers working on issues of landscape archaeology in Cambodia today are essentially following the course charted by Bernard-Philippe Groslier as early as the 1950s. Although the agenda he proposed was never completed in is lifetime, advances in remote sensing technologies and analytical techniques have made the fulfilment of his research objectives possible, even as studies of palaeoclimate and new theories about the nature of settlement growth re-orient the overall themes away from ideas of 'Oriental despotism' and power structures towards issues of landscape and the dynamic nature of human-environment interactions (Fletcher 1995, 2004). It should also be noted that, while virtually all of the criticisms of Groslier's 'hydraulic city' hypothesis can be dismissed on the basis of the new evidence, there is, on the other hand, no conclusive evidence at this stage that the hydraulic system was used for producing multiple crops of rice per year. Although we have demonstrated that it was certainly capable of doing so, intensive agricultural practises are extremely difficult to identify within the archaeological record. Work in this area is ongoing, and revolves primarily around analysis of the chemical characteristics of soils in the irrigable areas to the south (i.e. down slope) of the storage system at the centre of Angkor. Nonetheless, it is clear that, at the very least, the great reservoirs functioned as a kind of insurance policy against years of poor rainfall, and that they would have reduced the risk to basic subsistence inherent in short-term climatic variation (Lustig 2001). Long-term climate data for the region are currently unavailable but are in the process of being compiled; the important question, therefore, is now the issue of whether the system could have coped with the extended periods of drought that preliminary data now indicate were characteristic of the region during the classical Angkorian period (Stone 2009).

Work at sites beyond Angkor will help to bring some of the answers to these questions into sharper relief. We have shown that, on the current evidence, it now appears that Angkor was far from unique in its dependence on a massive and sophisticated water management system, and that such systems were in fact a very common, if not ubiquitous, feature of all major early Khmer settlements. Further research will concentrate on other Angkor-era complexes such as Preah Khan of Kompong Svay, as well as pre-Angkorian capitals such as Sambor Prei Kuk and post-Angkorian capitals in the vicinity of Phnom Penh, many of which—even on a very preliminary analysis—show compelling evidence for much the same urban and hydraulic structures that have been discovered at Angkor itself. In the immediate future, the increasing availability and falling costs of very high resolution multispectral datasets such as GeoEye-1 imagery, and the imminent availability of very high resolution elevation data from radar satellites such as DLR's TanDEM-X, will open up a range of interesting possibilities for archaeological research in Cambodia. As we have shown, some of this potential is already being realised under the auspices of a broad comparative study of Khmer urbanism, which, like GAP itself, will be fleshed out in the future by complementary ground-based methods such as excavation.

Acknowledgements The authors wish to thank the Australian Research Council (DP0880490, DP0211012, DP0558130, LX0882079); Faculty of Arts, School of Philosophical and Historical Inquiry, and Archaeological Computing Laboratory at the University of Sydney; Robert Christie Foundation; GeoEye Foundation; NASA; German Aerospace Centre (DLR); National Geographic; Carlyle Greenwell Bequest; Iain A Cameron Memorial Travel Grant; Royal Angkor Foundation Koh Ker Project; APSARA National Authority and Ministry of Culture and Fine Arts (Cambodia); Global Heritage Fund Banteay Chmar Project; Angkor Ultralight Survey; Bun Narith; Ros Borath; Soeung Kong; Heng Sophady; Ngaire Richards; Christophe Pottier; Roland Fletcher; Mitch Hendrickson; Martin King; Ian Johnson and Andrew Wilson.

References

Acker R (1998) New geographical tests of the hydraulic thesis at Angkor. South East Asia Res 6:5–47

Aymonier E (1900–04) Le Cambodge. Ernest Leroux, Paris

Baty P (2005) Extension de l'aéroport de Siem Reap 2004: Rapport de fouille archéologique. APSARA – INRAP, Siem Reap

Billman BR (1999) Settlement pattern research in the Americas: past, present and future. In: Billman BR, Feinman GM (eds) Settlement pattern studies in the Americas: fifty years since Viru. Smithsonian Institution Press, Washington, DC, pp 1–9

Bronson B (1978) Angkor, Anuradhapura, Prambanan, Tikal: Maya subsistence in an Asian perspective. In: Harrison PD, Turner BL (eds) Pre-Hispanic Maya agriculture. University of New Mexico Press, Albequerque, pp 255–300
Chandler DP (2000) A history of Cambodia. Silkworm Books, Chiang Mai
Crippen RE, Blom RG (2001) Unveiling the lithology of vegetated terrains in remotely sensed imagery. Photogramm Eng Remote Sens 67:935–943
Engelhardt RA (1995) Two thousand years of engineering genius on the Angkor plain. Expedition 37:18–29
Evans D (2002) Pixels, ponds and people: urban form at Angkor from radar imaging. Unpublished BA(Hons) thesis, University of Sydney, Sydney
Evans D (2007) Putting Angkor on the map: a new survey of a Khmer 'Hydraulic City' in historical and theoretical context. Unpublished PhD thesis, University of Sydney, Sydney
Evans D, Pottier C, Fletcher R, Hensley S, Tapley I, Milne A, Barbetti M (2007) A comprehensive archaeological map of the world's largest preindustrial settlement complex at Angkor, Cambodia. Proc Natl Acad Sci USA 104:14277–14282
FINNMAP (1992–1993) Aerial photography for resources mapping in Cambodia; 1:25,000. FM-International Oy FINNMAP (Cambodia), Phnom Penh
Fletcher R (1995) The limits of settlement growth: a theoretical outline. Cambridge University Press, Cambridge
Fletcher R (2000–2001) A.R. Davis memorial lecture. Seeing Angkor: new views of an old city. J Orient Soc Aust 32–33:1–25
Fletcher R (2004) Materiality, space, time and outcome. In: Bintliff JL (ed) A companion to archaeology. Blackwell Publishing, Malden
Fletcher R, Pottier C (2002) The Gossamer city: a new inquiry. Mus Int 54:23–7
Fletcher R, Barbetti M, Evans D, Than H, Sorithy I, Chan K, Penny D, Pottier C, Somaneath T (2003) Redefining Angkor: structure and environment in the largest, low density urban complex of the pre-industrial world. Udaya J Khmer Stud 4:107–121
Fletcher R, Penny D, Evans D, Pottier C, Barbetti M, Kummu M, Lustig T (2008) The water management network of Angkor, Cambodia. Antiquity 82:658–670
Gaughan AE, Binford MW, Southworth J (2008) Tourism, forest conversion, and land transformations in the Angkor basin, Cambodia. Appl Geogr 29:212–223
Goloubew V (1936) Reconnaissances aériennes au Cambodge. Bull École Fr Extrêm-Orient 36:465–478
Groslier B-P (1952) Milieu et évolution en Asie. Bull Soc Études Indoch 27:295–332
Groslier B-P (1960) Our knowledge of Khmer civilisation, a re-appraisal. J Siam Soc 48:1–28
Groslier B-P (1979) La Cité Hydraulique Angkorienne: Exploitation Ou Surexploitation Du Sol? Bull École Fr Extrêm-Orient 66:161–202
Hall KR (1992) Economic history of early Southeast Asia. In: Tarling N (ed) The Cambridge history of Southeast Asia: volume one, from early times to c.1800. Cambridge University Press, Cambridge
Hendrickson M (2007) Arteries of Empire: An operational study of transport and communication in Angkorian Southeast Asia. Unpublished PhD thesis. University of Sydney, Sydney
Jackson RD (1983) Spectral indices in n-space. Remote Sens Environ 13:409–421
Jacques C, Lafond P (2007) The Khmer empire: cities and sanctuaries, fifth to thirteenth century. River Books, Bangkok
Japan International Cooperation Agency (1999) Cambodia reconnaissance survey digital data. Japan International Cooperation Agency (JICA), Phnom Penh
Jensen JR (2000) Remote sensing of the environment: an earth resource perspective. Prentice Hall, Upper Saddle Rivers
Jordan CF (1969) Derivation of leaf area index from quality of light on the forest floor. Ecology 50:663–666
Kiernan B (2001) Myth, nationalism and genocide. J Genocide Res 3:187–206

Kummu M (2003) The natural environment and historical water management of Angkor, Cambodia. In: 5th world archaeological congress, Washington, DC. http://users.tkk.fi/~mkummu/publications/kummu_WAC_WashingtonDC_2003.pdf. Accessed Apr 2009

Lasaponara R, Masini N (2006) Identification of archaeological buried remains based on Normalized Difference Vegetation Index (NDVI) from QuickBird satellite data. IEEE Geosci Remote Sens Lett 3:325–328

Lasaponara R, Masini N (2007) Detection of archaeological crop marks by using satellite QuickBird multispectral imagery. J Archaeol Sci 34:214–221

Lillesand TM, Kiefer RW, Chipman JW (2004) Remote sensing and image interpretation, 5th edn. Wiley, New York

Logan TL, Strahler AH (1983) Optimal Landsat transforms for forest applications. In: McDonald DC, Morrison DB (eds) Machine processing of remotely sensed data with special emphasis on natural resources evaluation. Ninth international symposium, 21–23 June 1983. Purdue Research Foundation, Purdue

Lunet de Lajonquière E (1902–11) Inventaire déscriptif des monuments du Cambodge, 3 vols. École française d'Extrême-Orient, Paris

Lustig E (2001) Water and the transformation of power at Angkor, 10th to 13th centuries A.D. Unpublished BA (Hons) thesis, University of Sydney, Sydney

Ministère de la culture et des beaux-arts and École française d'Extrême-Orient (2007) Carte archéologique du Cambodge. Kim Long, Phnom Penh

Moore E (1995) The waters of Angkor. Asian Art Cult 8:36–51

Moore E (2000) Angkor water management, radar imaging, and the emergence of urban centres in Northern Cambodia. J Sophia Asian Stud 18:39–52

Moore EH (1989) Water management in early Cambodia: evidence from aerial photography. Geogr J 155:204–214

Nogi A, Weidong Sun, Takagi M (1993) An alternative correction of atmospheric effects for NDVI estimation. In: Geoscience and remote sensing symposium, 18–21 Aug 1993. IGARSS '93. Better Understanding of Earth Environment

Parmentier H (1939) L'art Khmer classique: monuments du quadrant nord-est. Les Éditions d'art et d'histoire, Paris

Penny D (2006) Vegetation and land-use at Angkor, Cambodia: a dated pollen sequence from the Bakong temple moat. Antiquity 80:599–614

Penny D, Pottier C, Kummu M, Fletcher R, Zoppi U, Barbetti M, Tous S (2007) Hydrological history of the West Baray, Angkor, revealed through palynological analysis of sediments from the West Mebon. Bull École Fr Extrêm-Orient 92:497–521

Perry CR, Lautenschlager LF (1984) Functional equivalence of spectral vegetation indices. Remote Sens Environ 14:169–182

Pottier C (1993) Préparation d'une carte archéologique de la région d'Angkor. Unpublished DEA thesis, Université Paris III—Sorbonne Nouvelle, Paris

Pottier C (1999) Carte Archéologique de la Région d'Angkor. Zone Sud. Unpublished PhD thesis, Université Paris III—Sorbonne Nouvelle, Paris

Pottier C (2000) Some evidence of an inter-relationship between hydraulic features and rice field patterns at Angkor during ancient times. J Sophia Asian Stud 18:99–120

Pottier C (2004) À propos du temple de Banteay Chmar. Aséanie 13:131–150

Pottier C, Chhem RK, Demeter F, Gabillault C, Guerin A, Heng T, Khieu C, Latinis K, Boer-Mah TS, Stulemeijer R, Watanasak M, Yam S (2004) Mission Archéologique Franco-Khmère sur l'Aménagement du Territoire Angkorien: Campagne 2004. École Française d'Extrême-Orient, Siem Reap

Pottier C, Bolle A, Llopis E, Soutif D, Tan C, Chevance JB, Kong V, Chea S, Sum S, Demeter F, Bacon A, Bouchet N, Souday C, Frelat M (2005) Mission Archéologique Franco-Khmère sur l'Aménagement du Territoire Angkorien: Campagne 2005. École Française d'Extrême-Orient, Siem Reap

Qi J, Chehbouni A, Huete AR, Kerr YH (1994) Modified Soil Adjusted Vegetation Index (MSAVI). Remote Sens Environ 48:119–126

Richards N (2007) Prehistoric and early historic settlement around Banteay Chmar, north-west Cambodia. Unpublished BA(Hons) thesis, University of Sydney, Sydney

Richardson AJ, Wiegand CL (1977) Distinguishing vegetation from soil background information (by gray mapping of Landsat MSS data). Photogramm Eng Remote Sens 43:1541–1552

Rouse JWJ (1974) Monitoring vegetation systems in the great plains with ERTS. In: Freden SC et al (eds) Third earth resources technology satellite-1 symposium- vol I: technical presentations, 10–14 Dec 1973. NASA, Washington, DC

Royal Angkor Foundation (1995) Angkor GIS v.1.0. Royal Angkor Foundation, Budapest

Siljestrom PA (1997) Technical note. The application of selective principal components analysis (SPCA) to a Thematic Mapper (TM) image for the recognition of geomorphologic features configuration. Int J Remote Sens 18: 3843–3852

Stone R (2009) Tree rings tell of Angkor's dying days. Science 323:999. doi:10.1126/science.323.5917.999b

Stott P (1992) Angkor: shifting the hydraulic paradigm. In: Rigg J (ed) The gift of water: water management, cosmology and the state in South East Asia. School of Oriental and African Studies, London

Traviglia A (2006) Archaeological usability of hyperspectral images: successes and failures of image processing techniques. In: Campana S and Forte M (eds) From space to place. Proceedings of the 2nd international conference on remote sensing in archaeology, Rome, 4–7 Dec 2006. Archaeopress, Oxford

Trouvé GA (1933) Chaussées et canaux autour d'Angkor Thom. Bull École Fr Extrêm-Orient 33:1120–1128

Tucker CJ (1979) Red and photographic infrared linear combinations for monitoring vegetation. Remote Sens Environ 8:127–150

Tucker CJ, Townshend RG, Goff TE (1985) African land-cover classification using satellite data. Science 227:369–375

Van Liere WJ (1980) Traditional water management in the lower Mekong Basin. World Archaeol 11:265–280

Chapter 10
Remote Sensing Study of the Ancient Jabali Silver Mines (Yemen): From Past to Present

Jean-Paul Deroin, Florian Téreygeol, and Jürgen Heckes

Abstract Archaeological research into the ancient Jabali silver mines in northern Yemen is a multidisciplinary project linking archaeologists, historians, and geologists, supported by remote satellite-based sensing. Mining companies are considering the development of the Jabali deposit for zinc ore. Some of the facilities, including a large open pit, will have a destructive impact on the old pits. Multispectral (Landsat TM, Terra ASTER, ALOS-AVNIR-2, etc.) and very high spatial resolution images such as QuickBird have been widely used for the geological setting and detail mapping of the different archaeological sites. The remote sensing data also provides a solid basis when it comes to detecting current operational sites.

Keywords Multispectral satellite image • Mining archaeology • Geoarchaeology • Yemen • Jabali

10.1 Introduction

Fundamentally, the objectives of mining archaeology are found to have antecedents in ancient works. As early as 1858, Louis Simonin, working in Tuscany, demonstrated the interest of a combined study of the ore deposits, the surficial

---

J.-P. Deroin (✉)
GEGENAA EA 3795, Université de Reims Champagne-Ardenne, 2 esplanade Roland Garros, 51100 Reims, France
e-mail: jean-paul.deroin@univ-reims.fr

F. Téreygeol
UMR 5060, Institut de Recherche sur les Archéomatériaux and CEA, Saclay, France
e-mail: florian.tereygeol@cea.fr

J. Heckes
Deutsches Bergbau Museum, Am Bergbaumuseum 28, 44791 Bochum, Germany
e-mail: info@dainst.de

R. Lasaponara and N. Masini (eds.), *Satellite Remote Sensing: A New Tool for Archaeology*, Remote Sensing and Digital Image Processing 16,
DOI 10.1007/978-90-481-8801-7_10,

and underground archaeological remains, and the old texts (Simonin 1858). However, mining archaeology is a relatively recent discipline. It deals with the study of ancient mines and their relation with the regional mineral economy. Within this context, however, shafts or entrances of the old networks are generally hidden under old mining dumps or colluvial deposits. Moreover, the underground remains are frequently filled up or flooded and some of them were intersected and damaged by nineteenth or early twentieth century works. These elements explain the limited interest of archaeologists in the old mines up until the 1970s.

Some disciplines can provide relevant data for mining archaeology. The Earth sciences, for instance, provide important data on the mineralogy of ore deposits, as well as on the overall geological setting. More recently, scientific analytical methods such as geochemistry have provided important information about the chemical composition of ore, slags, and artifacts, allowing archaeologists to built correlations or identify provenance. Although mining archaeology is inherently transdisciplinary in nature, it has only taken limited advantage of the potential of remote sensing, particularly its recent developments in spatial and spectral properties.

The ancient silver mines of Jabali, Yemen, are located in a mountainous and arid area of the southern Arabian peninsula, approx. 75 km north-east of Sana'a (Fig. 10.1). The Jabali deposit is hosted by dolomitized platform carbonates located at the southwestern edge of the oil-producing Wadi al Jawf basin (Al-Ganad et al. 1994). Jabali is a Zn-Pb-Ag deposit grading approx. 9.2% zinc, 1.2% lead and 68 g/t silver. Whereas Jabali was a well-known medieval silver mine, it has currently been explored only for zinc-ore. The archaeological missions carried out in 2006 and 2008 incorporated an important geological aspect supported by remote-sensing techniques. The first objective of this approach was an improvement in the geological background of the site, favoured by well-expressed morphologies (Deroin et al. 2011). At the same time, it allowed the study of the whole mining site across approx. 100 km^2 resembling a field, where displacements are difficult and the relationship with the local inhabitants is not without complications. Further to an overview of the sole Jabali mining site, the goals of our study include the comparison of the techniques used in the medieval Islamic world and those known at the same period in western Europe. The similarity with the Carolingian site of Melle (Poitou, France) is clear, as far as chronology, geology, and mineralogy are concerned (Téreygeol 2001), whereas the cultural and climatic aspects are obviously quite different.

The chronology of the archaeological mining site has been outlined during various field missions. Two main exploitation phases have been identified. The first phase took place between the last century of the Jahiliyya or pre-Islamic period (sixth to seventh century AD) and the second century *anno hegirae* AH (eighth to ninth century AD). The second phase extended from the fourth to the sixth century AH (twelfth to fourteenth century AD).

This chronology is based on a combination of field observations, document analyses and laboratory measurements. It provides thoroughly new insights into Jabali, knowledge of which had previously been based mainly on the analysis of the *kitab al-jawharatayn al-'atiqatayn* (Toll 1968), a book written by the so-called al-Hamdani, an Arab philosopher from the tenth century AD.

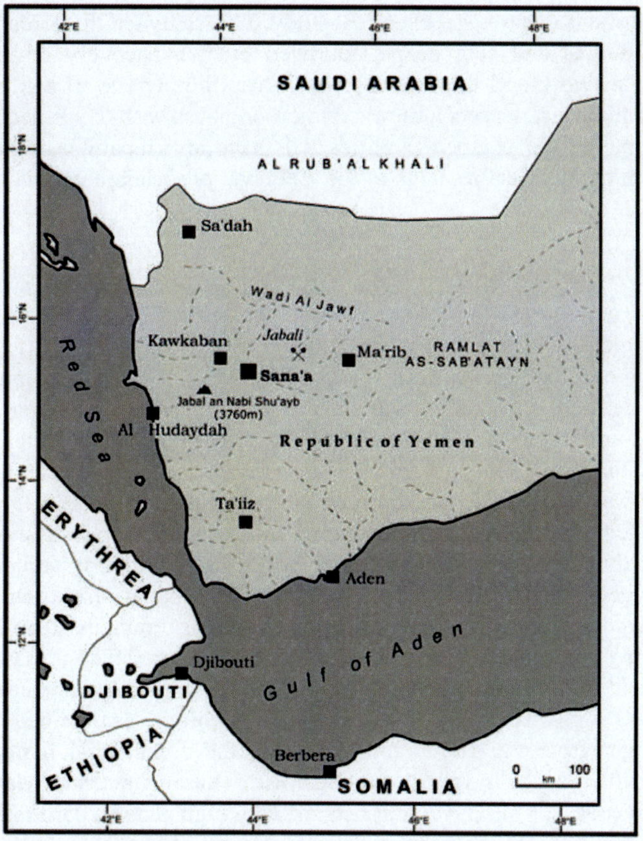

Fig. 10.1 Presentation of the study area

The text precisely describes the mines of *ar-Radrad*, which were related to the present Jabali site (Robin 1988). It defines the *terminus post quem* of the mining activities, because *al-Hamdani* indicates that silver exploitation suddenly ceased after the murder of Yu'firide Muhammad b. Yu'fir, a local sovereign in power in the year 270 AH (883–884 AD).

The miners of Persian origin were forced to flee during these troubled times. However, our observations indicate a significant renewal of the mining activities a few centuries later (twelfth to fourteenth century AD), during the Rasulide dynasty. During this revival, the mining site underwent considerable changes, including the dressing and crushing areas. Numismatic data combined with the archaeological data show a close link between this renewal and the coinage in Sana'a, the Rasulide capital city.

The history of the mining activities comprises alternating periods of exploitation and periods of relative abandonment. This also applies to the Jabali ore deposit, rediscovered by the French geological survey (BRGM) using archaeological information in the early 1980s (Christmann et al. 1983). It should be noted, that the rebirth of the site is due to a geochemical study. This study led the geologists to the old mining dumps and slag heaps, followed by the discovery of the ancient underground network and the orebody itself. The mining site of ar-Radrar, now Jabali, as a palimpsest, suffers a third exploitation phase with the project of a large open pit. Some of the planned facilities will have an irreparable impact on the mining and mineralurgical remains of the medieval pre-Islamic period.

10.2 Archaeological Remote Sensing

In archaeological remote sensing, the most conventional approach taken is the identification of crop marks from aerial photographs. Moreover, multispectral sensors can be used, particularly when dry conditions are encountered such as in arid or semi-arid regions. In the case of the remote Jabali area neither aerial photographs nor detailed topographic maps are available. However, satellite remote sensing data was available for the purpose of analyzing both the geological setting and the archaeological survey and prospection. Due to the arid conditions, interesting spectral domains such as medium or short wave infrareds are particularly relevant for geological mapping, whereas very high spatial resolution (less than 2 m) is needed for the identification of the archaeological remains. The analysis of remote-sensing data makes it possible to define and hierarchize areas of interest before the field missions, an approach that facilitates ground-based studies.

A relatively high spatial resolution is available with the declassified CORONA satellite images from the 1960s and now with the commercial remote-sensing imaging satellites such as Ikonos or QuickBird (Lasaponara and Masini 2005, 2006; Deroin et al. 2011). The comparison of very high and standard multispectral satellite imageries is rare (Schmid et al. 2008; Deroin et al. 2011).

We have used a wide range of satellite data. This data was mainly obtained by multispectral sensors (Landsat TM/ETM+, Terra-ASTER, ALOS-AVNIR-2), which are of particular interest because of their spectral response in the infrared range (700–2,400 nm) and in the red range (600–700 nm) and where iron caps or ferruginous soils are concerned (Table 10.1). Among this data, the Landsat ETM+ and ASTER

Table 10.1 Main satellite data used

Satellite	Sensor	GSD (m)	Spectral range (µm)	Period	Main data take
Landsat 1–5	MSS	80	0.50–1.10	1972–1992	7 scenes
Landsat 4–5	TM	30	0.45–2.35	Since 1982	26 scenes
Landsat 7	ETM+	30	0.45–2.35	Since 1999	18 scenes
Spot 1–3	PAN	10	0.50–0.73	1986–2009	April 21, 1995
Spot 5	PAN	5	0.50–0.73	Since 2002	August 29, 2002
EOS-Terra	ASTER	15 or 30 (SWIR)	0.52–2.36	Since 1999	10 scenes
ALOS	AVNIR-2	10	0.42–0.89	Since 2006	January 21, 2007 October 26, 2009
QuickBird	XS/P	0.61	0.45–0.90	since 2001	March 30, 2002

Fig. 10.2 3D view of the site. The QuickBird image acquired on 30 March 2002 is draped over the digital elevation model extracted from the 1:50,000 scale topographic map of Yemen. The area covered is about 15 km

data is of interest due to its spectral range extending toward the short wave infrareds (but with a medium spatial resolution), and that from QuickBird for the very high spatial resolution of this sensor. ALOS AVNIR-2 data, which has recently been made available, represents a good trade-off with its multispectral sensor (limited to the near infrared radiations) associated with a high resolution (10 m). This data was selected for the geological mapping of the area (Fig. 10.3). In addition to the set of image data, we also used digital elevation models (DEM). We built different DEMs using remote sensing data (Spot, SRTM, ASTER) and local topographic maps at 1:50,000. Satellite data could be draped over the DEM to obtain 3D-views.

Although relevant for geological mapping, the ground resolution from 10 to 30 m is clearly too rough for detailed field investigations required by the geoarchaeological survey. Therefore, we used the pansharpening of very high resolution multispectral images from QuickBird with a 0.67-m ground resolution. The QuickBird data draped over a high resolution DEM (5 m) allows quantification of the elementary watershed used for concentrating the water necessary to supply the ore-dressing areas (Fig. 10.2).

The work was complemented in the field by GPS measurements of the old canals and their connection with the ore-dressing areas. Some of them have been already destroyed, but it is expected that the satellite data could allow the detection of specific landforms.

The geological interpretation of this QuickBird data for the mine area results in a map representing 14 image facies corresponding to different limestone and dolomite facies (Deroin et al. 2006a). The latter were correlated with the lithologies as

Fig. 10.3 ALOS colour composite. The image was acquired on January 21, 2007. *Red* near infrared range, *Green* red range, *Blue* blue range. The area covered is about 10 × 9 km

observed in the field, using their spectral and textural characteristics. Thus, different kinds of limestone and dolomite were distinguished. QuickBird-based data represents the key image for the mapping of the archaeological remains.

10.3 The Geology of the Mining Site

The geology of the Jabali area is clearly put into light when using the ALOS image at an intermediary scale (Figs. 10.3 and 10.4). Other investigations have shown the interest of very high resolution data (Deroin et al. 2006b). The Jabali Zn-Pb-Ag

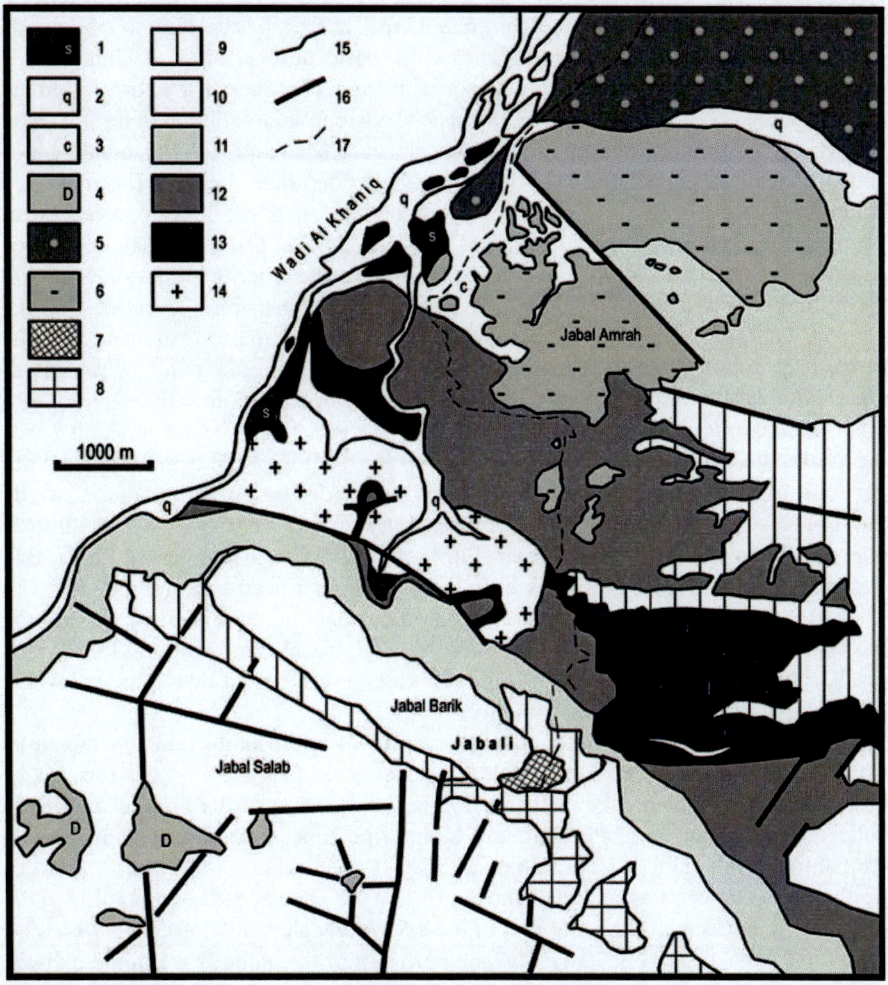

Fig. 10.4 Interpretation of the ALOS data represented in Fig. 10.3. *1* Main slag areas, *2* wadis deposits, *3* pediment deposits, *4* doline, *5* reg surface developed on marly facies, *6* volcanites, *7* Jabali iron cap, *8* late dedolomitization, *9* secondary dolomitization, *10* massive bioclastic limestone (U7) and black mudstone (U8), *11* greenish mudstone grading to limestone and marl (U6), *12* bryozoan calcarenite and oolitic limestone (U5), *13* sandstone, mudstone, and limestone (U1 to U4), *14* granite, *15* main lithological contour, *16* main fault, *17* track to the mine. NB. U1 to U8 (Triassic? to Upper Jurassic) are cited according to Al Ganad et al. (1994)

deposit lies within the topmost layers of the Amran Formation, which was deposited during a marine transgression that reached its maximum in the Late Oxfordian-Early Kimmeridgian (Upper Jurassic). This formation crops out along the south-western edge of the Wadi Al Jawf basin, the place of oil deposits, located in the 1980s. The lithological succession comprises a 300-m-thick sequence, the so-called

Amran Formation, divided into eight units, Unit 1 to Unit 8 according to Al-Ganad et al. (1994). Due to their low thickness, the basal units (Unit 1 to Unit 4) are generally difficult to separate using remote sensing techniques. They correspond to no. 13 in Fig. 10.4. Unit 1 is a 10 m-thick sandstone and conglomerate unit that occurs unconformably above the Proterozoic basement and, for the Jabali area, above Cenozoic granites (no 14). It could be older than Jurassic (Permian or Triassic?). Unit 2 is made of gypsiferous mudstone, marl and limestone reaching 25 m in thickness. Unit 3, dated to Callovian, is the first unit that could be stratigraphically dated owing to fossils present in the micritic limestone (about 50 m thick). Note that in the south-east area a micritic limestone, so-called Unit 3, is mainly outcropping. Unit 4 is also made of a micritic limestone associated with bedded dolomite of lagoonal facies (15 m). The field Unit 5 is a key unit, because it is a bryozoan calcarenite frequently overlain by coral-bearing oolitic limestone. This 40 m-thick unit is of Late Oxfordian-Early Kimmeridgian age. Unit 6 and Unit 7 are the thicker units with about 80 m each. Unit 6 (no 11, Fig. 10.4) is mainly made of gypsiferous mudstone at the base, grading to interbedded limestone and marl toward the top. Unit 7 (no 10, Fig. 10.4) comprises massive limestone but they are almost completely dolomitized. The Amran Formation contains a number of Pb-Zn-Ba occurrences. The Jabali deposit is hosted by the dolomitized limestone of Unit 7. Thus, the mineralization is associated with dolomitization controlled by WNW- to NNW-striking faults. Such a fault is clearly visible in the ALOS image (no 9, Fig. 10.4). Non-economic Pb-Ba occurrences are hosted by the lower layers of the Amran Formation (Unit 5 mainly).

The oxidized mineralization consists mainly of smithsonite (zinc carbonate), hydrozincite, cerussite (lead carbonate), and iron and manganese oxides forming a large iron cap or gossan at Jabali (no. 7, Fig. 10.4). Most of the mineralization is concentrated in massive bioclastic and biomicritic limestones, locally oolitic and containing corals (Unit 7, see unit no. 10, Fig. 10.4). This unit is dolomitized in its northern part, as well as in Jabali (no. 9, Fig.10.4). The secondary dolomitization phenomenon affects also a large part of the bryozoan calcarenite of Unit 5 which is represented by no. 12 (Fig.10.4). The southern part of the orebody is limited by late dedolomitization processes illustrated as a dark strip within the white carbonates (no. 8, Fig. 10.4). The top of the unit is deeply affected by fracturing and karstification in the highlands of the area (Jabal Salab), characterised by a number of sinkholes (dolines) (no. 4 and 10, Fig. 10.4).

Volcanic rocks are clearly visible in the north-eastern part of the image (no. 6, Fig. 10.4). It consists of trachytic rocks. Thermal metamorphism induced by the laccolith emplacement leads to transformation characterised by blue tones in Fig.10.3. The concerned units are Unit 5 and Unit 6 to the south-west and the east of Jabal Amrah, respectively. The magmatic rocks of Jabali are of particular importance to understand the geological setting of the mineralization (unit no. 6). They crop out in the large Jabal Amrah massif in the north of the mining area. A formation of small dykes of hypovolcanic rocks (trachyte) appears close to the gossan. From a geometrical point of view, a sill of volcanic rocks extends from the Jabal Amrah to the North, toward the ore deposit to the south (not visible in

Fig. 10.3). This sill has been cut off by drilling and its role in the emplacement of the orebody is clear.

Recent alluvial deposits are put into light owing to the large width of the wadis (no. 2). Some old terraces (not detailed in Fig. 10.4) have been identified with an amount of large blocks and boulders. Some of these blocks are metric in size, representing the hard rocks of the subtrate, granite, quartzite, migmatite, etc. Colluvial deposits are particularly dense on the western side of the Jabal Amrah (no. 3) due to the weathering (mainly kaolinization) of the volcanic rocks. A formation of slag areas mixed with alluvial materials have been identified in the field (no. 1). They cover at least 70 ha.

10.4 The Environment of the Mining Sites

The archaeological and geological study of the Jabali area using field observations supported by remote sensing data leads to an integrative view of the mining site over more than 20 km from the mines to the Wadi Al-Khaniq. Three main items should be emphasized: mineralization, water, and vegetation.

10.4.1 The Mineralization

The book written by Al Hamdani is named "*the book on gold and silver*" (Al Hamdani 1968). It could be considered as an actual treaty on silver and gold metallurgy, because it treats the technology and ancient chemistry of the metals from their extractions from the mines to the fabrication of coins. The Arab writer was well acquainted with Greek philosophers, particularly Aristotle, and with their ideas on the generation of substances from the four natural elements: air, earth, fire, and water. He mentioned a silver ore clearly associated with the lead: 'The evidence for silver ores is the "kohl" [. . .] the latter is represented by small quantities such as the quantity of moonlight is small regarding its size (Chapter V).' Al Hamdani's book also mentions the association of silver with galena and estimates the rate and geochemistry of the metals. However, Al Hamdani never describes the mine itself and the host rocks. He only mentions that miners from Jabali state and that there are no similar mines in the Khurasan, the old Persian province located in the North-eastern part of the present Islamic Republic of Iran.

Another study (Deroin et al. 2011) indicates that mapping, using remote sensing in the visible to short wave infrared range has great potential for the identification of iron caps possibly related to archaeological remains. The phases of works in Jabali have been clearly put into light by the study of a time series of Landsat data. Iron cap can be detected by the multispectral remote-sensing method and can also be used as a guide in the field for prospecting ore deposits.

Fig. 10.5 Detail of the Jabali mine area (3D view). Background: QuickBird data acquired on March 30, 2002. *1* Main canals according to Christmann et al. (1983), *2* main underground mine entries. Note the small watersheds. The shrubs appear as *red dots*

10.4.2 Water

Although dangerous for mining activities, water is obviously vital for a perennial life on site. Today, the area allows only a nomadic way-of-life. Water was possibly more present during the period of exploitation. In any case, there are a lot of canals running from the surrounding mountains as already mentioned in the first geological works (Christmann et al. 1983). Some other canals linked to small watersheds have been identified in the southern part of the Jabali iron cap (Fig. 10.5). The slag heaps also show evidence of major gullying.

10.4.3 Vegetation

Vegetation is closely related to water. It is necessary for life but also for the different phases of the metallurgical process. In the Jabali area, fuel can only be derived from woody vegetation, particularly from shrubs and trees. Wood intervenes as a reduction agent during the ore melting. It is also essential for cupellation, the process used to separate silver from base metals such as lead and zinc. The amount of wood available plays an essential part in reduction, cupellation and revivification, the reduction of a metal from a state of combination to its metallic state. Here charcoal is used only. Al Hamdani gives evidence for this practice: e.g. 'The best fuel for refining is charcoal', 'If you want to convert litharge

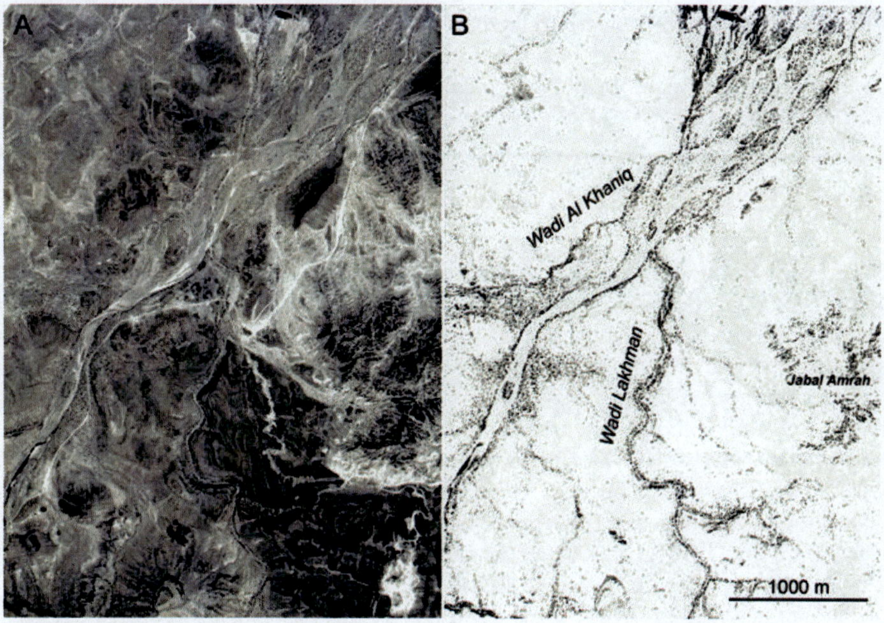

Fig. 10.6 (**a**) ALOS colour composite (see also Fig. 10.3). (**b**) Vegetation. Each black dot represents a small shrub. Note the concentration of vegetation along the Al Khaniq and Lakhman wadis

into lead, melt it twice', 'The village of the mine was a large village located in a watered valley (ghayl) with palm trees', 'There were 400 furnaces', etc. (Peli 2008).

Figure 10.6 shows the current vegetation extracted from the QuickBird image acquired on March 30, 2006. We have used the Normalized Difference Vegetation Index (NDVI). The NDVI is one of the most frequently used vegetation indices. The combination of its normalized difference formulation and use of the highest absorption region of chlorophyll (red range) and highest reflectance region of vegetation (near infrared range) make it robust over a wide range of climate conditions. For the multispectral QuickBird data, NDVI is defined by the equation

$$NDVI = \frac{(NIR - Red)}{(NIR + Red)} \tag{10.1}$$

Where NIR is near infrared band, Red is the red band.

The value of this index ranges from -1 to +1. The common range for green vegetation is 0.2–0.8. In the case of Jabali studied with the help of QuickBird data, 0.15 is the inferior threshold because of the low chlorophyll activity in this early spring image. By applying that threshold and conducting field checks, we were able to extract the main shrubs and trees. They are usually represented by black dots in

Fig. 10.6. Vegetation is clearly related to the banks of the two main rivers, the Al Khaniq and Lakhman wadis. It should be emphasized that currently there is a definite absence of palm trees in the local wadis, in fact there are small trees only.

10.5 Discussion: Jabali, a Moving Area

The area of Jabali was mined for silver as early as the pre-Islamic period. Basically, Jabali is known as one of the most important silver mines of the Abbasid world (Robin 1988). In the tenth century AD Al-Hamdani described this place then named al-Radrad. It was precised to have been exploited from a period before the coming of Islam until the end of the ninth century AD. However, new datings demonstrate that the mining activities occurred until the fourteenth century AD. According to Al-Hamdani, the village of al-Radrad (Jabali) was extensive and located in a watered valley where palm trees grew, the current wadi al-Khaniq. The mine was connected to southern Iraq by a road passing notably through al-Bahrayn and al-Basra (Peli and Téreygeol 2007). This site had been completely lost until the early 1980s, when geological prospections rediscovered the ancient silver mine and opened up new perspectives for the mining activities of Yemen.

At Jabali, the old workings cover up to 10 ha. Other archaeological places have been identified, such as Shutayba, Souq, plateau or in the wadi Al Khaniq (Fig. 10.2). At Jabali, the irregular landforms follow the ore concentration morphology. The latter is crossed by many shafts, drifts, low chambers and large tunnels, which in some cases exceed 100 m in length (Robin 1988). Apart from silver, the only exploited metal, Arab medieval metalworkers identified a zinc oxide, the so-called *tutyia*, deposited on the walls of the furnaces during the smelting process (Peli and Téreygeol 2007). The Jabali area is currently explored for zinc-ore. The deposit contains a geological resource estimated at 12.6 million tonnes of oxide ore, grading about 9% zinc. The mineralization located within Jurassic limestones and dolomites includes carbonates and oxides (smithsonite, hydrozincite, etc.). It should be noticed that silver is mainly related to sulfides, which are also represented, even less than oxides, by rare sphalerite and especially galena (Al Ganad et al. 1994).

The area is the focus of attention of archaeologists, historians and geologists. Basically, for these three topics Jabali offers an exceptional setting. However, the continuance of the site is now in question. Some of the planned facilities, particularly a large open pit located close to the major site, will have a destructive impact on the old pits, the ore-dressing and ore-crushing areas, and especially on the old canals coming from the Jabal Salab, already damaged by the preliminary mining works carried out since the 1980s. Other archaeological remains and slag heaps located far from the mine itself, especially in the wadi Al Khaniq will be unaffected.

The most recent impact has been characterized using a pair of ALOS AVNIR-2 data (Fig. 10.7). The 2007 ALOS image (already illustrated in Fig. 10.3) is compared to the last acquisition obtained by the same sensor on October 26,

Fig. 10.7 Comparison of the recent evolution of the Jabali site using ALOS AVNIR-2 data. Each image represents about 6 × 7 km (subimage of Fig. 10.3). Both colour composites correspond to the following composite: *Red* near infrared range, *Green* red range, *Blue* blue range. Numbers *1–5* refer to the text

2009. The main track is larger (1), there are new tracks in the Souq area (2) and to the iron cap (3), the area close to the lower military camp has been reworked (4). The main difference is at Jabali itself with works for the new open pit and for building the new plant (5). Thus, the mineralurgical area is the most affected by the new mining activities. Unfortunately, this is also the less known aspect of the Jabali silver mines. Indeed, Al Hamdani only describes a few elements of the process. For example he writes about the grain size of the ore: '*one crushes (the ore) to obtain the size of large grapes. The fine part is mixed with water and yellow clay*'. The preparation of the ore was probably limited to crushing and manual sorting.

Al Hamdani's account is particularly vivid where metallurgical activities are concerned: '*When birds came near the village of the mine, they dropped dead because of the fire from the furnaces*'. In some ways he is right, because field prospections confirmed the presence of a large amount of slags on about 70 ha. However, the quantity of slags is relatively low because the thickness of the slag layer never exceeds 20 cm. In addition, the detailed chronology of the deposit is missing. The metallurgical wastes may have been created between the sixth and the fourteenth century AD. Although slags are reprocessed by crushing and sorting out, it is frequent to find lead billet as large as 5 mm. There is obviously no shortage and the ore production was probably large enough to feed the furnaces.

10.6 Conclusions

Teleanalytical geological mapping is a powerful method to analyse and interpret the geological setting of a poorly-known region such as the Jabali area. During this study, new geological topics have been pointed out, such as the setting of the Jabal Amrah magmatic body, the extent of Zn-Pb-Ag mineralizations (iron caps), the mapping of Quaternary terraces along the Wadi al-Khaniq or the detailed cartography of (de)dolomitization phenomena. Among the set of available sensors, the ALOS-AVNIR-2 sensor with its 10 m-ground resolution and its visible-near infrared range appears as a good trade-off for classical geological mapping at 1:50,000, complementary to the widespread large-scale LANDSAT imagery and to the higher resolution data (for example QuickBird or SPOT 5).

One major issue of the study of the ancient Jabali silver mines concerns the climate of the past: Where did the trees grow, necessary for the melting of the ore? Al Hamdani mentioned 400 furnaces and the slag heaps cover more than 70 ha. Moreover, the mineral extracted from the mine was washed to separate ore from the gangue, an operation that requires a large amount of water drained from the local watersheds. Therefore, the hydrological regime (precipitation, run-off, palaeocourses of the wadis) as well as the landscape were clearly different from the present state. Thus, field data and the detailed analysis of ALOS and QuickBird images opened up promising perspectives. New places for future field prospections have been identified using remote sensing data. We have particularly recognized new gossans (iron caps) to the south-east of Jabali owing to the specific spectral reflectance of the iron oxides in the red range.

According to Jansen et al. (2007) the Medieval Warm Period (MWP) occurred from about AD 800–1300. The MWP is well illustrated during the European Middle Ages using information such as evidence of treeline and vegetation changes, or records of the cultivation of cereals and vines. This information is particularly extensive for western Europe. It was initially believed that the temperature changes were global. However, this view has been questioned. Unfortunately, there is insufficient documentation as to the existence of this warm period in the Arabian peninsula. The detailed study of Jabali can probably provide new data for metallurgical studies, as well as for the palaeoclimate studies.

Acknowledgements The field missions have been facilitated by the Yemen Geological Survey and Mineral Board (Dr Ismail N. Al Ganad), the Jabal Salab Company (Yemen) ltd. (General Manager Brett Grist), and the French Ministry for Foreign Affairs. We are particularly indebted to Major Khaled Al Naphani and the National Guard of Yemen for assistance with field work. The ALOS data have been acquired in the Pilot-Project ADEN-ALOS no. 3463 (*Geological Mapping of Sensitive Environments*) supported by the Japanese Space Agency (JAXA) and the European Space Agency (ESA).

References

Al Ganad I, Lagny P, Lescuyer JL, Ramboz C, Touray JC (1994) Jabali, a Zn-Pb-(Ag) carbonate-hosted deposit associated with Late Jurassic rifting in Yemen. Mineralium Deposita 29:44–56

Al Hamdani A (1968) Kitab al-Jawharatayn al-'atiqatayn (the book on gold and silver). Almqvist & Wiksell, Uppsala

Christmann P, Lagny P, Lescuyer JL, Sharaf ad-Din A (1983) Résultats de trois années de prospection en République arabe du Yémen. Découverte du gisement de Jabali (Zn-Pb-Ag) dans la couverture jurassique. Chron rech min 473:25–38

Deroin JP, Al-Ganad I, Al-Thari M, Benoit P, Téreygeol F, Heckes J, Hornschuch A (2006a) The effect of spatial resolution for teleanalytical geological mapping, example of the Kawkaban-Sana'a-Ma'rib region, Republic of Yemen. Photo-Interprétation 42(3):3–18

Deroin JP, Téreygeol F, Benoit P, Al-Thari M, Al-Ganad I, Heckes J, Hornschuch A, Peli A, Pillault S, Florsch N (2006b) Archaeological remote sensing in Yemen, the Jabali test site. From large-scale survey to field investigation. Brit Archaeol Rep, BAR Int Ser 1568:7–12

Deroin JP, Téreygeol F, Heckes J (2011) Evaluation of very high to medium resolution multispectral satellite imagery for geoarchaeology in arid regions, case study from Jabali, Yemen. J Archaeol Sci 38:101–114

Jansen E, Overpeck J, Briffa KR, Duplessy JC, Joos F, Masson-Delmotte V, Olago D, Otto-Bliesner B, Peltier WR, Rahmstorf S, Ramesh R, Raynaud D, Rind D, Solomina O, Villalba R, Zhang D (2007) Palaeoclimate. In: Solomon S, Qin D, Manning M, Chen Z, Marquis M, Averyt KB, Tignor M, Miller HL (eds) Climate change 2007: The physical science basis. Contribution of working group I to the fourth assessment report of the intergovernmental panel on climate change. Cambridge University Press, Cambridge/New York

Lasaponara R, Masini N (2005) QuickBird-based analysis for the spatial characterization of archaeological sites: case studies of the Monte Serico medieval village. Geophys Res Lett 32:L12313

Lasaponara R, Masini N (2006) On the potential of QuickBird data for archaeological prospection. Int J Rem Sens 27(6):3607–3614

Peli A (2008) Monnaies, métal et pouvoir. Frappes et techniques monétaires au Yémen (IIe-VIe/VIIIe-XIIe siècles). Thesis, Université Paris 1 – Panthéon-Sorbonne, 358 pp

Peli A, Téreygeol F (2007) Al-Radrad (al-Jabali): a Yemeni silver mine, first results of the French mission (2006). Proc Semin Arabian Stud 37:187–200

Robin C (1988) The mine of ar-Radrar: al-Hamdani and the silver of the Yemen. In: Daum W (ed) Yemen, 3000 years of art and civilization in Arabia Felix. Pinguin Editor, Innsbruck, pp 123–124

Schmid T, Koch M, DiBlasi M, Hagos M (2008) Spatial and spectral analysis of soil surface properties for an archaeological area in Aksum, Ethiopia, applying high and medium resolution data. Catena 75:93–101

Simonin L (1858) De l'exploitation des mines et de la métallurgie en Toscane pendant l'Antiquité et le Moyen Age. Ann Mines 14:557–815

Téreygeol F (2001) Les mines d'argent carolingiennes de Melle. Thesis, Université Paris 1 – Panthéon-Sorbonne, 3 vol, 615 pp

Toll C (1968) Al-Hamdani, Kitab al-Gauharatain Al-'aliqatain Al-Mâ'i'atain As-Sâfra' Wa't Barda (Die beiden Edelmetalle Gold und Silber). studia Semetica Upsaliensa 1, Acta Université, Uppsala, 394 pp

Chapter 11
Irrigation Is Forever: A Study of the Post-destruction Movement of Water Across the Ancient Site of Sri Ksetra, Central Burma

Janice Stargardt, Gabriel Amable, and Bernard Devereux

Abstract Sri Ksetra is a very large urban site created by the Pyu people, ca. fourth century AD, on the southern edge of the Dry Zone of Central Burma. Ancient irrigation works, discovered by Stargardt in 1985–1988 and first published in 1990, were essential to its urban development and, though rarely visible as surface features today, still profoundly affect the post-destruction settlements on the site and their rice harvests. This paper documents the sub-surface and surface movement of moisture across the site after the end of the monsoon season, i.e. in the period from September to February. It is based on the study of combined resources, including spaceborne multi-spectral and multi-temporal satellite imagery, aerial photographs and archaeological maps. It reveals how surface and sub-surface water still flows along the ancient irrigation channels and collection tanks of the site and demonstrates that an ancient system, which has been in an advanced state of sedimentation for over 1,000 years, still functions mainly at a subsurface level, creating differences in soil productivity and drainage today. The study explores the potential value of freely available imagery of medium resolution for archaeological research on other ancient landscapes embedded in modern ones where differences of wetness and vegetation are significant. Finally, these satellite images record the presence of ancient ritual ponds and associated burial terraces in the southern extramural area in a degree of detail that, surprisingly, compares well with the aerial photographs taken over 50 years ago when surface disturbance was at a lower level.

J. Stargardt (✉)
Department of Geography and Sidney Sussex College, University of Cambridge, Cambridge CB2 3HU, UK
e-mail: js119@cam.ac.uk

G. Amable • B. Devereux
Unit for Landscape Modelling, Department of Geography, University of Cambridge, Cambridge CB2 1QB, UK
e-mail: gsa1000@cam.ac.uk; bjd1@cam.ac.uk

R. Lasaponara and N. Masini (eds.), *Satellite Remote Sensing: A New Tool for Archaeology*, Remote Sensing and Digital Image Processing 16,
DOI 10.1007/978-90-481-8801-7_11, © Springer Science+Business Media B.V. 2012

Keywords Multi-temporal and multi-spectral satellite imagery • Burmese urban archaeological site • Ancient irrigation • Enduring effects • Modern landscape

11.1 Introduction

Sri Ksetra is one of a very small number of urban sites in Southeast Asia as a whole which existed for most of the first millennium AD. In the 1980s, Stargardt uncovered a major system of irrigation works threading through the site integrating it into the surrounding plains and villages (Stargardt 1990, 45–107, 343, 349–69, Fig. 14, Pl 11; Figs. 11.2 and 11.3a here). She concluded that this system explained the longevity of the site, but that local levels of knowledge and management (together with the responsiveness and high productivity of traditional rice species) provided the flexibility of the ancient irrigation system. It expanded and contracted rapidly in scale and level of productivity in response to changing needs and opportunities, supporting Pyu urban culture at Sri Ksetra over many centuries of occupation. These findings challenge the concepts of despotic force proposed by Wittfogel under *Oriental Despotism* and *Hydraulic Society* (Wittfogel 1957). The satellite images examined here provide new data on the astounding capacity of the ancient irrigation works of Sri Ksetra to continue to operate largely spontaneously, mainly below the modern ground level, for over 1,000 years after the end of urban levels of settlement and overall maintenance of the whole irrigation system. It continues to exert differential effects on agricultural productivity and the livelihoods of the inhabitants up to the present. Thus irrigation is forever; it can be locally managed to provide a highly flexible, but strong and durable crop plant for society quite unlike the despotic trajectories proposed by Wittfogel.

Sri Ksetra is located in the eastern plain of the Irrawaddi basin in central Burma (18°48′N × 95°17′, Fig. 11.1), near the Nawin River, a highly seasonal but perennial tributary flowing into the Irrawaddi near Pyay [Prome] (mistakenly labeled "Pye" on Google Earth and on some modern maps). Archaeological research began at Sri Ksetra in 1904 and continues up to the present through the work of the Archaeological Training School at Sri Ksetra (Director U Thein Lwin and his colleagues) and Dr Bob Hudson of the University of Sydney. A series of site maps have been made, of which only the two most recent will be discussed here for their differing data sources and interpretations (Stargardt 1990; Hudson 2007, Figs. 11.2 and 11.6 here). Sri Ksetra is located in the zone of transition between the heavy rainfall coastal belt of Burma and the large Dry Zone of central Burma. Rainfed rice cultivated here in the nineteenth to twenty-first centuries fails in 2 out of every 3 years, causing long-running controversies among past scholars over how the ancient city sustained itself. Stargardt's first studies of aerial photographs and surface surveys in the 1980s (good remote sensing cover not then being available), showed that Sri Ksetra's development depended on water harvesting in canals, tanks and moats, which had not been previously detected but formed an integral part of the urban format: e.g. multiple moats defined the boundaries of the city; a great storage tank originally closed its eastern side.

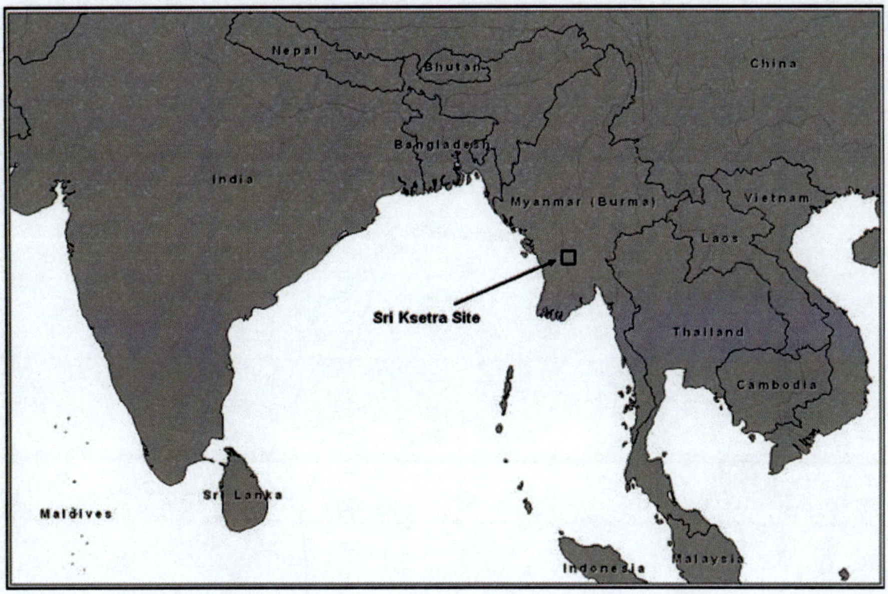

Fig. 11.1 Southern Asia showing the location of the ancient city of Sri Ksetra, Central Burma

The slope factors affecting Sri Ksetra are complex and the location of the site reveals the precise understanding of its ancient inhabitants of how to extract maximum benefit from them. The prevailing slope is from the ranges to the west and southwest, where the crests reach 121 m amsl. Within a short distance the elevation drops to 60 m amsl and this contour skirts the western moats and walls of the site. From that point, the slope continues towards the northeast corner at a more gradual rate (Fig. 11.2, contour lines). Just outside the northeast corner at 34 m amsl it encounters another yet more gradual slope coming down from the northeast. This slope affects the eastern edge of the city (its lowest area being 32 m amsl) and continues to drain away to the southeast. A third slope factor runs outside the northern edge of the city, east–west, and permits the main body of the Nawin Stream to flow through a gap in the ranges to enter the Irrawaddy River at Pyay [Prome]. These multiple slopes were mastered to bring water in a series of canals and tanks to converge on Sri Ksetra: from the western ranges a heavy waterforce rushed down the culverts during the rainy season (May–September). Its destructive impact was immediately reduced by the circular alignments of the outer moats backed by strong multiple brick walls. Months later this water could be safely channeled along further circular canals inside the walls and later still distributed along canals converging on the citadel or inner city and then be allowed to flow across the entire urban terrain. At the northeast corner of the city this water encountered another canal system fed by barrages across the Nawin River, and exploiting minimal northeast/southwest slopes to flow towards the city while

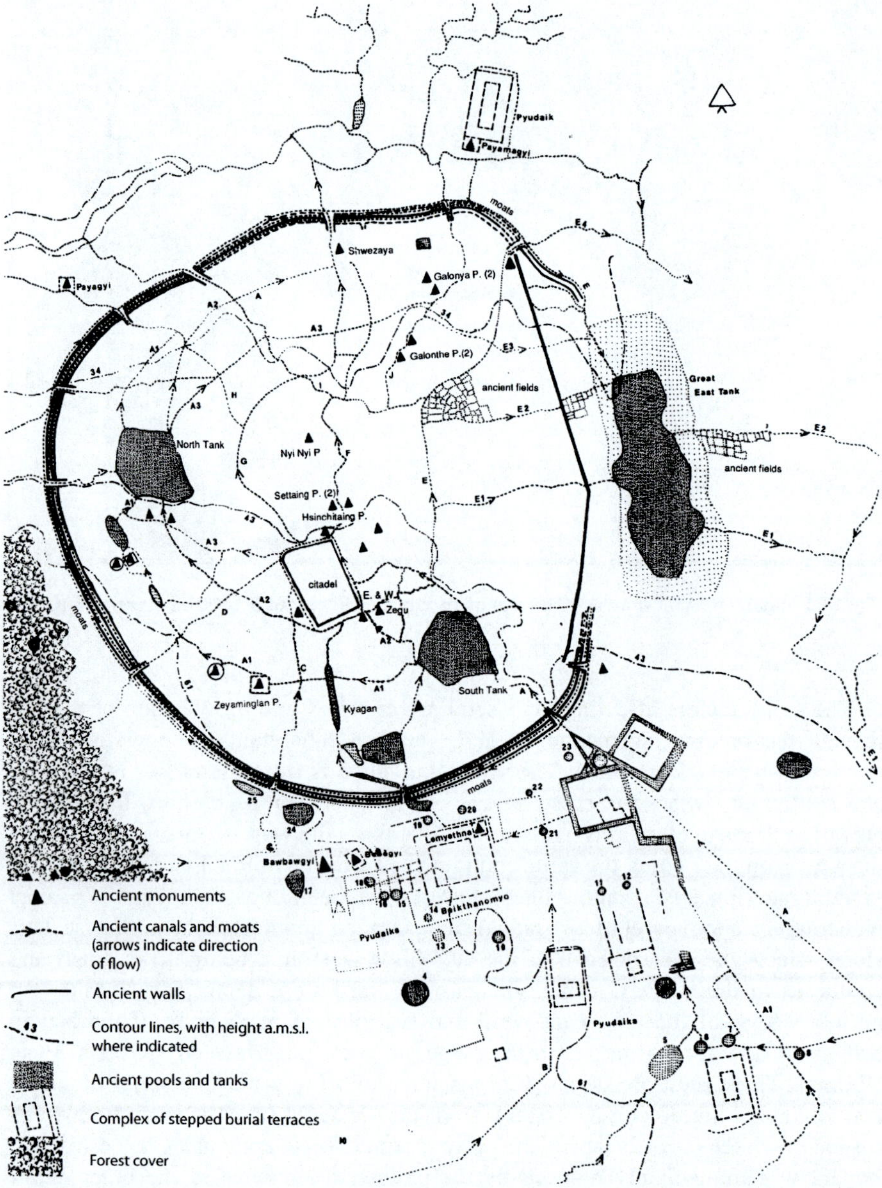

Fig. 11.2 Archaeological map of Sri Ksetra compiled by Stargardt, 1986–1989 (*Source*: Stargardt 1990, Fig. 11)

the main body of this stream continued, as mentioned, in a westerly direction to join the Irrawaddy.

The canals from the west and southwest fed the outer moats, the south and north tanks, contributed to the great east tank and fed all the canals inside the urban perimeter.

The Nawin canals contributed significantly to the contents of the east tank and the canals running from it towards the southeast. The superb skills of the Pyu in surveying and water control are demonstrated at Sri Ksetra where canals flow from the southwest into the city while others flow almost parallel c. 3 km away but in the opposite direction and c. 30 m lower. Together, these two irrigation systems with separate origins supported rice cultivation and horticulture both within and outside the city walls (Stargardt 1990, 54–65, 84–107, Pl. 11, Fig. 14; Fig. 11.2 here). When operating fully, irrigation water prolonged the impact of the rainy season by c. 4 or 5 months, transforming the reliability and volume of the rice harvests, c. fourth to ninth century AD. Within that period there were phases when the constituent canals contracted and expanded and when the great east tank experienced such an extended period of aridity that it was no longer able to enclose the eastern side of the site. A small lightly constructed east wall was then built along a strikingly different alignment to close the site on this side. This chapter presents a range of satellite images and considers what they reveal about the continued operation today of parts of the ancient irrigation systems from September (the end of the rains) to the following February, thereby prolonging the effects of the rainy season by no less than 5 months.

Although foreign coins and artefacts have been found at Sri Ksetra, the main foundations of its prosperity lay in its irrigated agriculture rather than trade. The satellite images examined below show that, operating largely below the modern land surface, the ancient irrigation works still affect the livelihoods of the villages occupying the site today.

The enclosed city area covers more than 17 sq. km inside the external rings of moats and walls. Beyond them, extramural areas were linked by continuous, homogeneous patterns of land use forming a nuclear area of at least 100 sq km in the Nawin Valley itself, and probably much more. It was defined by a continuous irrigation system serving the same combination of agriculture, horticulture and major clusters of monuments, burials and settlement areas both inside and outside the walls. This extended, multi-functional urban format differs sharply from the compact ancient urban settlements of Europe and the Near East, but represents a type of ancient royal city found frequently in South, Southeast and East Asia (cf. Stargardt 1990, 45–104; 343–6 on Pyu sites; Moore 2003 on Pyu and Mon sites, Hudson 2004 on Pagan and Fletcher et al. 2008 on Angkor). While South and East Asia shared the extended format of royal cities, there the culturally and economically significant zones of ancient urban space were typically delineated by walls and pathways. In the royal cities of Southeast Asia, this role was traditionally played by man-made bodies of water in the form of moats, canals, ponds and larger tanks – fragments of such works are still visible in the inner city at Bangkok, Ayutthya and Sukhothai. These ancient irrigation systems (as at Sri Ksetra) not only served vitally important practical functions but also, it is argued, conferred symbolic meanings on the urban space and its constituent parts (Wheatley 1971; Stargardt 2002–2003, 156–67; Willis 2009 among many others).

The present day landscape supports a mixture of land uses, dominated by mixed (partly dry) riziculture and scattered low-density settlements. The northwestern part of Sri Ksetra now supports a small extension of the nearby city of Pyay [Prome].

As most of the historic man-made water-bearing structures have not been maintained for many centuries, few remain visible at ground level. In the following sections, we show how many of these early historic structures can still be detected by satellite remote sensing study and how they impact on the spatial and temporal patterns of surface and ground water resources in the landscape to produce local differences in soil productivity.

11.2 Data and Methods

The use of moderate resolution multispectral imagery from space in archaeological mapping hitherto has been limited. This is mainly because their spatial resolution has often proved unsuitable for the identification in sufficient detail and effective mapping of features at the field or site level. High-resolution surface elevation data of the kind widely in use in recent mapping of surface archaeology at UK and European sites (see for example, Devereux et al. 2005, 2008; Bewley et al. 2005; Doneus and Briese 2006) are not yet available for the Sri Ksetra site. However, as this study will show, the extended urban format of these site-types, just described, means that moderate resolution multispectral images can provide useful overviews at the landscape level as well as information on the broader scale context and landscape associations. In some respects, noted below, they capture impressively detailed data.

The detailed archaeological mapping of the Sri Ksetra site discussed here was first compiled in the late 1980s by Stargardt (Stargardt 1990, Fig. 14, Pl 11; Stargardt 2002–2003, 150, Fig. 3) employing field mapping techniques and detailed interpretation of air photos flown by the Royal Air Force (RAF) in 1953.

With emendations in 2009 and 2011 these provide the key baseline data for the present study. An extract of this map is shown in Fig. 11.2. Additionally, Stargardt compiled an air photo mosaic from the RAF imagery (Fig. 11.3a), which was checked by surface survey and used for site-level detail. In addition to these data sets we employed spaceborne multispectral LANDSAT Enhanced Thematic Mapper (ETM) data available from the United States Geological Survey. The fourth century Sri Ksetra landscape was dominated by the network of irrigation channels, tanks and moats, which have contributed to the character of the present-day landscape. Thus, the remote sensing methods adopted here were aimed at analysing spatial and temporal patterns in the distribution of surface and sub-surface moisture across the landscape. In order to be able to track moisture movement across the landscape the imagery selected were acquired in the period between the end of the rainy season and the start of the dry season. This corresponds roughly to the period between September and February. Unfortunately due to persistent cloud cover in this region it was not possible to acquire cloud-free time series imagery for the entire period. In order to minimize the effects of seasonal variations in rainfall patterns (both in amounts and seasonality) only images acquired in the same season were selected. Thus three images were used, acquired in September 2002, January 2003 and February 2003, corresponding to the parts of the season when extensive variations in moisture conditions occur. The images are shown in Fig. 11.3b–d.

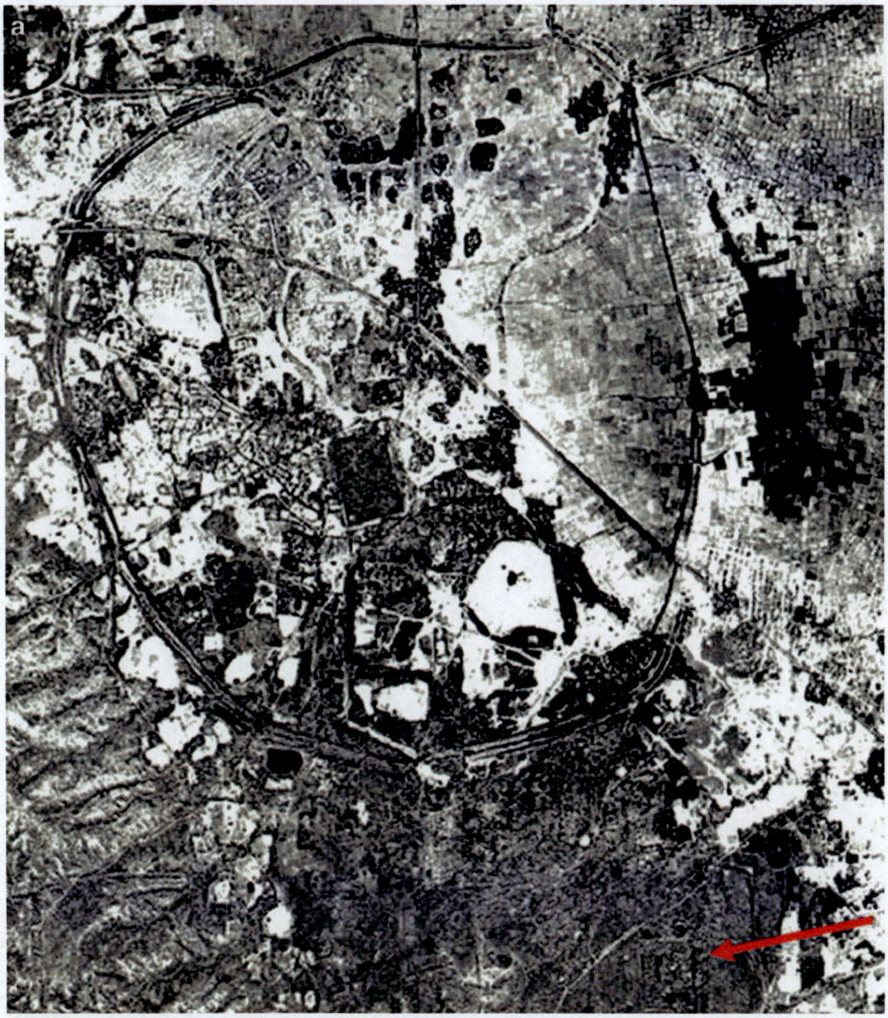

Fig. 11.3a Air photo mosaic of Sri Ksetra compiled from RAF aerial survey of 1953. *Rectangular-shaped* burial terraces can be seen in the south, each with two circular ritual ponds at one end (*example arrowed*) *Source*: Royal Air Force; mosaic Janice Stargardt

Many techniques have been developed for extracting landscape, soil and vegetation moisture content from remotely sensed imagery. A number of these have been developed for mapping surface water and moisture content from multi-spectral optical imagery. Gao (1996) described a Normalised Difference Water Index (NDWI) for detecting and mapping the liquid water content of vegetation from spaceborne multispectral imagery. The index is a modulation ratio of the reflectance in near infrared and short wave infrared bands. It has been used in

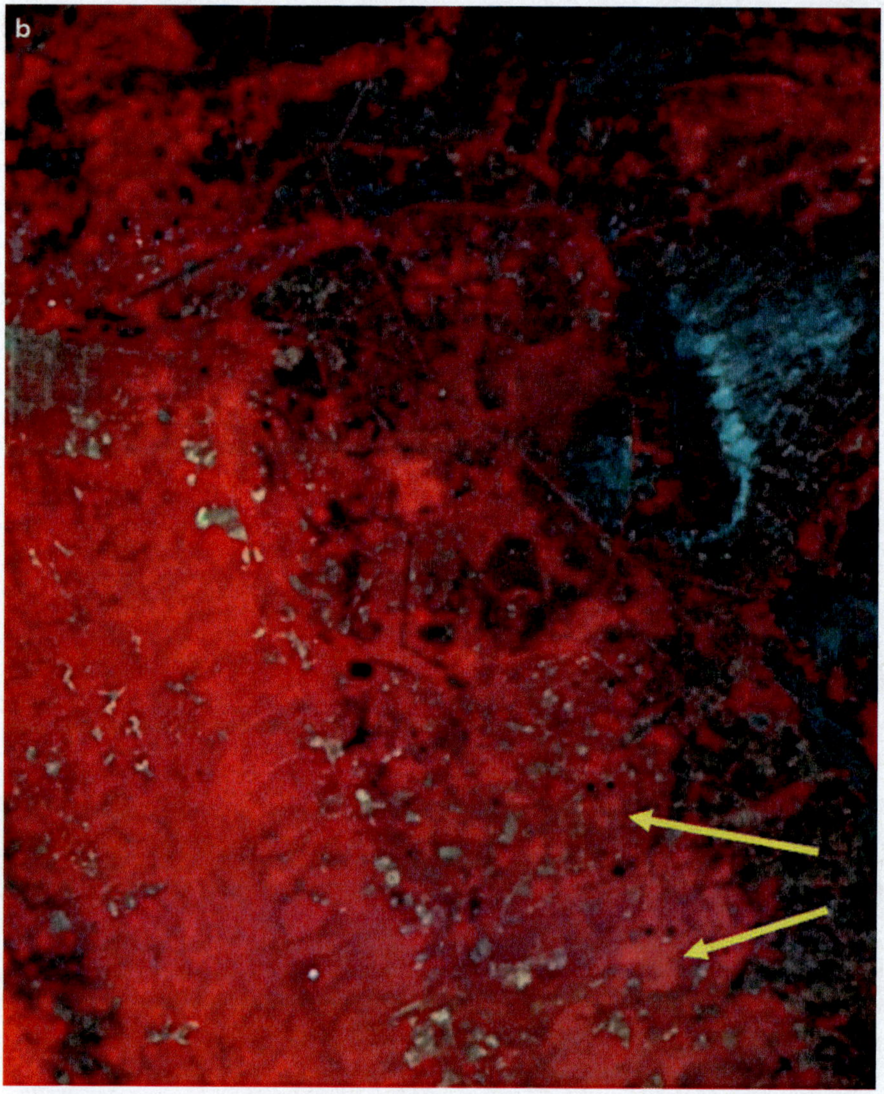

Fig. 11.3b Infrared colour composite of LANDSAT ETM+ imagery, September 2002. The dark features in the image are open water features. *Rectangular-shaped* burial terraces can be seen in the south, each with two circular ritual pools at one end

detecting vegetation water content (Jackson et al. 2004), vegetation water stress (Eitel et al. 2006) and for determining and mapping the dates of spring phonological greening (Delbart et al. 2006). McFeeters (1996) described a slightly different version of the NDWI which was later modified by Xu (2006) for use in mapping open water features. In order to improve its performance, by maximizing

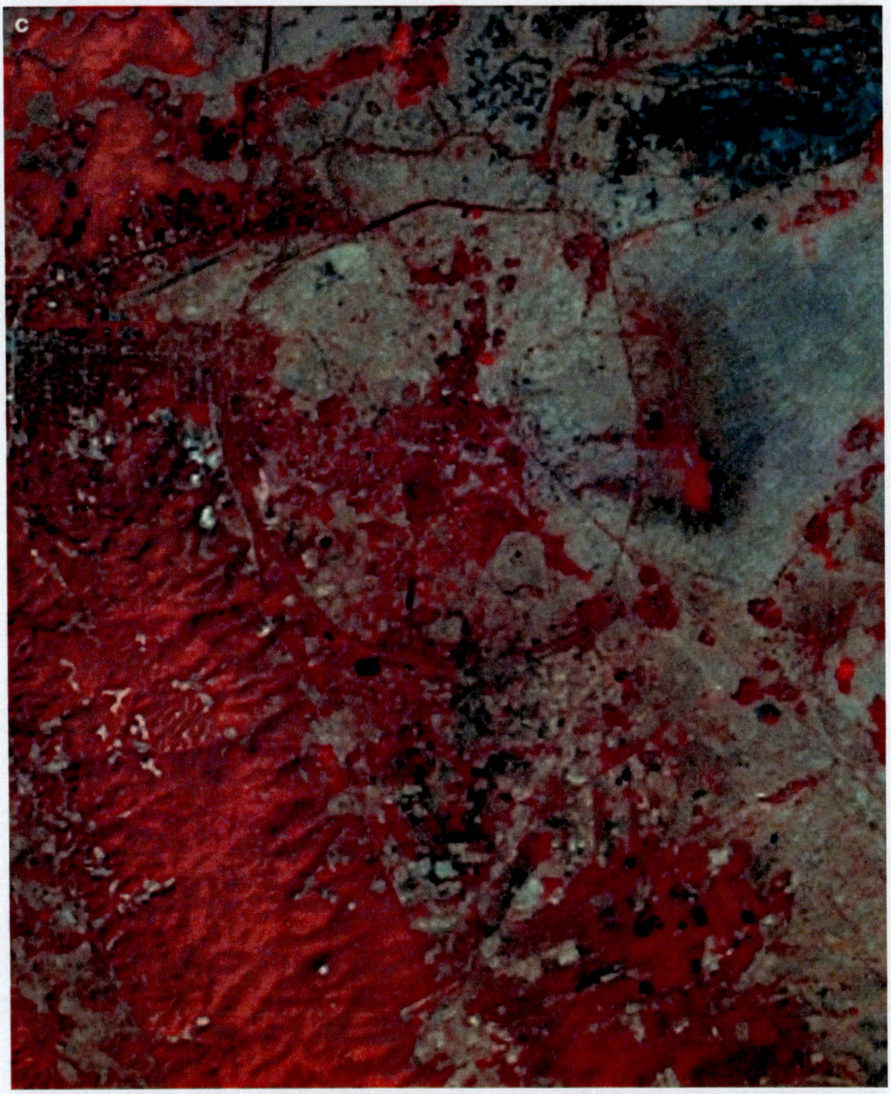

Fig. 11.3c Infrared colour composite of LANDSAT ETM+ imagery, January 2003. The dark features in the image are open water features. *Rectangular-shaped* burial terraces can be seen in the south, each with two circular ritual pools at one end

reflectance of water whilst minimizing the influence of vegetation, it employs a normalized ratio of the reflectance in the green and short wave infrared bands as follows:

$$NDWI = \frac{ETM2 - ETM5}{ETM2 + ETM5} \tag{11.1}$$

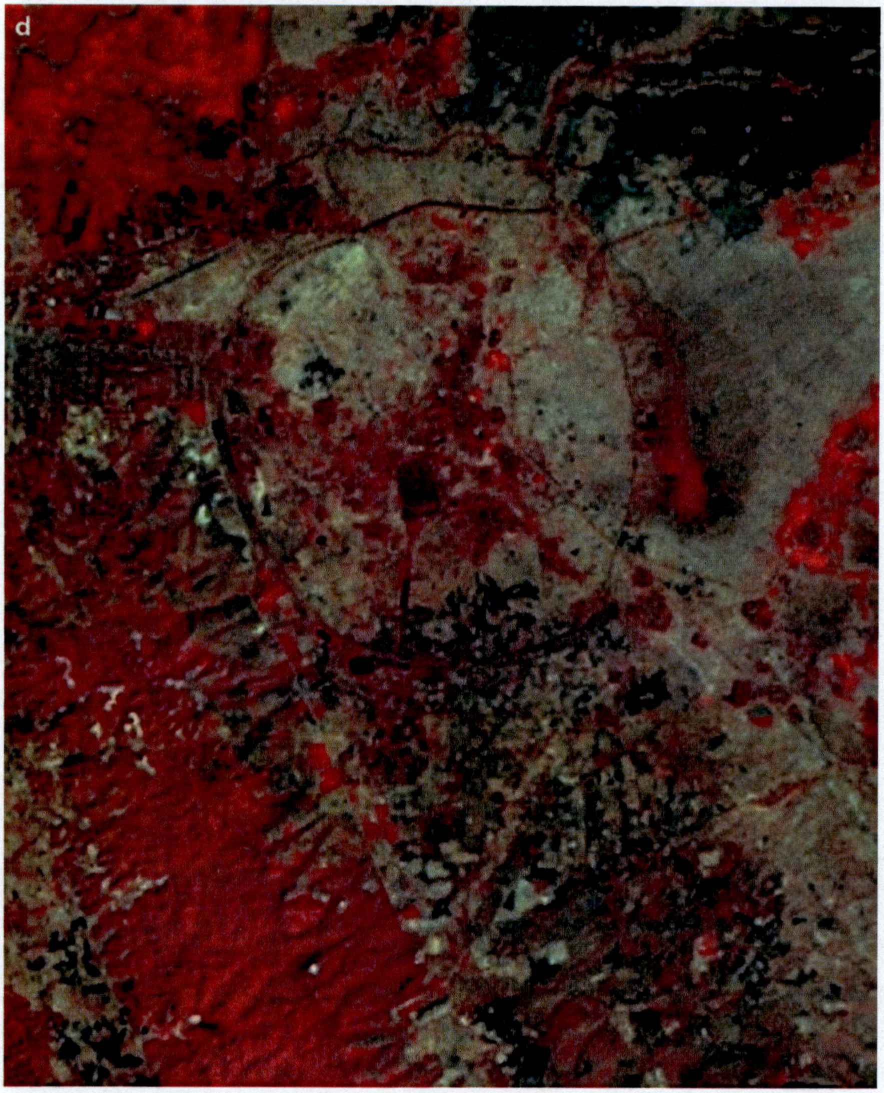

Fig. 11.3d Infrared colour composite of LANDSAT ETM+ imagery, February 2003. The dark features in the image are open water features. *Rectangular-shaped* burial terraces can be seen in the south, each with two circular ritual pools at one end

ETM2 and ETM5 are respectively the reflectance in green and mid-infrared bands of LANDSAT ETM. In this study we have used the NDWI to map the distribution and variation of surface water as well as vegetation and soil moisture for subsurface water at the Sri Ksetra site.

The NDWI images were computed using the equation above. The resulting images for the three dates are shown in Fig. 11.4a–c. NDWI values range from +1,

Fig. 11.4a *NDWI* image for Sri Ksetra, September 2002. *Lighter shades* indicate surface water or areas of higher moisture content. *Arrowed* features are again rectangular burial terraces with a pair of water-bearing pools at the northern end

indicating high moisture content or surface water, to −1 for dry or low moisture conditions. In practice NDWI values commonly range between −0.5 and +0.5. In this study the absolute values of the NDWI are not of interest but their spatial and temporal variations are.

In particular, we examined these spatio-temporal variations in the areas of historic water-bearing structures. In the NDWI images shown in Fig. 11.4 higher

Fig. 11.4b *NDWI* image for Sri Ksetra, January 2003. *Lighter shades* indicate surface water or areas of higher moisture content. *Arrowed* features are burial terraces and pools as in Fig. 11.4a

moisture levels are indicated by lighter shades of gray. These images allowed for the study of the changing spatial patterns and dynamics of surface and ground water regimes in the period between the wet and dry seasons. They also highlight the enduring impact of the historic irrigation system on the present-day landscape. Later in this paper, the Sri Ksetra map compiled by Hudson in 2005–2007 from different sources (Fig. 11.6) is compared and contrasted.

Fig. 11.4c *NDWI* for Sri Ksetra, February 2003. *Lighter shades* indicate surface water or areas of higher water content. Burial terraces and pools are still visible

11.3 Results and Discussions

The NDWI for the three dates shown in Figures 11.4a–c portray a clear pattern of ground water movement across the site. As expected the general pattern of water movement through the season follows the prevailing slope of the site from the ridge outside the city in the west and southwest to the eastern and northeast directions. In addition to showing surface water features (bright white) and other low reflectivity surfaces (such as the tarmac on the airport runway – the straight linear feature in the northwest), the NDWI for the three dates clearly portray the strong differential patterns of water retention, at a sub-surface level – also showing as white or grey. These are consistently concentrated at the sites of the c.1,700-year old irrigation structures (cf. Fig. 11.2) as the following account of a selection of the features visible on these images shows.

Taking Fig. 11.4a and starting in the southern extramural area (c. 59 m downslope to 58 m amsl, the second highest level of the site), one can clearly see no fewer than 15 small, circular ritual pools scattered over this part of the site, often in pairs. A comparison with the mosaic of aerial photographs of approximately the same area (Fig. 11.3a) where 25 such pools were recorded, reveals the striking fact that a great amount of detail about these pools was captured by the medium resolution satellite images. They were originally dug at the northern end of ancient, rectangular, stepped burial terraces constructed mainly from rammed earth and sometimes faced with bricks (see arrows on Fig. 11.4a, b). These terraces were the repositories of large numbers of cremated human burials in terra cotta urns. Some sheltered hundreds, others thousands of urns, which in turn contained human bones, ash, clay and sometimes Buddhist artifacts. The burial terraces of Sri Ksetra probably contain the largest number of pre-Islamic burials in South Asia as a whole. It has already been noted (Stargardt 2002–2003, 151–154) that these terraces appear to have been constructed over a long period because instances of partial superposition occurred as well as differences in their state of preservation (cf. Figs. 11.2 and 11.3a).

On Fig. 11.4a, the latest burial terrace with two pools on its northern edge shows up quite clearly while a pair of pools belonging to another terrace (now obscured) can be seen at its southeast corner and the corner of an older terrace can just be distinguished immediately to the west (see arrows). Turning to Fig. 11.4b, c, most of the small pools can still be distinguished even though some of their water content has diminished through subsurface percolation to lower levels. On the other hand, the right-angle corners of three burial terraces can now be distinguished, one partly under the most recent terrace with two others immediately to the west. What is being recorded in Fig. 11.4b by the NDWI, now that the general environment is less saturated, is the water content of ancient canals which (as recorded on Fig. 11.2) originally flanked these burial terraces before flowing towards the city and joining the moats. The pools must have served some ritual purposes connected with the burials originally, but are still seasonally important water bodies today and thus show up white on Fig. 11.4a, b and to a lesser extent on Fig. 11.4c. The low intensity cropping of this area today undoubtedly benefits differentially from the superior water retention of these pools and canal beds well beyond the end of the wet season. Approaching the walled area of the site from the southwest, one distinguishes on both Fig. 11.4a, b the multiple outer moats surrounding the site, now largely silted up but exploited for ponds, wells, fields and gardens. Figure 11.4b, capturing the situation in late January 2003 over 8 weeks after the end of the wet season, clearly shows the existence of high levels of moisture in the ancient courses of the double and sometimes triple moats flanking the outer walls on the southeast, south, southwest and west of the site. Figures 11.4a–c also show a big single alignment of moisture concentrated along the north wall and turning southeast into the site of the ancient east tank.

This northern band of moisture again follows the course of the ancient north moat – the north and east being the lowest levels of the site originally at c. 34–32 m amsl (cf. Fig. 11.2 with contour lines and ancient canals, with Figures 11.3a, b, 11.4a–c). Outside the walls on the north and northeast of Sri Ksetra, Fig. 11.4a (September) shows a very widespread retention of water both at the surface and sub-surface levels, but significant water patterns can nonetheless be distinguished, especially in the eastern sector of the site. For instance, here the small streams originating from the Nawin River appear on the image from the northeast corner and flow in general towards the west. But one branch exploits minimal slope factors to flow southeast along the eastern perimeter of the site. That they were originally canals is revealed by their geometric angles. In the less saturated conditions captured by Fig. 11.4b in January 2003, the same pattern is visible. The tortuous course of this northeast canal is typical of Pyu techniques to lower water forces immediately after the wet season and thus reduce their erosive capacities (Stargardt 1990, 54–65). This moisture reaching the extreme north-east of the site January-February onwards originates in non-local rain falling on the Pegu Range to the east and percolating downslope to Sri Ksetra site along the beds of ancient canals branching out of the Nawin River. It accurately but indirectly follows the slight northeast/southeastwest slope to flow into the east tank. Further features in the northeast area, as revealed by Fig. 11.4a, b, also contribute significant information on the continued operation of the ancient irrigation system and its modern impacts. A strong band of moisture in the northeast canal can be seen on both images, branching out of the Nawin stream at a sharp angle and flowing to the northeast "corner" of the ancient city wall. It then flows south along the eastern edge of the site. In September (Fig. 11.4a) this band is wider and at its southern end less defined because of the heavy general moisture content prevailing in the depression it has entered (i.e. the ancient east tank, Fig. 11.2). By January (Fig. 11.4b), when moisture has become both spatially more concentrated and somewhat reduced in volume, these features are also visible. The Stargardt-Nicolazzi surface surveys in December 2007–January 2008 showed this canal still contained surface water along a very limited part of its course. Figures 11.4b and c, however, reveal that at a subsurface level, this feature extends over a further 4 km, delivering a significant amount of water to the site of the ancient east tank. It does so along a well-defined subsurface channel created by its own water force through the permeable sediments occupying the site of the tank (see the grey area spreading along the east side of the urban site, Fig. 11.4b, c). It is worth emphasizing that none of this water flow is visible from the surface today, but the area is still known to the local population as *pinle*, the sea, in reference to the great tank of c. 7 × 3 km that once filled this area. Under pressure from the Agriculture Department, the ground level here has been raised and is now under what appears to be dry rice cultivation in recently created large fields. The crop yields will, however, benefit differentially depending on their proximity to the bodies of moisture conducted into this area as a result of the continued, invisible operation of sections of the ancient irrigation system.

Figure 11.5 presents wetness profiles for September and January respectively, across the southern part of Sri Ksetra following the transect indicated on Fig. 11.5a. From left to right (in the direction of slope), the peaks on Fig. 11.5b, c are labelled in relation to features visible on Figs. 11.3 and 11.4 and named on Fig. 11.2. These features are: the two outer western moats (labelled A), the Kyagan (B) – a broad band of surface water running south-north (discussed below), the northern part of the ancient south tank (C), and a large area of moisture collected in the area around and abutting the east wall all the way into the great east tank (D), which was the major collection and redistribution point of the ancient irrigation system. The situation in January, revealed by Fig. 11.5c shows that while much of that water has flowed away significant differences remained. But significant patches of red (intense photosynthesizing vegetation) in January–February (cf. Fig. 11.3c, d) show that this water had done its work in producing markedly concentrated patches of dense vegetation, especially visible in the area of the east tank, the burial terraces and the outer moats as well as other site canals.

Thus these different aerial and satellite sources offer mutually supporting evidence of the existence and basic features of the ancient irrigation system in and around Sri Ksetra, on whose existence Stark (2006, 415–146) and Hudson (2007 – Fig. 11.6 discussed below) appear to have some reservations. The multispectral sources also demonstrate the enduring effects of that irrigation system in that the strong presence of subsurface humidity at Sri Ksetra today occurs in a patterned way for as long as 4 months after the rainy season, and its benefits can be seen in the clusters of dense plant growth for at least 1 further month.

The differential impacts demonstrated here for only a selection of the ancient water works of Sri Ksetra affect all the present field and garden crops over an area much greater than the one discussed here. The reasons for their enduring wetness impacts are fourfold: firstly ancient canals and tanks were located so accurately along the slopes that water continues to flow today as it did when the canals and tanks were originally dug during the first millennium; secondly the sediments that accumulated in the ancient irrigation works over the course of the second millennium are more fertile than the natural local clays of marine origin; thirdly these sediments are more permeable than the clays and thus provide favourable conditions for the movement of water across the site; fourthly, the blocky geological structure of the clays ensures that the moisture remains within those sediments in the ancient irrigation channels. A partial exception, the Kyagan, will be discussed below. That the local inhabitants are well aware of this differential impact to be gained from the ancient water bodies is not only revealed in the survival of toponyms like *pinle*, but also in the way they sink ponds and wells into ancient moats, tanks and canals to exploit, at a low level of intensity, the old Pyu traditions of water harvesting and irrigation.

Hudson's 2007 map of Sri Ksetra (Fig. 11.6) is based on intensive surface surveys with GPS and using the MapInfo GIS package. The main differences between Figs. 11.2 and 11.6 concern irrigation works. Firstly, the only major water body

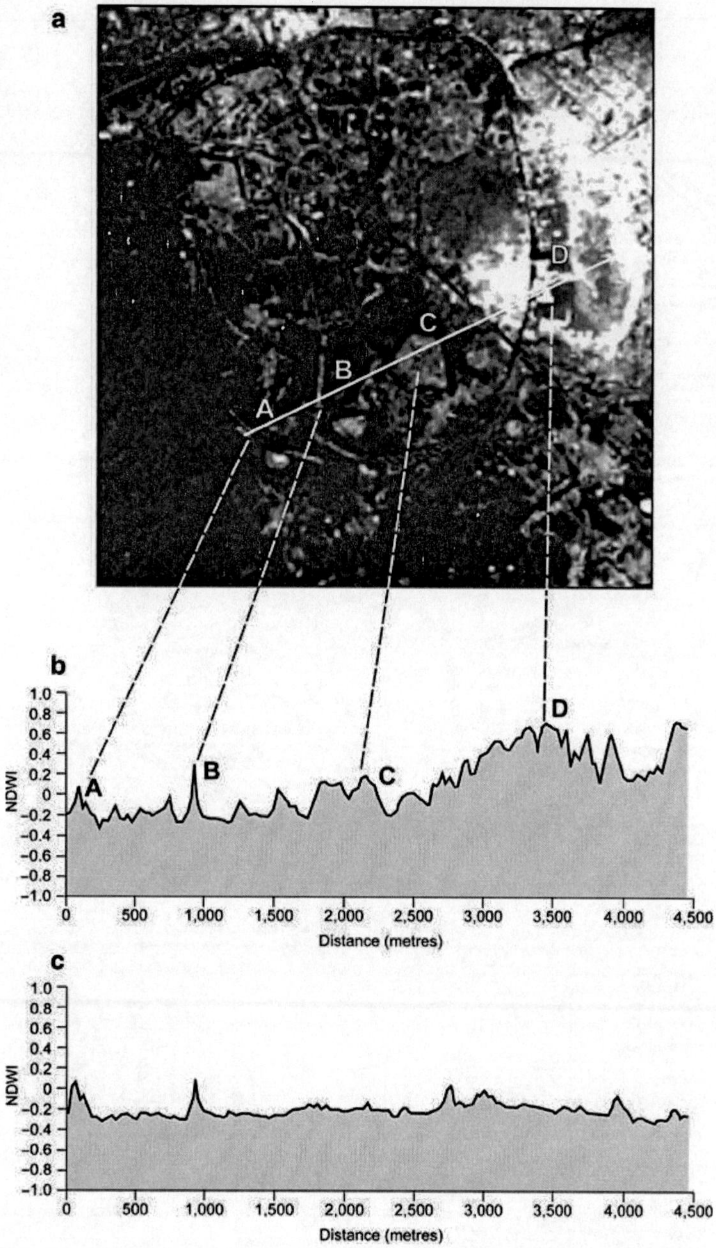

Fig. 11.5 Profiles of surface and sub-surface moisture variation along a transect of the southern part of the Sri Ksetra site 2002–3. (**a**) *NDWI* image for September 2002. As expected the September 2002 moisture profile (**b**) shows relatively higher moisture levels with strong variations. The January 2003 profile (**c**) still reveals significant variations in water retention but levels are generally lower (see discussion on previous page)

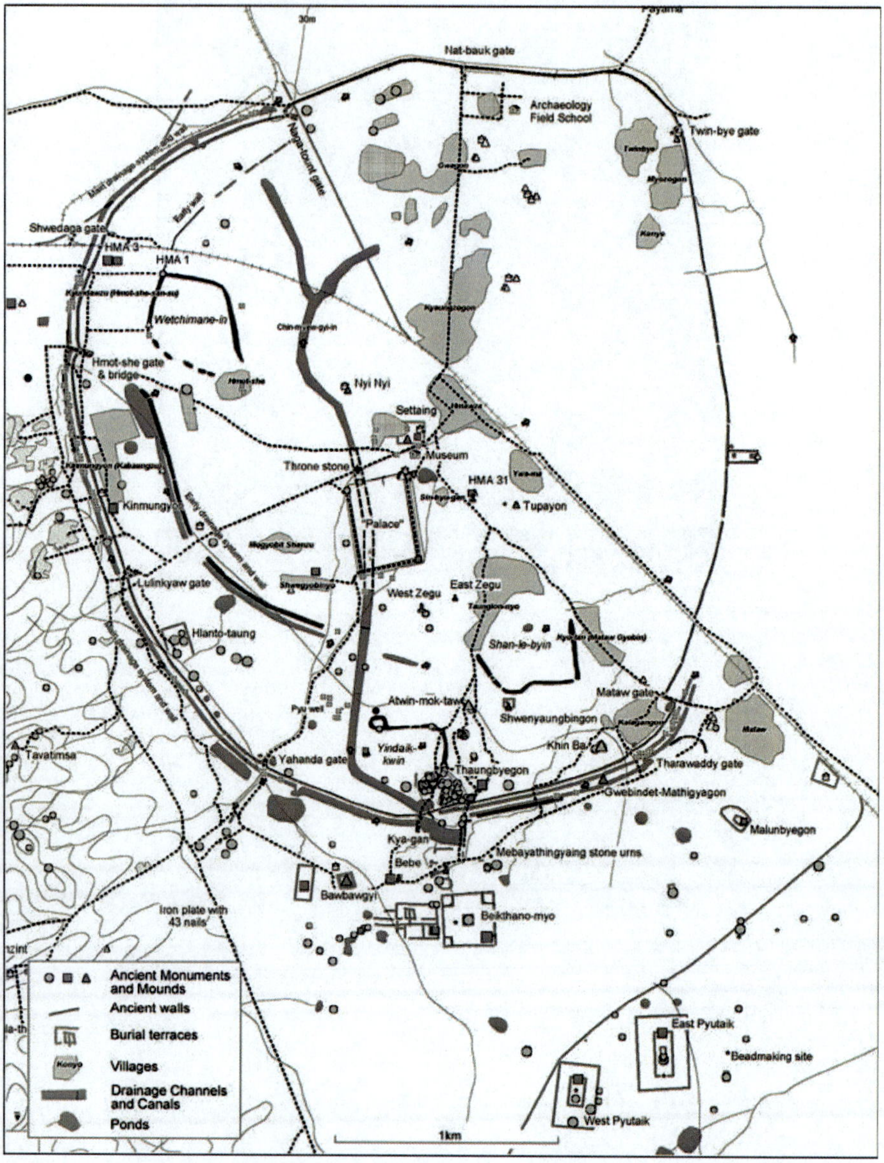

Fig. 11.6 Archaeological map of Sri Ksetra compiled from different sources by Hudson in 2005–2007 (*Source*: after Hudson 2007)

recorded on Fig. 11.6 is the broad Kyagan channel running c. south-north across the site and apparently passing under or through the area of the royal citadel or palace. Although contour heights are not supplied, they can be found on Fig. 11.2. From these it is clear that this channel passes by a direct route from c. 60 to 43 then to 34 m amsl in a distance of c. 3.5 km. This is the partial exception to the continuities of directions of water flow mentioned above. From satellite, aerial and surface reconnaissance, it seems to us that this channel *in its present form* results from post-destruction erosion especially recent and rapid. Otherwise it would, in the life-time of the ancient royal city, have threatened its central structures as well as the operation of the main irrigation system.

Moreover, surface surveys by Stargardt in the 1980s recorded Kyagan as c. 20 m wide (cf. Fig. 11.2), while Stargardt and Nicolazzi in 2007–2008 found that this channel had continued to erode to a width of over 50 m, being very shallow and full of sand (U Thein Lwin, personal communication). Thus Kyagan exhibits the characteristics of an active erosional channel today but may indeed have ancient origins as a canal of the same dimensions as the other convergent canals inside the walls registered in Fig. 11.2, that could be seasonally activated when water forces were low.

Evidence presented above on the Pyu irrigation techniques consistently show awareness of the need to reduce the erosive force of water flows by means of circular-tending canal courses along the contours or meandering canal courses (cf. Figs. 11.2, 11.3, 11.4). Other major differences between Figs. 11.2, 11.3, 11.4 and 11.6 are as follows: Fig. 11.6 does not record the network of ancient canals within and outside the walled site, the great east tank, or the lesser south and north tanks but the earthen banks that once enclosed them are plotted on Fig. 11.6 and the local toponyms, both of which end in *–in* [lake] are given. Further differences exist in the number of outer walls and moats recorded. (cf. Figs. 11.6, 11.2, 11.3a, 11.4a–c and 11.5b, c). On the other hand, Fig. 11.6 contains many valuable data on find sites and does record parts of the course of the first circular canal inside the walled area (Al on Fig. 11.2). It also presents for the first time an important new discovery made by U Thein Lwin, his colleagues and Hudson: namely an ancient earthen wall running just inside that canal. Tribute must also be paid to Hudson's valuable work on hand-held, detailed GPS measurements (not available at the time of Stargardt's original mapping in the 1980s). To be conclusive, however, altitude measurements need confirmation by mounted differential GPS records. The differences between these maps may be summarized as the differences between surface survey – however admirable – and aerial or remotely sensed perspectives. The latter are an indispensable complement to surface surveys for the detection of traces of the ancient irrigation works anywhere. The main thrust of this paper has thus been to demonstrate the enduring economic impact of the irrigation systems at Sri Ksetra and the adequacy of the Remote Sensing methods used to document those processes.

11.4 Conclusions

Through the case study of Sri Ksetra, we have shown that the geology and extended format of this urban type allows the existence of ancient water works to be amply demonstrated and their continued impact on modern landscapes to be detected from multi-spectral remotely sensed imagery and derived indices. In spite of the technical limitations of the medium resolution of these images, mentioned at the outset, this study has shown them to be sensitive instruments for recording the enduring effects of ancient irrigation works embedded in the present landscape in ways that are not visible in surface surveys. The techniques employed in this study may well provide new insights into the dialogue between past and present in other archaeological sites with similar geology, extended format and suspected or known ancient irrigation works in South, Southeast and East Asia. They may even be appropriate to the search for ancient water works in urban sites of this type where no surface traces remain and – as here – even compare well with aerial photographs in their ability to capture certain features and their temporal changes in the course of the year. Environmental factors which have contributed to the success of this study are these: the site is located where there are enduring, if complex water flows; the city and its irrigation works were constructed on stable clays of marine origin which have retained the clear imprints of human activities since their origin; the lighter sediments washed into the irrigation canals and tanks obliterated many of their surface traces but also served as a permeable medium for water to flow beneath the surface along well-defined channels thus creating differentiated patterns of wetness and vegetation. The monsoon regime, though subject to some variations, delivers seasonally defined weather conditions to this area. The existence of such seasonal variations makes it desirable to study images that form a time sequence.

References

Bewley RH, Crutchley S, Shell C (2005) New light on an ancient landscape: LiDAR survey in the Stonehenge World Heritage Site. Antiquity 79:636–647

Delbart N, Le Toan T, Kergoat L, Fedotova V (2006) Remote Sensing of spring phenology in boreal regions: a free of snow-effect method using NOAA-AVHRR and SPOT-VGT data (1982–2004). Remote Sens Environ 101:52–62

Devereux BJ, Amable G, Crow P, Cliff A (2005) The potential of airborne LiDAR for detection of archaeological features under woodland canopies. Antiquity 79:660

Devereux BJ, Amable G, Crow P (2008) Visualisation of LiDAR terrain models for archaeological feature detection. Antiquity 82:470–479

Doneus M, Briese C (2006) Full-waveform airborne laser scanning as a tool for archaeological reconnaissance. In: Campana S, Forte M (eds) From space to place. Proceedings of the 2nd international conference on remote sensing in archaeology. BAR Int Ser 1568, Oxford, pp 99–106

Eitel JUH, Gessler PE, Smith AMS, Robberecht R (2006) Suitability of existing and novel spectral indices to remotely detect water stress in Populus spp. For Ecol Manag 229:170–182

Fletcher R et al (2008) The water management network of Angkor, Cambodia. Antiquity 82 (317):658–670

Gao BC (1996) NDWI – a Normalised Difference Water Index for remote sensing of vegetation liquid water from space. Remote Sens Environ 58:257–266

Higham CFW (2002) Archaeology in Myanmar: past present and future. Asian Perspect 40:127–134

Hudson B (2004) The origins of Pagan; the archaeological landscape of Upper Burma to AD 1300. PhD thesis, Department of Archaeology, University of Sydney

Hudson B (2007) Sriksetra survey map 2005–2007. SOAS Bull Burma Res 5:1

Jackson TJ, Chen D, Cosh M et al (2004) Vegetation water content mapping using Landsat data derived normalized difference water index for corn and soybean. Remote Sens Environ 92:475–482

McFeeters SK (1996) The use of the Normalized Difference Water Index (NDWI) in the delineation of open water features. Int J Remote Sens 17(7):1425–1432

Moore E (2003) Bronze and iron age sites in upper Myanmar: Chindwin, Samon and Pyu. SOAS Bull Burma Res 1:1–25

Ouma Y, Tateishi R (2006) A water index for rapid mapping of shoreline changes of five East African Rift Valley lakes: an empirical analysis using Landsat TM and ETM+ data. Int J Remote Sens 27(15):3153–3181

Stargardt J (1990) The ancient Pyu of Burma, vol I, Early Pyu cities in a man-made landscape. PACSEA Cambridge in assoc ISEAS, Singapore

Stargardt J (2001) The historical geography of Burma; the creation of enduring patterns in the Pyu period. Newsl Int Inst Asian Studs [IIAS] Leiden July Special Burma Issue

Stargardt J (2002–2003) City of the wheel, city of the ancestors: spatial symbolism in a Pyu royal city of Burma. Indo-asiatische Zeitschrift [Berlin] 6/7:144–167

Stark MT (2006) Early mainland Southeast Asian landscapes in the first millennium AD. Ann Rev Anthropol 35:407–432

Wheatley P (1971) Pivot of the four quarters: a preliminary enquiry into the origins and character of the ancient Chinese city. Aldine Pub. Co., Chicago

Willis M (2009) The archaeology of Hindu ritual. Cambridge University Press, Cambridge

Wittfogel KA (1957) Oriental despotism; a comparative study of total power. Yale University Press, New Haven

Xu H (2006) Modification of normalised difference water index (NDWI) to enhance open water features in remotely sensed imagery. Int J Remote Sens 27(14):3025–3033

Chapter 12
Following the Ancient Nasca Puquios from Space

Rosa Lasaponara and Nicola Masini

Abstract Precious information to reconstruct ancient environmental changes, still fossilized in the present landscape, may be captured from multispectral satellite images from medium to high spatial resolution. In particular, satellite derived moisture content may facilitate the identification of areas involved in early environmental manipulation mainly addressed to set up irrigation and artificial wet agro-ecosystems where the natural rainfall was insufficient to support agriculture. Up to now, only a few number of archaeological studies on spatial patterns of moisture have been carried out through the world using satellite optical data. In this chapter, Landsat and ASTER data were analyzed for some areas near Nasca river within the drainage basin of the Rio Grande, densely settled over the centuries and millennia even if the physical environment presented serious obstacles to human occupation. This region is one of the most arid areas of the world, so that the pluvial precipitations are so scarce that they can not be measured. To face this critical and extreme environmental conditions, ancient populations of the Nasca River valley, devised an underground aqueducts called *puquios*, some of which are still used today. Archaeologists suggest that during the Nasca flourishing period, certainly the number and spatial distribution of *puquios* was larger than today. We used satellite data to identify areas to be further investigated to assess if and where therein *puquios* were constructed for water control and retrieval.

R. Lasaponara
Institute of Methodologies for Environmental Analysis, CNR-IMAA,
C. da S. Loya, 85100 Tito Scalo, PZ, Italy
e-mail: lasaponara@imaa.cnr.it

N. Masini (✉)
Institute of Archaeological and Architectural Heritage, CNR-IBAM,
C. da S. Loya, 85050 Tito Scalo, PZ, Italy
e-mail: n.masini@ibam.cnr.it

R. Lasaponara and N. Masini (eds.), *Satellite Remote Sensing: A New Tool for Archaeology*, Remote Sensing and Digital Image Processing 16,
DOI 10.1007/978-90-481-8801-7_12,

Keywords Satellite remote sensing • Hydrography • Puquios • Vegetation indices • NDWI • Nasca • Peru

12.1 Introduction

Archaeological investigations on ancient societies are addressed to cast light on important basic questions such as: how civilizations came about, how they adapted effective strategies to face adverse environment conditions (drought, flood), and why they ceased to exist. To address these questions, different approaches and perspectives have been employed through the years, including socio-economic (Diamond 2009, 2010) and socio environmental analysis (Folan et al. 2000; Adams and Jones 1981), predictive models (Vaughn and Crawford 2009), fractal (Brown and Witschey 2003) and investigation from in situ material and data analysis.

Archaeologists have faced the necessity to enrich information content about ancient settlements and improve the knowledge of historical local environmental changes not only using material from excavations but also environmental and palaeo-environmental parameters. This is required especially in areas for which little information is generally available, as in the case of South America, where the majority of ancient populations did not use writing and therefore, no documentary sources are available.

Palaeo-environmental studies may take benefit from the use of multispectral satellite images from medium to high spatial resolution along with digital terrain models.

Satellite derived parameters, such as temperature and moisture with their spatial patterns and variations, can help us to extract precious information to reconstruct ancient environmental changes still fossilized in the present landscape. One of the most significant parameters is the moisture content, because historically, civilizations have predominately located their communities close to water ways for their basic survival, agriculture, ritual and domestic use. Changes in the ecosystems and landscapes can obscure the ancient water ways and hidden important information on the distribution and organization of populations. Palaeo-environmental characteristics may help not only to discover new and unknown ancient settlements, but also to improve the knowledge about the social organization, agricultural production, survival strategies and settlement distribution.

Archaeological evidences of early environmental manipulation such as, irrigation and artificial wet agro-ecosystems, have been found, throughout the world, in areas where the natural rainfall was insufficient to support agriculture, as in Ancient Egypt, Mesopotamia, and Ancient Persia (modern Iran). Therein population still use today the *Qanats* (about 800 BC), considered one of the oldest known irrigation methods. The *Qanats* systems are based on a network of vertical wells and sloping tunnels to retrieve groundwater.

In North India, the Indus Valley Civilization devised sophisticated irrigation made of an extensive network of canals and storage systems to practice large scale agriculture.

In Central and South America, including parts of what are now Belize, Guatemala, northern Honduras and southern Mexico, the Maya civilization utilized the karst groundwater resources by means of springs and caves. Archaeological evidence reveals that, in the semiarid zones, Maya pattern settlements occurred in areas with greater access to the groundwater as expected given that Maya urban, rural, and religious life was strongly linked to water exploitation. Nevertheless, compared to other Maya technological achievements, groundwater retrieval methods were quite primitive, inefficient, and strongly affected by human contamination. According to some authors, see for example (Veni 1990) and reference therein quoted, this caused widespread disease and may have contributed to Maya downfall.

In northern Peru, at least 4,500 years ago, early civilizations addressed great efforts to construct small-scale gravity canals in higher-elevated coastal valleys to 'dominate' the arid environment which limited the development of local food production necessary for the increase population (Dillehay et al. 2005). In Southern Peru, ancient populations of the Nasca River valley, devised an efficient system for the control and retrieval of water, based on an underground system of aqueducts called puquios, which made it possible to transform one of the most arid area of the world in a densely settled zone, as archaeological record and findings clearly testify.

Up to now, only a few number of archaeological studies on spatial patterns of moisture have been carried out through the world using satellite optical data.

Effective methodologies for the retrieval of these valuable information are still to be developed. Actually, moisture content may be estimated from remote sensing techniques by using active and/or passive sensors, such as, (i) Radar microwave (Synthetic Aperture Radar (SAR)) and/or (ii) multispectral optical remote sensing technique. In this study, we will focus on the use of optical data, such as Landsat MSS and ASTER.

12.2 Satellite Based Estimation of Soil Moisture Variation

12.2.1 Satellite Based Estimation of Soil Moisture Variation: Brief Overview

An accurate knowledge of the state of vegetation and soil moisture and their spatial and temporal patterns is a key issue for a wide range of applications including, agriculture, water management, drought monitoring and forecasting. Over the last decades, numerous studies have been carried out to assess vegetation/soil parameters using remote sensing to simulate soil moisture (Ragab 1995; Walker et al. 2001).

Tucker (1980) used the Normalized Difference Vegetation Index (NDVI) (see formula12.1), developed by Rouse et al. (1973), to estimate leaf water content and other physiological variables for herbaceous cover. Actually, for more than three decades the NDVI has been used as a surrogate to estimate vegetation water content even with strong limitations mainly encountered when vegetation coverage is dense and the index is close to the saturation level.

$$NDVI = \frac{(NIR - R)}{(NIR + R)} \tag{12.1}$$

The arithmetic combination of the red and Near infrared (NIR) channels enables us to exploit the different spectral behaviour of vegetation cover in the different bands. NDVI provides a dimensionless numerical value. The formula is designed as a ratio, in order to normalize its variability field between -1 and $+1$. NDVI assumes values less than 0 for water, slightly higher than 0 for bare soils, higher than 0.4 for vegetation, can exceed 0.8 in presence of dense vegetation or be close to saturation (NDVI $\approx$ 1) for a rainforest.

Tucker (1979) recognized that NDVI may be a useful "surrogate" for the estimation of vegetation water content for grassland but, as a general rule, the relationship between this index and the vegetation moisture content is strongly linked with the amount of vegetation. NDVI provides information closer to the amount and greenness of vegetation rather than moisture content and it is generally limited by soil reflection. Moreover, its effectiveness is reduced due to its sensitivity to atmosphere.

The advantage of using NDVI for water content estimation is that this index is simple and available routinely and globally, but a number of limitations (Ceccato et al. 2002a, b) can be summarized as follows:

(i) NDVI saturates at intermediate values of leaf area index (LAI), therefore it is not responsive to the full range of the canopy.
(ii) Each plant species has its own relationship of chlorophyll and moisture content;
(iii) A decrease in chlorophyll content does not imply a decrease in moisture content;
(iv) A decrease in moisture content does not imply a decrease in chlorophyll content.

Nevertheless, the NDVI has been used as an indicator of vegetation moisture content mainly because it has been the only information available.

More recently, the potentiality of using satellite SWIR spectral bands for moisture content estimation has been supported by both modelling (Ceccato et al. 2001; Fourty and Baret 1998; Ustin et al. 1998; Zarco-Tejada et al. 2003) and experimental studies based on the available multispectral satellite datasets (Ceccato et al. 2002b; Peñuelas et al. 1993; Peñuelas and Inoue 1999; Ripple 1986; Roberts et al. 1997; Chen et al. 2003; Jackson et al. 2004; Yu et al. 2000).

Using multispectral satellite data, the estimation of moisture content into soil and vegetation may be improved using spectral indices based upon NIR and SWIR and in general on the longer wavelength reflective infrared range (1,240–3,000 nm), for example, the short-wave infrared (SWIR) reflectance (1,300–2,500 nm).

Several spectral indices, such as Normalized Vegetation Moisture Index (NVMI) or Normalized Difference Water Index (NDWI), mainly based on SWIR bands, can be computed to estimate moisture content for both soil and vegetation. The mathematical formulation of these indices (see formulas 12.2 and 12.3) is very similar to the NDVI, but based on specific bands of water absorption (Rogers and Kearney 2004).

$$NVMI = \frac{(NIR - SWIR)}{(NIR + SWIR)} \tag{12.2}$$

$$NDWI = \frac{(RED - SWIR)}{(RED + SWIR)} \tag{12.3}$$

Both of these two indices NVMI and NDWI are sensitive to water content in vegetation and soil, respectively; being that the absorption of water content of vegetation close to NIR band (and that of soil close to red band) is negligible, whereas a small absorption is present into the SWIR spectral range. Moreover, in comparison with NDVI both NVMI and NDWI are less sensitive to the effects of the atmosphere, but, the effects of soil reflection are still present.

There are two main satellite sensors which offer information in these spectral bands from medium to low spatial resolution: they are Landsat TM/ETM+ (available at 30 m) and ASTER images (with a spatial resolution ranging from 15 to 90 m). For Landsat TM/ETM+, NIR and SWIR correspond to bands 4 (780–900 nm) and 5 (1,550–1,750 nm), respectively. ASTER offers more bands than TM/ETM+ useful for computing both NVMI and NDWI. Finally, Landsat MSS data have been used as useful data for vegetation monitoring, and in turn for vegetation, using mainly NDVI.

Nevertheless, these indices have not received much attention until recently mainly because the systematic coverage offered by TM and ETM+ bi-weekly is not considered adequate for many applications based on near-real time acquisitions. Technical details of Landsat Mss, TM and ASTER are in Sect. 12.2.2.

12.2.2 Medium Spatial Resolution Satellite Data for Estimating Moisture Variations

The Landsat satellite program was designed mainly for vegetation monitoring. In particular the five spectral bands of MSS sensor (see Table 12.1) were selected: (i) to study green reflectance from healthy vegetation (band 1), (ii) to analyze chlorophyll absorption in vegetation (band 2), (iii) for recording near-infrared

Table 12.1 MSS bands: data. For practical use, it is important to remind us that for technical reasons MSS bands on the first three Landsat satellites were labelled as 4, 5, 6, and 7. Later, these bands were relabelled to 1, 2, 3, and 4 for MSS onboard Landsat 4 and 5 satellites (see Table 12.2)

Band	Resolution (m)	Wavelength (μm)	Description
1	79	0.5–0.6	Green
2	79	0.7–0.7	Red
3	79	0.7–0.8	Near infrared
4	79	0.8–1.1	Near infrared
5	240	10.4–12.6	Thermal infrared

Table 12.2 Satellite platforms carrying the MSS Sensor

Satellite	Launched	Decommissioned
Landsat 1	July 23, 1972	January 6, 1978
Landsat 2	January 22, 1975	February 25, 1982
Landsat 3	March 5, 1978	March 31, 1983
Landsat 4	July 16, 1982	June 15, 2001
Landsat 5	March 1, 1984	MSS sensor failed in 1992, TM sensor still operational

reflectance peaks in healthy green vegetation and for detecting water-land interfaces (bands 3 and 4, respectively).

The four bands from visible to near infrared have a spatial resolution of 79 × 79 m, whereas the thermal infrared (only present on Landsat 3) has a spatial resolution of 240 × 240 m.

Landsat's satellite program was designed to have regular acquisition schedule (revisits each spot on the earth every 16–18 days) and long-term data archive with data available from 1972 to 1992 (Table 12.2). The MSS has been one of the first sensors at high spatial resolution onboard Landsat satellites, but of course today it is classified as moderate-resolution image source and its spatial resolution is considered its major limitation.

MSS has been operative until 1992 when it stopped acquiring images because it was replaced by Thematic Mapper (TM) with improved technical characteristics namely higher spectral and spatial capability.

LANDSAT TM multispectral data are acquired from a nominal altitude of 705 km (438 miles) in a near-circular, sun-synchronous orbit at an inclination of 98.2°, imaging the same 185-km (115-mile) swath of the Earth's surface every 16 days. Seven are the channels, among them six at 30 m and one in the thermal range at 90 m of resolution (for additional information see Table 12.3). All of the remote sensed data were georeferenced in the UTM projection.

ASTER is a high resolution imaging instrument that is flying on the Earth Observing System (EOS) Terra satellite. It has the highest spatial resolution of all five sensors on Terra and collects data in the visible/near infrared (VNIR), short wave infrared (SWIR), and thermal infrared bands (TIR). Each subsystem is

Table 12.3 Thematic Mapper (TM) Landsat 4–5 bands: spectral range and spatial resolution

Band	Wavelength (μm)	Resolution (m)
Band 1	0.45–0.52	30
Band 2	0.52–0.60	30
Band 3	0.63–0.69	30
Band 4	0.76–0.90	30
Band 5	1.55–1.75	30
Band 6	10.40–12.50	120
Band 7	2.08–2.35	30

Table 12.4 ASTER imagery bands: spectral range and spatial resolution

Band	Wavelength (μm)	Resolution (m)
Band 1	0.52–0.60	15
Band 2	0.63–0.69	15
Band 3N	0.76–0.86	15
Band 3B	0.76–0.86	15
Band 4	1.60–1.75	30
Band 5	2.145–2.185	30
Band 6	2.185–2.225	30
Band 7	2.235–2.285	30
Band 8	2.295–2.365	30
Band 9	2.36–2.43	30
Band 10	8.125–8.475	90
Band 11	8.475–8.825	90
Band 12	8.925–9.275	90

pointable in the crosstrack direction. The VNIR subsystem of ASTER is quite unique. One telescope of the VNIR system is nadir looking and two are backward looking, allowing for the construction of 3-dimensional digital elevation models (DEM) due to the stereo capability of the different look angles. ASTER has a revisit period of 16 days, to any one location on the globe, with a revisit time at the equator of every 4 days. ASTER collects approximately 8 min of data per orbit (rather than continuously). All the 14 ASTER spectral bands are listed in Table 12.4.

The utilization of Landsat TM/ETM and ASTER images for moisture mapping has been generally limited by their costs and infrequent temporal coverage. Nevertheless, today, TM and ASTER images are free available from NASA web site and they offer suitable data sets for studies which do not require real or near real time data availability, as in the case of archaeological investigations. The TM/ETM and ASTER generally provide a large historical data archive along with update acquisitions very useful for extracting information useful for environmental and palaeo-environmental studies. It should be noted that, actually the ASTER SWIR bands are not available due to some serious acquisition errors encountered since the year 2008.

12.3 Satellite Based Estimation of Moisture Content: Case Study in the Drainage Basin of the Río Grande

12.3.1 Study Area and Aims

The geomorphology of the coastal desert of southern Peru, has been shaped by the drainage basin of the *Río Grande*, which empties into the Pacific Ocean after passing through the coastal range and collecting water from nine tributaries (Fig. 12.1).

From top to bottom, the Rio Grande first collects the waters of the Vizcas, Ingenio and Palpa rivers, then those of the Santa Cruz River, and afterwards joins the Nasca River, whose water flow is contributed by the Aja, Tierras Blancas, Taruga, Tunga and Las Trancas rivers, to which are added the Atarco, Carrizales and Usaca narrow ravines (Fig. 12.2). Fed with the seasonal precipitation coming from the Andes, at an altitude higher than 2,000 m above sea level, the southern tributaries of the Nasca River are substantially smaller than those of the northern tributaries and also those of other coastal valleys. Since about 5000 BC, the ecosystem of the valley of the Nasca River and its main tributaries has been essential in forming the first complex societies. The physical environment presents serious obstacles to human occupation mainly linked with the lack of water due to scarce pluvial precipitation, so scarce that it can not be measured. It is widely recognized that this is one of the most arid areas of the word and these complex environmental conditions have characterized the drainage basin of the Río Grande over the centuries and millennia. The scarcity of precipitation is not the only issue

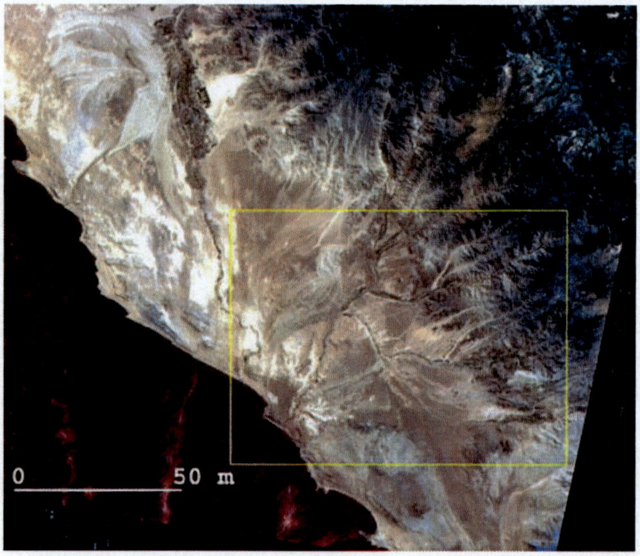

Fig. 12.1 RGB Landsat ETM 7 image (2003): study area location indicated by the *yellow box*

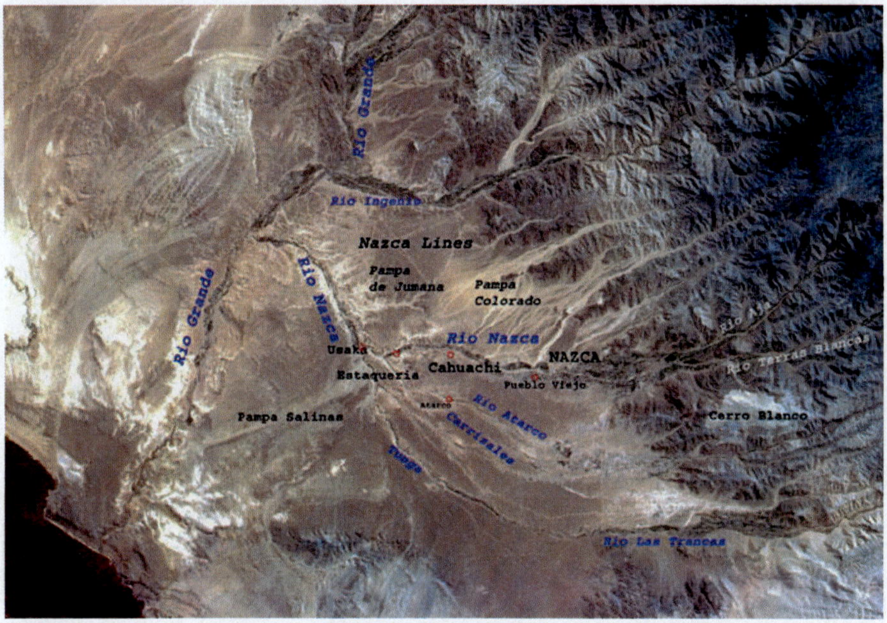

Fig. 12.2 Zoom of Fig. 12.1 which depicts Rio Grande Basin: from top to bottom, the Rio Grande first collects the waters of the Vizcas, Ingenio and Palpa rivers, then those of the Santa Cruz River, and afterwards joins the Nasca River, whose water flow is contributed by the Aja, Tierras Blancas, Taruga and Las Trancas rivers, to which are added the Atarco, Carrizales and Usaca narrow ravines

limiting the hydric resource availability but also the infiltration capacity which determines a substantial reduction of the water transmission causing the underground flow of water. As a result there is no water available for irrigation and for domestic purposes. Therefore, we can expect this to result in sparse occupation, but, on the contrary, archaeological evidence clearly testifies that the area was densely settled in pre-columbian times (Orefici and Drusini 2003).

Historically, the development of cultures was only possible thanks to the efficient control of the water, based on an underground system of aqueducts called puquios, devised since Early Intermediate Period (known also as Middle Nasca) as proved by archaeological findings associated to *puquios* found by archaeologists (Schreiber and Lancho Rojas 2009).

In other words, for thousands of years, populations have successfully adapted to the low rainfalls in this area, aided by the fact that underground water was likely very close to the surface and accessible by constructing wells and developing a new water source, reaching to the water table through a number of underground aqueducts. Throughout the years, several researchers have studied *puquios*: Alfred Kroeber and Toribio Mejía Xesspe between the 1920s and 1930s (Kroeber and Collier 1998; Mejía Xesspe 1942), Francisco González García (1934), a scientific mission of Tokyo University (Kobori 1960), and researchers of the former

Fig. 12.3 Puquios with the shape of open trenches accordingly to the way the galleries were built

Peruvian development agency CORDEICA in the 1980s (Solar La Cruz 1997). Studies from Lancho Rojas and Schreiber started up back since 1985 and continue as of today (Lancho Rojas 1986; Schreiber 1995, 2003; Schreiber and Lancho Rojas 1988, 2009).

Puquio is a horizontal water well, a trench and/or underground gallery that connects a place on the surface with the ground water source (Fig. 12.3). Such system of aqueducts was the most ambitious hydraulic project in the Nasca area and it made possible to have water availability for the whole year not only for irrigation but also for domestic needs. In fact, the water filters inside the puquios and flows throughout it toward a small reservoir (*cocha*) or directly to a stream or canal.

Results from recent investigations carried out by Lancho Rojas (2009), have pointed out that nowadays there are 36 *puquios* still in use in these valleys. As suggested by archaeological evidence, during the Nasca period, there was a number of such galleries certainly greater than today and also their spatial distribution spread out over a larger area. Of course, some *puquios* have been substantially altered in the last centuries and others have been abandoned or destroyed. The original *puquios* devised by ancient Nasca civilizations can be categorized into two types according to their shape and size: (i) *puquios* with the shape of open trenches and (ii) *puquios* trench galleries characterized by greater length and depth compared to the open trenches.

Fig. 12.4 Openings, also called chimneys, or more commonly, eyes which allow the access to the galleries for their annual cleansing. The shape and dimensions of the eyes vary

Puquios with the shape of open trenches (see, Fig. 12.3), with borders usually covered by pebbles, have generally a base of approximately 1-m width but they spread on the surface around 10 m or more. Some trench-shaped *puquios* present multiple branches, but their common feature is having just one canal. Figure 12.3 shows that pebbles form a sort of contention wall which requires a systematic cleansing and actually were repaired annually and changed year by year.

The *puquios* built with shape of tunnels can measure up to 10 m of width, and they generally present a large part of their branches covered, thus creating "filled trench galleries" (Schreiber and Lancho Rojas 1988). All along the galleries and at different points there are openings, also called chimneys, or more commonly, *ojos* (eyes) which allow the access to the galleries for their annual cleaning as well as the entrance of air and sunlight (an example is shown in Fig. 12.4). The shape and dimensions of the *ojos* vary accordingly to the way the galleries were built. Finally, most *puquios* flow to small reservoirs or *cochas* in their lowest extremes, right where water is directed to the irrigation streams or canals.

Archaeologists found evidence that the area could not have been inhabited without the existence of these aqueducts. The survey of ancient *puquios* is crucial not only to detect unknown ancient settlement remains as well as to understand how the Nascas – also by a suitable choice of ecological niches – were able to develop effective and complex agro-ecosystems to meet the demands of a large population, as part of a cosmo-vision of the natural environment, which comprised the total respect of their resources.

To investigate and search for the whole drainage basin, our analysis focused on the moisture content extracted from a multi-temporal data set made up of medium-resolution satellite Landsat MSS, TM and Aster images.

12.3.2 Analysis and Results

The investigations based on remote sensing have been carried out using a multitemporal data set made up of Landsat MSS (1974, 1990), Landsat TM (1990), Landsat 7 (2003) and Aster (2003, 2004, 2007). More detailed features have been extracted by the available VHR satellite dataset used for archaeological prospection (for additional information see Chap. 14).

Herein, we will focus on the Landsat MSS acquired on month 1974, ASTER datasets acquired on June of 2003, 2004 and 2007 and GeoEye images on 2011 February.

These data were processed using the most adequate algorithms according to the available spectral bands. For the MSS data we computed the NDVI (see formula 12.1) and the Tasseled Cup Transformation (TCT) to better discriminate the vegetation from desert areas and to emphasize the spatial variation of moisture content from the most drought regions.

The TCT is a fixed coefficient linear transformation, based on empirical analysis of physical features space, proposed by Kauth and Thomson (1976) mainly for agricultural monitoring. The TCT, based on the four MSS bands, provides a new reference system composed of: (i) "soil brightness" defined by the signature of non vegetated areas, (ii) "greenness" obtained from vegetation signatures, (iii) and other two new axis orthogonal to the first two axes, which indicate changes in atmospheric haze conditions (for additional details on TCT the reader is referred to Chap. 2).

Figure 12.5 shows the 1974 NDVI map which clearly enhances brightness areas characterized by the presence of vegetation. The same figure clearly points out that the desert area between the Nasca river and Pampa de Chauchilla is characterized by a large dry hydrographic reticulum, within which we can distinguish underground rivers. The latter are characterized by an intermittent water flow due to soil infiltration capacity which causes that some parts of their course disappear under the surface.

The first two components of TCT visualize greenness and soil brightness, respectively. TCT2 confirms and integrate information provided by NDVI regarding the vegetated area. TCT1, shown in Fig. 12.6, emphasizes ancient mudslides (known as *huaycos*), due to flash flood caused by torrential rains occurring in the mountains. Moreover, TCT1 enhances the presence of linear geoglyphs which cross

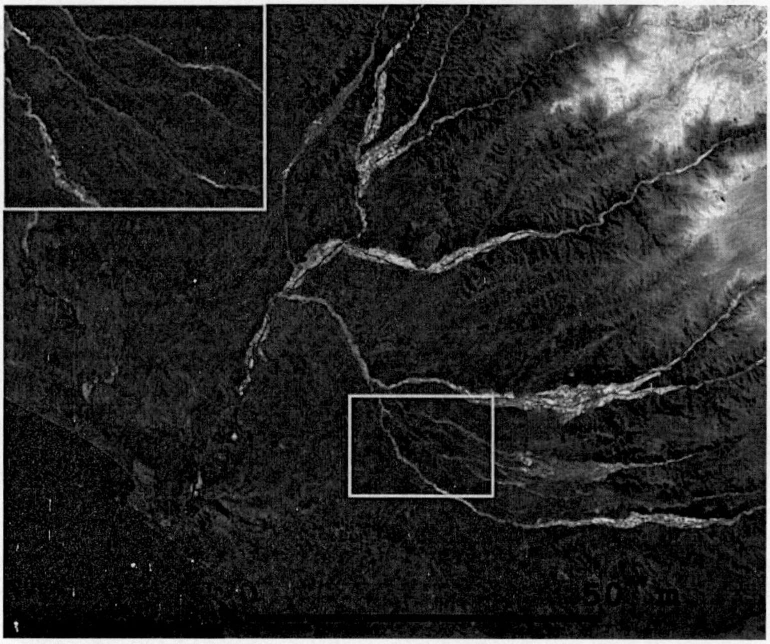

Fig. 12.5 MSS NDVI map (1974) at 80 m spatial resolution. The map enhances brightness' areas characterized by higher NDVI values which indicate the presence of vegetation. The desert area between the Nasca river and Pampa de Chauchilla is characterized by a large dry hydrographical reticulum, within which it is possible to distinguish underground rivers (*upper left*)

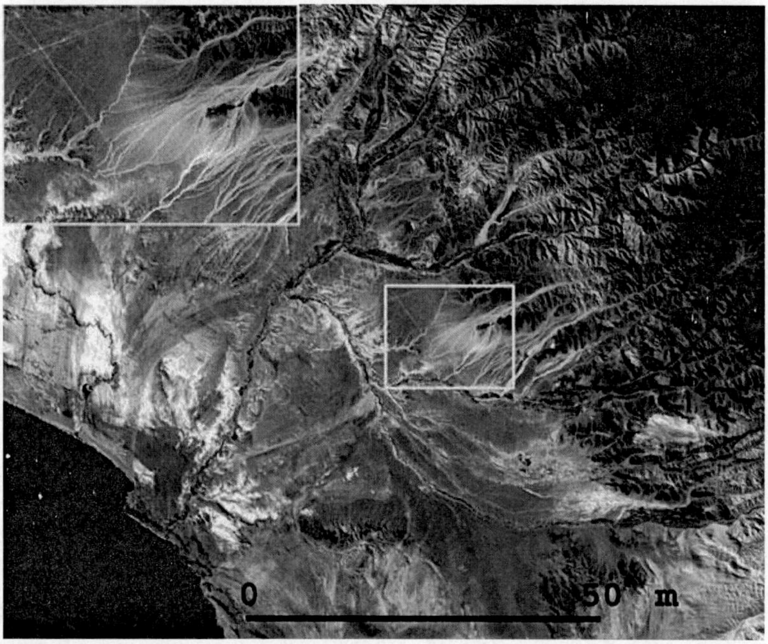

Fig. 12.6 MSS TCT1 map (1974). The soil brightness map enhances the reflectance of the Nasca lines and the beds of *huaycos* (mudslides)

the Pampa de Jumana and Pampa Colorado (see Fig. 12.6 and zoom in the same figure, upper left).

The multitemporal data analysis shows that the most interesting results in terms of variation of vegetation and moisture content were recorded in areas which were further investigated using ASTER images which offer higher spatial and spectral resolution compared to Landsat MSS and TM.

For each Aster data set (2003, 2004, 2007) we computed NDVI and NDWI maps shown in Figs. 12.7, 12.9 and 12.10. NDVI maps (Fig. 12.7) put in evidence some variations related to vegetation cover changes during the period considered.

They are easily identifiable by the chromatic visualization offered by RGB composition. In particular: (i) constant low values over time are identified by dark-grey to black tone; (ii) constant high values are visualized by lighter tones from white to lighter grey; (iii) low non constant values over time produce a colour whose different intensity levels of red, green and blue depends on the NDVI values for the years 2003, 2004 and 2007, respectively. For example, a pixel with a higher intensity value of red respect to green and blue is characterized by higher 2003 NDVI value than those in 2004 and 2007.

It is worth to note that the investigated area is mostly desert and therefore visualized in RGB with black colour (Fig. 12.7d). However from Fig. 12.7d, we can identify a large area characterized by changes of NDVI over time in the fluvial oasis of the Nasca river. Whereas, smaller areas indicated as v1, v2, v3 and v4 in Fig. 12.7a–d are characterized by NDVI variations located quite close to the tributaries Taruga, Carrizales, Tunga and Atarco. Just Atarco stream Fig. 12.8 referred to: it shows a picture (2010 August) of the typical vegetation close to the stream which from our satellite data set appears as NDVI higher values compared to the surrounding desert areas. The presence of vegetation is only possible for the presence of underground water.

With regard to the above mentioned variations v1-v4, it is worth to note that three of them (namely v1, v2 and v3 in Fig. 12.7d) are recorded in correspondence to the parts of the rivers where water usually flows underground; thus providing information about the fluctuation over time of water level. This is confirmed by other elaborations carried out by using PCA of the three NDWI maps of ASTER dataset (Fig. 12.9). In the current work, NDWI is computed by formula 12.2 using band 6 as SWIR and assuming negative sign (low moisture content should be characterized by high NDWI value and vice versa).

NDWI maps also allow us to extract information about flow characteristics of the rivers at South of the Nasca basin. In particular, by comparing the above mentioned PC2 (Fig. 12.9) and the RGB composition (Fig. 12.7d) of NDWI ($R = 2003$, $G = 2004$, $B = 2007$) we can distinguish three different spectral behaviours.

One is related to pixels characterized by constantly low NDWI values (from -0.1 to 0, see for example the profile in Figs. 12.10b and 12.11a) over time associated with perennial rivers (visualized by darker grey to black tone in RGB image of Fig. 12.10a). The size of these water courses is much more less than the ASTER SWIR pixel (30 m). In the investigated area the largest riverbed is in the Rio Nasca.

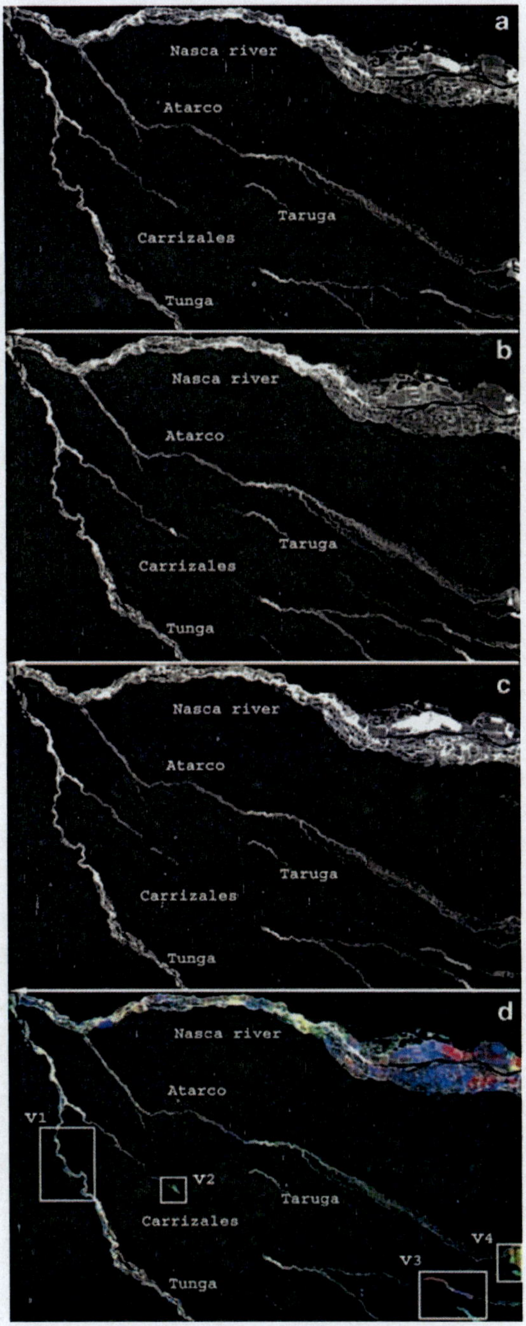

Fig. 12.7 (**a–c**) NDVI maps of ASTER data acquired in June of the years 2003, 2004 and 2007. (**d**) RGB composition of the above said NDVI maps (*R* NDVI2003; *G* NDVI2004; *B* NDVI2007). The rectangular white boxes denote significant NDVI variations in the considered period

Fig. 12.8 Picture taken on 2010 August near the Atarco stream. The image displays the typical vegetation which grows close streams of the Rio Nasca basin which from our satellite data set appear as NDVI higher values compared to the surrounding desert area

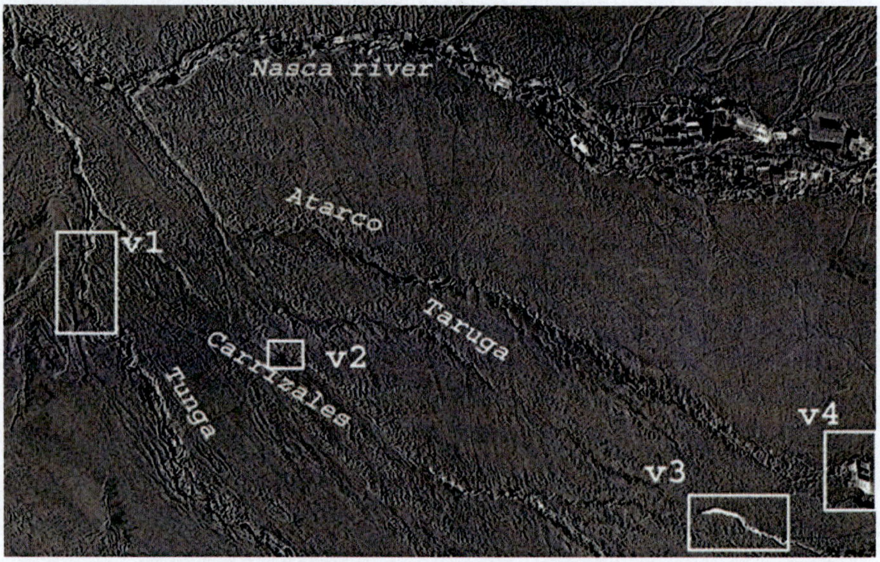

Fig. 12.9 PCA2 of NDWI maps which evidence the temporal variation of moisture content in agree with the changes of NDVI shown in Fig. 12.7d

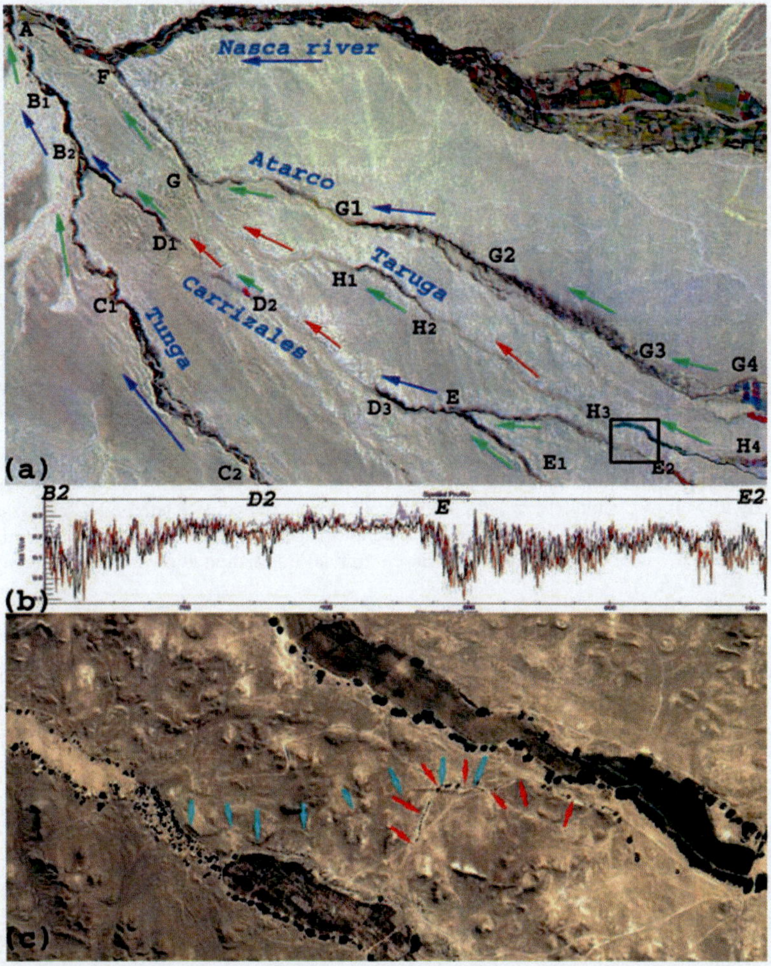

Fig. 12.10 (**a**) RGB composition of NDWI maps computed for the years 2003, 2004 and 2007. It provides information on the hydraulic regime (perennial, ephemeral, dry) of the four stream tributaries of the Nasca river. The black rectangular box denotes an area where two disused *puquios* have been detected. (**b**) Spatio/temporal profile of NDWI ASTER Index of the Carrizales which allowed us the different hydraulic characteristics. (**c**) Two disused *puquios* identified from a GeoEye scene in an area located between Carrizales and Taruga where the hydraulic regime is ephemeral, as shown by the analysis of ASTER data

The wet edge of the river ranges from 2 to 5 m and the depth is less than 1 m. At South the riverbed sizes are less than 2 m, whereas the depths are less than 0.50 m. Therefore, what we focus is a small part of the pixel whose signal is enough to be appreciated but remarkably lower than the typical values expected in presence of rivers with a significantly greater water flow.

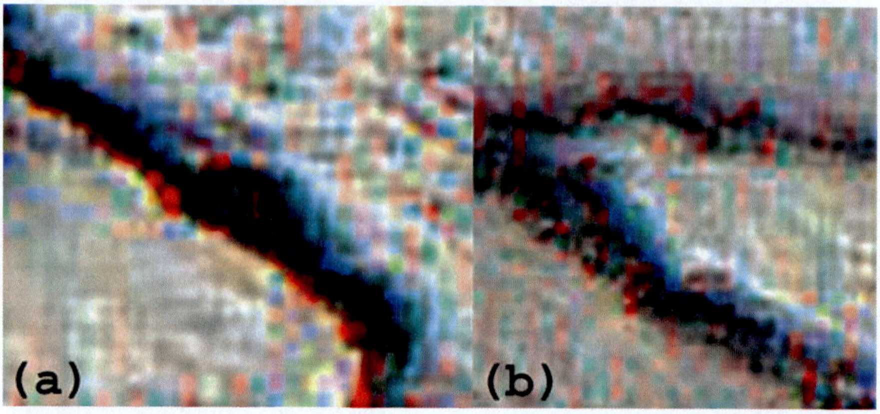

Fig. 12.11 Two different patterns observed from the RGB composition of NDWI index computed for the years 2003, 2004 and 2007. The pattern shown in (**a**) is typical of a stream characterized by constantly low NDWI values (from −0.1 to 0) over time associated with perennial water courses. The pattern in (**b**) evidences a spatial and temporal "dispersion" in NDWI values referable to underground water courses whose seasonal and/or annual fluctuation allow to feed vegetation as shown in Fig. 12.8

The second spatial/temporal pattern we identified is related to pixels with the highest NDWI values (0.25–0.30) and, therefore, we associate them with underground rivers in accordance with our in situ analyses and maps available in Lancho Rojas and Schreiber (2009).

Finally, the multitemporal analysis enabled us to identify the third feature pattern characterized by a spatial and temporal "dispersion" in NDWI values. In other words, as shown in Fig. 12.11b, from 1 year to another the NDWI values change within a buffer which denotes that water table is not very deep (visualized by a colour with different intensity value of red, green and blue). So, inter year water level oscillations cause significant variations in soil moisture and presence of vegetation, as detected by ASTER multitemporal dataset considered in this work.

The processing of ASTER data allowed us to characterize the tributaries of Nasca rivers. Such water courses have different flow characteristics which change seasonally and from 1 year to another, as evidenced by a NDWI profile for the considered period extracted from the *Carrizales* river (Fig. 12.10a). AS a whole, Fig. 12.10b shows the flow characteristics of the Nasca basin. In detail, blue arrows denote stretches of streams with perennial regime, red arrows indicate underground streams and finally green arrows denote an ephemeral regime of the streams.

The areas crossed by stretches of rivers characterized by significant fluctuation of water table are extremely important for the study of historical landscape with particular reference to the palaeo-hydrography and the ancient hydraulic systems, such as the Nasca *puquios*. This task needs the use of very high resolution satellite imagery. Figure 12.10c shows two disused *puquios* observed by QuickBird image, one more ancient than the other, as evident by the fact that the latter is characterized by the presence of vegetation (denoted by red arrows) fed by water which still flows through the *puquio*.

As a whole, the most important result of our investigation has been the possibility to detail the flow characteristics of the hydrography, thus providing information about possible ancient aqueduct systems.

12.4 Conclusion

Multispectral satellite images from medium to high spatial resolution can offer precious information useful to reconstruct ancient environs and environmental changes, still fossilized in the present landscape.

In our investigations, Landsat MSS, ASTER and VHR data sets were analyzed for some areas selected from within the drainage basin of the Rio Grande. This region is considered one of the most arid areas of the world, so that the pluvial precipitations are so scarce that they can not be measured.

Considering this extreme drought, which characterizes this area today as several centuries ago, ancient populations of the Nasca River valley devised an efficient system for retrieval water and to face these critical environmental conditions. This system was based on underground aqueducts called *puquios*, which in part are still used today. Archaeological record put in evidence that during the Nasca flourishing period, the number and spatial distribution of *puquios* were larger than today

In this chapter, vegetation and moisture indices computed from Landsat and Aster multitemporal data allowed us to identify three different spatial/temporal patterns each of them referable to a distinct hydraulic regime: perennial, ephemeral and dry. Most of them are completely dry, due to soil infiltration capacity which causes a substantial loss of water flow of the rivers or are characterized by an intermittent water flow (ephemeral) or they. The latters flow partially on the surface and in some parts of their course disappear in the subsoil. The rainfall volume and geological characteristics determine the point at which the rivers disappear under the surface.

In the past, the intermittent characteristic of water flow did not prevent the population in farming the fluvial oasis, thanks to the water retrieved by the *puquios*. Today the agriculture is mainly practised near stretches of rivers characterized by perennial hydraulic regime.

The results of our investigations show that the integrated use of satellite data (from medium to very high resolution) and in situ analyses can fruitfully support the detection and the survey of old aqueduct systems, such as the ancient disused *puquios*.

In future, it is desirable to make operative the use of earth satellite observation technologies for mapping unknown Nasca aqueducts thus contributing: (1) to improve the knowledge about the social organization, agricultural production, survival strategies and settlement distribution over time; (2) to revitalize the agriculture in the fluvial oases; (3) and to valorize from the touristic point of view this important cultural heritage.

References

Adams REW, Jones RC (1981) Spatial patterns and regional growth among classic Maya cities. Am Antiquity 46(2):301

Brown CT, Witschey WRT (2003) The fractal geometry of ancient Maya settlement. J Archaeol Sci 30:1619–1632

Ceccato P, Flasse S, Tarantola S, Jacquemoud S, Gregoire JM (2001) Detecting vegetation leaf water content using reflectance in the optical domain. Remote Sens Environ 77(1):22–33

Ceccato P, Flasse S, Gregoire J-M (2002a) Designing a spectral index to estimate vegetation water content from remote sensing data: Part 2: validation and applications. Remote Sens Environ 82 (2–3):198–207

Ceccato P, Gobron N, Flasse S, Pinty B, Tarantola S (2002b) Designing a spectral index to estimate vegetation water content from remote sensing data: Part 1: theoretical approach. Remote Sens Environ 82(2–3):188–197

Chen CM, Hepner GF, Forster RR (2003) Fusion of hyperspectral and radar data using the IHS transformation to enhance urban surface features. J Photogramm Remote Sens 58:19–30

Diamond J (2009) Maya, Khmer and Inca. Nature 461:479–480

Diamond J (2010) Two view of Collapse. Nature 463:880–881

Dillehay TD, Eling HH, Rossen J (2005) Preceramic irrigation canals in the Peruvian Andes. Proc Natl Acad Sci USA 102(47):17241–17244. doi:10.1073/pnas.0508583102, 2005 November 22

Folan WJ, Faust B, Lutz W, Gunn JD (2000) Social and environmental factors in the classic Maya collapse. In: Lutz W, Prieto L, Sanderson W (eds) Population, development, and environment on the Yucatan Peninsula: from ancient Maya to 2030. Research report RR-00–14, IIASA, Laxenburg

Fourty T, Baret F (1998) On spectral estimates of fresh leaf biochemistry. Int J Remote Sens 19 (7):1283–1297

Gonzalez Garcia MF (1934) Los acueductos incaicos de Nazca. Aguas e Irrigación 2(2):207–222

Jackson TJ, Chen D, Cosh M et al (2004) Vegetation water content mapping using Landsat data derived normalized difference water index for corn and soybean. Remote Sens Environ 92:475–482

Kauth RJ, Thomas GS (1976) The Tasseled Cap – a graphical description of the spectral-temporal development of agricultural crops as seen by Landsat. In: Proceedings of the symposium on machine processing of remotely sensed data, Purdue University, West Lafayette, pp 4B41–4B51

Kobori I (1960) Human geography of methods of irrigation in the Central Andes. In: Andesu, Andes, the report of the University of Tokyo scientific expedition to the Andes in 1958, Bijitsu Shuppan sha, Tokyo, pp 74–97, 417–420

Kroeber A, Collier D (1998) The archaeology and pottery of Nazca, Peru: Alfred L. Kroeber's 1926 expedition. Altamira Press, Walnut Creek

Lancho Rojas J (1986) Descripción y problemas de mantenimiento y rehabilitación de los acueductos de Nasca. Informe entregado al CONCYTEC (Consejo Nacional de Ciencia y Tecnología)

Mejia Xesspe T (1942) Acueductos y Caminos Antiguos de la Hoya del Río Grande de Nasca. In: Actas y Trabajos científicos del XXVII Congreso Internacional de Americanistas, Lima. Librería e Imprenta GIL, Lima, vol 1, pp 559–569

Orefici G, Drusini A (2003) Nasca: hipótesis y evidencias de su desarrollo cultural. Documentos e Investigaciones 2. Ediciones CISRAP, Brescia

Peñuelas J, Inoue Y (1999) Reflectance indices indicative of changes in water and pigment contents of peanut and weat leaves. Photosynthetica 36(3):355–360

Peñuelas J, Filella I, Biel C, Serrano L, Savé R (1993) The reflectance at the 950–970 nm region as an indicator of plant water status. Int J Remote Sens 14(10):1887–1905

Ragab R (1995) Towards a continuous operational system to estimate the root-zone soil moisture from intermittent remotely sensed surface moisture. J Hydrol 173:1–25

Ripple WJ (1986) Spectral reflectance relationships to leaf water stress. Photogramm Eng Rem S 52(10):1669–1675
Roberts DA, Green RO, Adams JB (1997) Temporal and spatial patterns in vegetation and atmospheric properties form AVIRIS. Remote Sens Environ 62:223–240
Rogers AS, Kearney MS (2004) Reducing signature variability in unmixing coastal marsh Thematic Mapper scenes using spectral indices. Int J Remote Sens 25(12): 2317–2335
Rouse JW, Haas RH, Schell JA, Deering DW (1973) Monitoring vegetation systems in the Great Plains with ERTS. In: Third ERTS symposium, vol I, NASA SP-351, pp 309–317
Schreiber KH (1995) The puquios of Nasca. Latin Am Antiquity 6(3):229–254
Schreiber KH (2003) Irrigation and society in the Peruvian Desert: the puquios of Nasca. Lexington Books, Lanham
Schreiber KH, Lancho Rojas J (1988) Los puquios de Nasca: un sistema de galerías filtrantes. Boletín de Lima 59:51–62, Editorial Los Pinos, Lima
Schreiber KH, Lancho Rojas J (2009) El control del agua y los puquios de Nasca. In: Nasca. El desierto del los Dioses de Cahuachi. Graph, Lima, pp 132–151
Solar La Cruz F (1997) Nasca filtering galleries; galerías filtrantes. Universidad Abraham Valdelomar, Lima
Tucker CJ (1979) Red and photographic infrared linear combinations for monitoring vegetation. Remote Sens Environ 8:127–150
Tucker CJ (1980) Remote sensing of leaf water content in the near infrared. Remote Sens Environ 10:23–32
Ustin SL, Roberts DA, Jacquemoud S, Pinzon J, Gardner M, Scheer GJ, Castaneda CM, Palacios A (1998) Estimating canopy water content of chaparral shrubs using optical methods. Remote Sens Environ 65:280–291
Vaughn S, Crawford T (2009) A predictive model of archaeological potential: an example from northwestern Belize. Appl Geogr 29(4):542–555
Veni G (1990) Maya utilization of karst groundwater resources. Environ Geol 16(1):63–66. doi:10.1007/BF01702224
Walker JP, Willgoose GR, Kalma JD (2001) One-dimensional soil moisture profile retrieval by assimilation of near-surface observations: a comparison of retrieval algorithms. Adv Water Res 24:631–650
Yu GR, Miwa T, Nakayama K, Matsuoka N, Kon H (2000) A proposal for universal formulas for estimating leaf water status of herbaceous and woody plants based on spectral reflectance properties. Plant Soil 227(1–2):47–58
Zarco-Tejada PJ, Rueda CA, Ustin SL (2003) Water content estimation in vegetation with MODIS reflectance data and model inversion methods. Remote Sens Environ 85:109–124

Chapter 13
High-Resolution Satellite Imagery and the Detection of Buried Archaeological Features in Ploughed Landscapes

Ioana A. Oltean and Lauren L. Abell

Abstract Traditionally, archaeological aerial surveyors fly routes selected on the probability of site visibility and record only what they perceive as significant. This approach to aerial reconnaissance has come under criticism in recent years with respect to productivity, bias and cost. Through panchromatic imagery with sub-metre resolution and high-resolution multispectral products, modern commercial satellites are capable of recording fine details of both extant sites and archaeological cropmarks. They are not limited by fuel costs, airspace laws, pilot time or conflicting regions of interest. In addition, the multiple wavelength images and powerful image analysis software available offer the facility to manipulate images in order to maximise the visibility of surface details and to detect any variations in the non-visible spectrum. The present paper employs a comparative analysis of above-ground remotely-sensed data recovered on various dates and with various methods of acquisition (QuickBird and Corona satellite imagery; WWII vertical aerial photography and recent oblique aerial photographs), in order to discuss the potential of satellite imagery to reveal buried archaeological features in the area of southern Dobrogea (Romania) and to assist in the reconstruction of past landscapes. It outlines some of the methodological approaches employed, highlights areas for further research and presents some preliminary outcomes.

Keywords Landscape archaeology • Aerial reconnaissance • Satellite imagery • QuickBird • Corona • Cropmarks • Romania

---

I.A. Oltean (✉)
Department of Archaeology, University of Exeter, Exeter EX4 4QE, UK
e-mail: i.a.oltean@exeter.ac.uk

L.L. Abell
Remote Sensing Center, Naval Postgraduate School, Monterey, CA 93943, USA
e-mail: llabell@nps.edu

R. Lasaponara and N. Masini (eds.), *Satellite Remote Sensing: A New Tool for Archaeology*, Remote Sensing and Digital Image Processing 16,
DOI 10.1007/978-90-481-8801-7_13,

13.1 Introduction

Throughout much of the twentieth century aerial archaeology has greatly enriched our knowledge of past human activity, as sites and landscapes across Europe and beyond have been revealed from the air. In countries such as Britain, since its implementation over 60 years ago, sites discovered through aerial photographic reconnaissance have increased the number of known archaeological sites tenfold. Some of the sites visible from the air are extant and are easily detectable on the ground and from the air. However, the success of aerial archaeology is most evident in ploughed landscapes, where archaeological features are buried under modern cultivation. In these areas archaeological remains are not visible from the ground and have therefore been overlooked by early surveys; nevertheless, they can be revealed by variations in vegetation development, reflecting variations in the availability of soil moisture and nutrients to crops, which produce recognizable patterns of differential growth known as cropmarks (e.g. Wilson 2000, 67–86; Hanson 2008). Considerable sums are invested every year in aerial reconnaissance, a method of data collection which utilises primarily light aircraft as aerial platforms to take oblique photographs of known or newly identified archaeological sites. In the United Kingdom, aerial reconnaissance amounted to some 450 h of flying *per annum*, at a cost of £87,500 in aircraft hire alone when the data was last collated a decade ago (British Academy 2001, 13).

Unfortunately, such a situation is not comfortably paralleled elsewhere. By 2000, for example, in most European countries some archaeological aerial reconnaissance was taking place annually. However, the extent, funding and usage (e.g. academic research, heritage protection and management) varied significantly from country to country, with none of them reaching a level comparable to that in Britain (Bewley and Raczkowski 2001). Moreover, there are still European countries such as Greece, Bulgaria and Turkey, where for legal or other reasons no such archaeological aerial reconnaissance is being undertaken. An additional problem is that funding for such activities is not always constant. In Romania, where aerial reconnaissance was discouraged until well after the fall of the communist regime, for a decade after its beginning in 1998, aerial reconnaissance remained confined to the resources offered by funding bodies such as the British Academy and the Leverhulme Trust to British-based research projects in parts of Transylvania and Dobrogea (e.g. Hanson and Oltean 2001; Oltean and Hanson 2007b). Accordingly, given the susceptibility of this method of data acquisition to annual variations in climate and crop cultivation, gaps in data collection due to lack of funding over one or several seasons are bound to create significant bias in the resulting reconstruction of the archaeological landscape. Finally, despite its importance for the understanding and management of archaeological landscapes, few countries use aerial photographs for archaeological mapping, with only England having a coherent national programme in place (British Academy 2001, 16).

Traditional aerial reconnaissance methodology presents archaeologists with a number of advantages. Given the size of archaeological contexts and sites, there is always a limitation to what can be detected from the air compared to ground-based remote sensing and, indeed, excavation. From a low height of 500 m it is possible to achieve a significantly better ground resolution compared to the products of medium or high altitude coverage. In addition, aerial reconnaissance offers the opportunity to photograph sites from various viewing angles and, as cropmark formation is heavily dependent on climatic conditions, cultivation patterns and soil types, to schedule data acquisition more easily at the best possible time in crop development to reveal cropmarks. Detection of features by other means (e.g. soilmarks, thaw/frost marks, etc. – Wilson 2000, 43–67; Hanson 2007) is possible, but tends to be less preferred – and equally reliant on climatic and geological conditions. Unfortunately this means that most reconnaissance flights are seasonally restricted. Moreover, given the funding limitations, there is a tendency for flights to focus on areas most likely to produce good results at the expense of large areas with heavier soils and/or wetter climates, supposedly less favourable conditions for data recording (see various contributions in Mills and Palmer 2007). Unfortunately, this generates an inbuilt bias in data recovery (e.g. Hanson 2005), dangerously easily transferred into wider archaeological knowledge and becoming the basis of archaeological and historical interpretations. It is therefore unsurprising that the serendipitous and subjective coverage of wider landscapes has been criticised repeatedly, particularly in the recent years, as not being the most productive and cost-effective method (e.g. Palmer 2005).

Arguments in favour of block vertical aerial photographic coverage, normally used for the purposes of photogrammetry, have usually been dismissed by most aerial archaeology practitioners as impractical, with an unproven or poor cost-benefit ratio. The debate, however, has largely overlooked the significant potential contribution of satellite imagery. Although satellite remotely-sensed image data has been employed by archaeologists over the past decades (most notably LANDSAT and CORONA products; e.g. Hamlin 1977; Fowler and Fowler 2005) its wider use has been confined by issues of ground resolution, cloud coverage, access to and cost of data, and availability of processing skills to studies of ancient environments. There has been less use for site examination, and that predominantly in areas where for various reasons aerial photographs were not accessible (Parcak 2009, 13–31).

However, satellite imagery is becoming increasingly accessible to specialists and the general public alike. A cost reduction trend is already apparent, indicating increased ease of purchase over time. For the detection of buried archaeological remains the traditional advantages associated with observer-directed aerial data collection listed above are increasingly becoming less unique, with only the ground resolution remaining superior, and the unique advantages that satellite data are offering has already greatly improved their cost-benefit ratio. The currently available commercial satellite imagery offers significantly improved ground resolution (the highest is available for panchromatic datasets at 0.5 m), off-nadir viewing options, and increasing availability of data acquisition over time for continuous global site tasking (De Laet et al. 2007b, 831–2). In addition, image data is easier to handle and process with current digital technology, but negotiation of its meaning in order to allow wider integration within archaeological knowledge cannot yet be divorced completely from subjectivity (Hanson 2008). Automated pattern detection data processing by means of mathematical algorithms and statistical analyses is a technology still very much in progress and does not yet always allow confident interpretation of archaeological features, but promises to do so in the future at least in relation to morphologically uncomplicated, fairly standardised structures (see Aurdal et al. 2006). However, the visual recognition of potentially significant patterns and employment of photographic interpretation methods may allow for more detailed site plan reconstruction (Lasaponara and Masini 2006; Lasaponara and Masini 2007; De Laet et al. 2007a).

13.2 Study Area and Previous Research

The sites discussed in this paper are located in Southern Dobrogea (Romania) not far from the western coast of the Black Sea and the Bulgarian border (Fig. 13.1). The first site discussed here (site A – UTM-WGS84 35N 625000E/853400N) is positioned immediately to the west of the modern town of Mangalia; site B (UTM-WGS84 35N 622247E/4848137N) is in the middle of the cultivated plateau

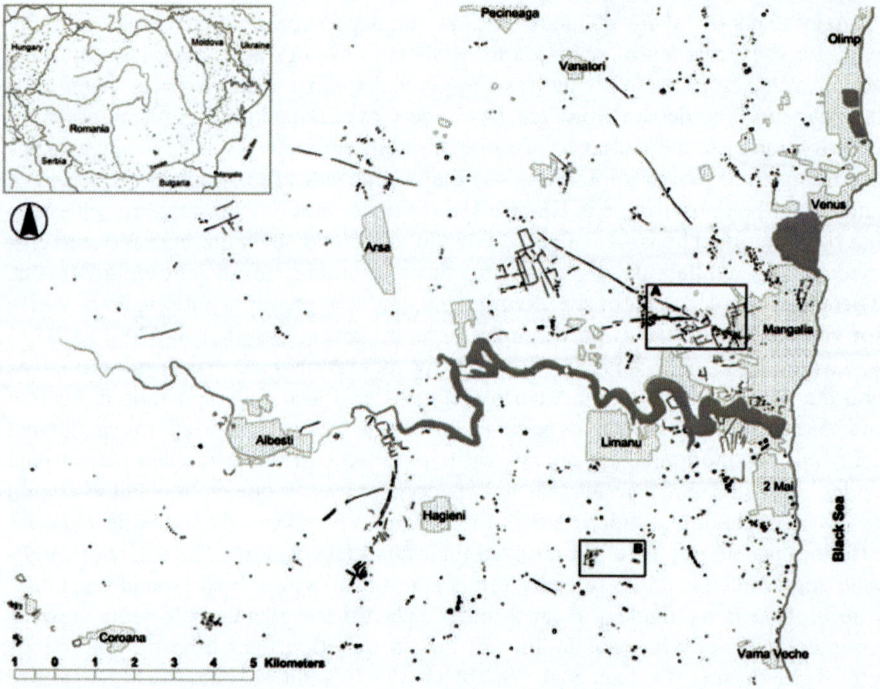

Fig. 13.1 Areas of *sites A* and *B* within the archaeological landscape around the modern town of Mangalia (Romania)

to the south of Mangalia, some distance from all the modern villages in the area (Hagieni, Limanu and Vama Veche).

Given the local climate, geology and cultivation pattern, the conditions for the formation of cropmarks are very favourable in this region. Dobrogea is Romania's warmest and driest region, with predominantly intense and short-lived rainfall, much of it in the spring, and occasional hot dry winds emanating from the Russian steppe. Furthermore, the geology allows rapid drainage of moisture. The soil from both areas is levigated chernozem. In comparison with site B, which is located on loess at some distance from the Black Sea or any other water bodies, Site A may benefit from more humidity through the presence of clays in combination with limestones and sands or through the proximity of a silted up estuary (Oltean and Hanson 2007a).

Nevertheless, given the local geology and climate, moisture stress can be a natural occurrence in the area, particularly around site B. The modern landscape is predominantly given over to arable cultivation focused mostly on cereals, with pastureland occurring where water shortage restricts cultivation. Based on these facts the peak period in crop development for revealing cropmarks should be throughout June and very early in July. However, excesses of precipitation over the spring months combined with the natural conditions may produce local variations: excessive rainfall may delay the cropmark formation by a few weeks, but drought could prevent germination altogether.

Southern Dobrogea has long-lived evidence of human occupation since the Neolithic. An area of constant ebb and flow of populations and cultural interference between the classical Mediterranean and the prehistoric Eurasian cultures, Dobrogea experienced early exposure to the Greek colonisation of the Black Sea shores and subsequent Roman occupation, which eventually dissolved into Byzantine, then Ottoman imperialism. A wide array of man-made features can be found in the archaeological landscape, from those of larger dimensions such as settlement enclosures, roads, large funerary barrows dating from at least the sixth century BC to the third century AD, to smaller features including small sunken-floored structures or buildings utilising stone walls of sub-metre widths in unenclosed settlements.

Archaeological preservation in the area until the twentieth century has been fairly good. Although Mangalia has been continuously occupied since its foundation as the Greek colony of Kallatis in the sixth century BC, ruins have been traditionally subjected only to occasional robbery and accreted settlement. More recently, the intensification of arable agriculture and the expansion of urban areas and infrastructure development present clear dangers to archaeological preservation. In both cases discussed here, ploughing erosion is currently the main danger, but given its location on the fringes of the modern town of Mangalia, which has seen significant urban expansion throughout the twentieth century, site A may soon be engulfed by development.

The wider area has been the focus of research by the first named author during 2004–2007 (*Contextualizing change on the Lower Danube: Roman impact on Daco-Getic landscapes*), funded by the British Academy, which aimed to

Fig. 13.2 Burial mounds and ancient roads west of Mangalia (*site A*) as visible on Corona imagery DZB00402700026H018001, 21st April 1966 (USGS); North to the right

evaluate the nature and impact of Roman conquest and colonisation on the transformation of native settlement patterns. It involved aerial reconnaissance (June 2005 and 2006) in conjunction with the examination of WWII aerial vertical photographs (April–May 1944) and Cold War Corona satellite imagery (ID: DZB00402700026H018001; KH-7 2–4 ft High Resolution Surveillance Camera; 21st April 1966 – Oltean and Hanson 2007b; Fig. 13.2). The methodology employed at the time was successful in identifying a large number of archaeological remains in the area, most of them now buried, and enabled the reconstruction of large parts of the archaeological landscape (see Fig. 13.1). However, several issues were also raised. For example, dealing with an area where aerial reconnaissance had never been applied before meant that estimation of the right time in terms of crop development was constrained from the beginning, and further hindered by the damp seasons in both 2005 and 2006 (in contrast to 2004). As a result, cropmarks were still at their early stage of formation at the time of aerial reconnaissance and, although larger structures (e.g. funerary barrows; enclosures) were recognisable, fewer than expected stone-built open settlements characteristic of the Hellenistic and Roman periods have been identified. Moreover, the acquisition season of the historical aerial and satellite imagery consulted makes them very compatible both in terms of visibility of archaeological remains (as soilmarks) and the types of features recoverable (mostly large, solid buried remains such as burial mounds and ancient roads), but unfortunately makes them less useful in revealing those types of sites severely under-represented in the aerial reconnaissance (e.g. open settlements).

The sites selected for a closer analysis in this study are located within areas covered by these various datasets, which allows for easy comparison with previous results obtained through alternative methodologies. The scale and nature of the archaeological remains detected at sites A and B is also contrasting. In the area of site A, several large archaeological features have previously been located and photographed from the air in 2005 and 2006, though they were also visible on the earlier image data sets from 1944 and 1966. They include funerary barrows and roads belonging to the very large area of burials buffering the ancient town of Kallatis, with a number of roads connecting the town with the surrounding territory best visible as soilmarks/germination marks (Figs. 13.2 and 13.3).

Aerial reconnaissance over 2005 and 2006 recorded cropmarks in the area of site B indicating the presence of buried remains of a stone-built settlement, the only example noted in the area under survey (Fig. 13.4). However, apart from the methodological comparison intended in this study, this offered the opportunity to analyse further several aspects of interpretation needing further attention for both sites. On site A for example, no surrounding ditch and/or traces of burial chambers or robbing trenches have yet been detected, although other barrows in the vicinity revealed such details, and a series of smaller circular features visible as cropmarks in 2005 were missing from the earlier, historical imagery. Also, a further cluster of buried archaeological structures with morphological similarities identified in 2006 some 700 m to the East of site B indicates the strong possibility that the settlement was much larger than was originally suspected (Oltean and Hanson 2007b, 674906-5).

Fig. 13.3 Oblique aerial photograph showing the area of *site A* (seen from NNE) made during aerial reconnaissance in June 2005 indicating the presence of large archaeological objects such as burial mounds and ancient roads (WS Hanson and IA Oltean)

Fig. 13.4 Oblique aerial photograph of *site B* (seen from the W) made during aerial reconnaissance in June 2006 indicating the presence of buried remains of stone buildings of an ancient open settlement (WS Hanson and IA Oltean)

13.3 Data Processing

In the area under study here, the initial data sets from QuickBird (DigitalGlobe) included two collects from the same season (2004), one from 25 June (catalogue ID 10100100030CB803) and the second on 18 July (catalogue ID 1010010031E5101). For traditional aerial archaeology, the date of the June collection window would be estimated to be ideal for aerial reconnaissance targeting cropmarks, while the July date, when most cereal crops are already being harvested, would normally be estimated as slightly too late in the season. The spectral characteristics and apparent texture indicate that the vegetation covering the main fields under detailed scrutiny here was consistent for both sites (probably corn/maize), if far from ideal for good cropmark conditions. Both collection windows included multispectral (four bands: blue 479.5, green 546.5, red 654 and near-infrared 814.5) and panchromatic imaging, enabling us to manipulate these datasets and analyse them in more detail in order to explore their potential for revealing buried features of archaeological significance and compare them with results from the oblique and vertical aerial photographs and the 1966 satellite imagery.

The processing and analysis of the data is at its beginning, with the immediate objective of selecting the most relevant methods of artificially enhancing feature detection within the multispectral signature, particularly for cropmarks. The results from both datasets have been analysed in parallel, by utilising both traditional aerial photo-interpretation and standard remote sensing approaches.

All images were superficially scanned visually using ENVI 4.3 software. On site A, the RGB "true colour" multispectral images were subjected to a linear 2% enhancement to increase contrast. In the second scan, Band 4 (IR) was run in the green. Several areas were revealed as having a higher concentration of object figures and lines, noted as contrasting vegetation marks. This secondary processing included a variety of band combinations to ascertain highest contrast and field depth. Band 4 in the red and Band 2 in both the blue and green, with a sharpening mask and linear 2% stretch, initially proved to highlight the features of interest most consistently. A contour line filter was further applied to distinguish real surface variation from colour features. However, the dense background clutter (plough marks) could not be reliably extracted and the resulting image was difficult to interpret and potentially misleading.

A Normalized Difference Vegetation Index (NDVI) analysis was run to highlight real vegetation differences and determine which feature patterns did not register (Fig. 13.5, upper). Further analyses were run with ENVI 4.5 and ENVI Zoom 4.5 feature extraction processing tool. An attempt was made on site A to cluster and segment the features visible in the RGB image, but plough mark interference (as with the contour lines) made segmentation and merging nearly impossible. Anomaly detection performed better, but provided no new information. On site B however, archaeological features were too obscured to easily use the feature extraction tool, requiring manual segmentation of visible objects.

Processed images of site A were run through the feature extractor also with little success, still apparently due to clutter. A scatterplot analysis helped to determine the boundaries of apparent features. Two very clear features of large dimensions appeared on site A, but no distinct features were apparent on site B, further confirming the visible contrast.

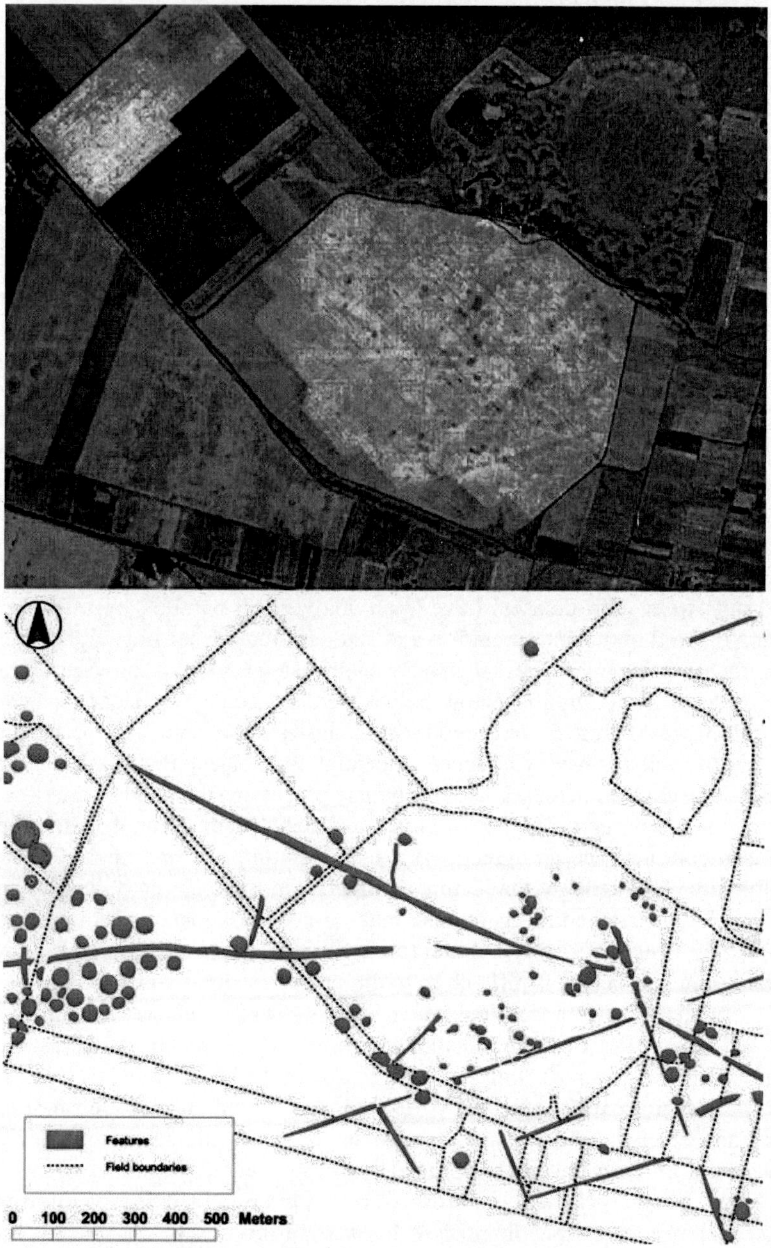

Fig. 13.5 Normalized Difference Vegetation Index (NDVI) analysis enhanced by Gaussian stretch (*upper*) and a cumulative plot of features with archaeological potential (*lower*) detected on site A based on the visual interpretation of all available datasets

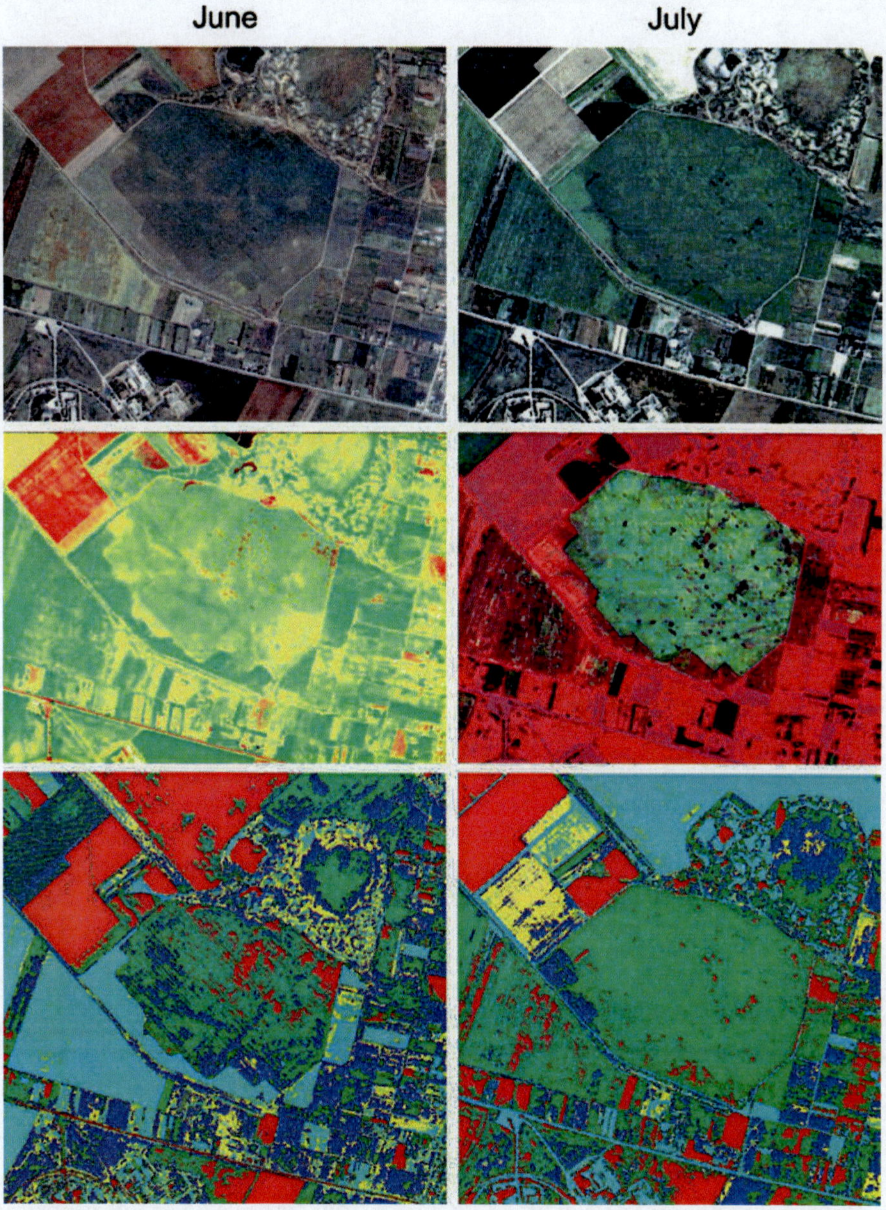

Fig. 13.6 Multispectral data processing on *site A* in June and July: vegetation suppression (*above*); false colouring interactive stretch in *yellow* (*middle*); and K-means tercat (*below*)

Several further processing techniques using standard ENVI internal algorithms were applied with extremely consistent results on both sites: k-means classification (Richards and Jia 2005), vegetation suppression and colour mapping (Fig. 13.6). NDVI with a Gaussian stretch enhancement provided the best grayscale contrast (see Fig. 13.5, upper). Individual band histogram adjustment (interactive stretching) provided the best small feature distinctions as well as a depth of field showing differentiation of previously-identified homogenous features – mostly linear.

Given that the clear discrepancy of results between sites A and B is likely to be connected to the ground resolution of the multispectral datasets, data fusion has been employed to artificially increase the ground resolution of the processed multispectral data. Efforts at data fusion to improve ground resolution had varied results. Panchromatic images and multi-spectral images were warped together using ENVI image-to-image registration or "warping" to create hybrid images. Typically image to image registration is done with data sets from different sensor vehicles to match geographic features and orient or register the images. Our approach was intended to increase the resolution by combining the higher resolution panchromatic data with multispectral data. The resampling method with the sharpest results was the "nearest neighbor" method. Bilinear interpolation and cubic convolution both reduced resolution significantly. Variations were run using Rotation, Scaling and Translation (RST) warps and 1st degree Polynomial warps. Ground control points ranged from 6 to 25.

13.4 Discussion

For independent analysis based exclusively on this methodology, the results produced by the processing methods discussed above are fairly limited and, especially in the case of sites B and C, insufficient for any confident selection of features. Moreover, the occasional interference from plough marks or other disturbances prevents for the time being any automatic feature extraction. Data fusion aided interpretation of features based on visual pattern recognition, but was still insufficient for developing automated feature extraction.

However, when compared with data provided by aerial reconnaissance and the Corona satellite, it is possible to identify several archaeological features in the QuickBird imagery and even to refine and complete the site plans reconstructed previously. Although in the area of site A the panchromatic dataset from 25th June was not sufficiently contrasting to enable identification of cropmarks, the multispectral set was able to reveal the presence of differential vegetation growth signalling larger buried features such as Roman roads, ancient funerary barrows and field boundaries previously noted on aerial photographs and Corona imagery. These were enhanced by both the application of the vegetation suppression algorithm and the NDVI. Given the nature of the crop (corn/maize) and its characteristic stages of development, sufficient contrast in vegetation suppression was not maintained in July. However, the NDVI algorithm was still able to reveal a further set possible features of smaller dimensions which are consistent with a set of possible features revealed by oblique aerial photographs 1 year after in 2005, though not by Corona images of the site in 1966 (Figs. 13.3 and 13.5). Nevertheless, additional aerial or ground-based survey will be needed in order to fully clarify their nature.

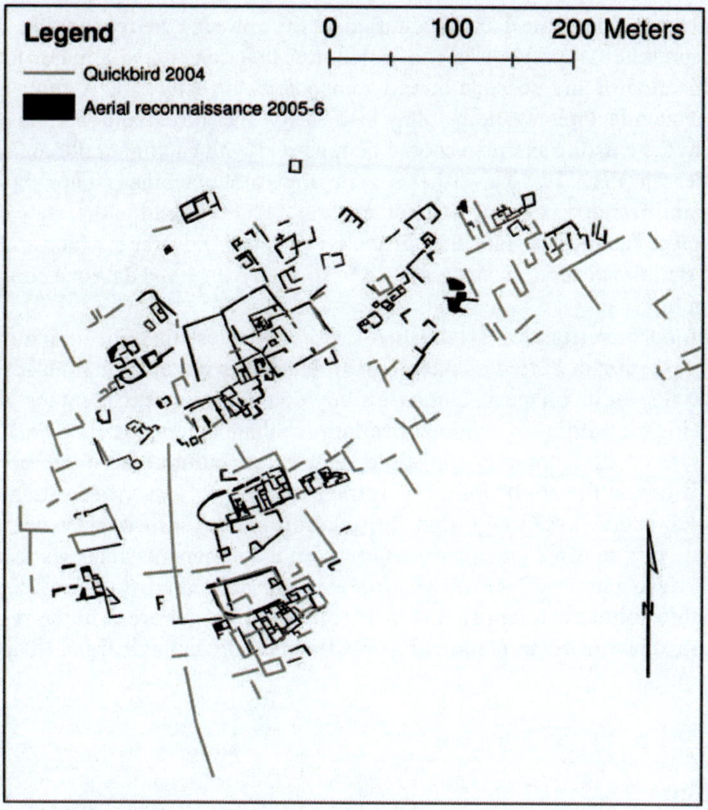

Fig. 13.7 Cumulative interpretative plot of buried archaeological features on *site B*: recovered from aerial photographs of 2005 and 2006 and from QuickBird imagery of 2004

Interpretation of the panchromatic imagery available for area B was difficult, given that both were also covered at the time of data acquisition by vegetation not conducive to cropmark formation, and the multispectral data sets produced more limited results in terms of independent recognition of buried archaeological features (probably because of the nature of the site, consisting of buried remains of ancient buildings with sub-metre stone walls). However, comparison of the k-means classification and of the colour mapping applied to the multispectral dataset with the set of features already mapped at site B from aerial photographs enabled the recognition of a few of these features (Fig. 13.7). Of particular importance was the observation that potential features spread outside the areas previously mapped, which lends further weight to the possibility that site B extends beyond the limits to which its site plan has already been reconstructed.

In general, visual pattern recognition of potential features provided more confident and consistent interpretations in both case studies presented here. However, if at site A the ground resolution of the datasets is sufficient to secure feature recognition, at site B it was the presence of previously-known features that constituted a major factor for the identification of any possible buried archaeological features. Unfortunately, this means that such features would be otherwise easily overlooked and that the possible additions to the existing site plan should be regarded with caution until confirmed by further research (Fig. 13.7). Nevertheless, one important outcome is that in both areas of sites A and B, the processing of multispectral data revealed the presence of some archaeological features despite the difficult vegetation coverage which may have prevented traditional aerial reconnaissance from delivering good data at a comparable acquisition time.

Overall, our investigation established that even in less than ideal conditions of visibility, detection of buried archaeological features is possible on high-resolution multispectral satellite imagery. Our case study could provide excellent test datasets for improving algorithms of artificial resolution enhancement based on deconstruction models, or of automated object detection on cultural heritage sites. Our intention is to test the applicability of further methods of analysis for this data or similar, such as the VARI algorithm (vegetation index sensitive to water content and indicative of relative greenness) and principle components analyses to attempt to filter surface clutter. Once these problems are addressed, we aim to attempt further viable solutions to apply this methodology to larger areas in the region, in order to assist reconstruction and analysis of wider archaeological landscapes.

References

Aurdal L, Eikvil L, Koren H, Loska A (2006) Semi-automatic search for cultural heritage sites in satellite images. In: Campana S, Forte M (eds) From space to place. Proceedings of the 2nd international conference on remote sensing in archaeology/2nd international workshop, CNR, Rome, 4–7 Dec 2006, BAR Int Ser 1568, Oxford, pp 1–6

Bewley RH, Raczkowski W (eds) (2001) Aerial archaeology – developing future practice. NATO Science Series, Amsterdam

British Academy (2001) Aerial survey for archaeology, Report of a British Academy working party 1999, London

De Laet V, Paulissen E, Meuleman K, Waelkens M (2007a) The effect of pixel resolution and spectral characteristics on the extraction of archaeological features from very high-resolution remote sensing imagery: Sagalassos, southwest Turkey. In: Ehlers M and Michel U (eds), 6749 0A

De Laet V, Paulissen E, Waelkens M (2007b) Methods for the extraction of archaeological features from very high-resolution Ikonos-2 remote sensing imagery, Hisar (southwest Turkey). J Archaeol Sci 34:830–841

Ehlers M, Michel U (eds) (2007) Remote sensing for environmental monitoring, GIS applications, and geology VII, 17–20 Sept 2007, Florence. Proceedings of SPIE no. 6749, Bellingham

Fowler MJF, Fowler YM (2005) Detection of archaeological crop marks on declassified CORONA KH-4B intelligence satellite photography of southern England. Archaeol Prospect 12:257–264

Hamlin CL (1977) Machine processing of LANDSAT data: an introduction for anthropologists and archaeologists. MASCA Newsl 13:1–11

Hanson WS (2005) Sun, sand and see: creating bias in the archaeological record. In: Brophy K, Cowley D (eds) From the air: understanding aerial archaeology. Tempus, Stroud, pp 73–85

Hanson WS (2007) Site discovery: remote sensing approaches, aerial. In: Pearsall DM (ed) Encyclopedia of archaeology. Academic, New York/London, pp 1907–1912

Hanson WS (2008) The future of aerial archaeology in Europe (or Are algorithms the answer?). In: Lasaponara R, Masini N (eds) Advances in remote sensing for archaeology and cultural heritage management. Aracne, Rome, pp 47–50

Hanson WS, Oltean IA (2001) Recent aerial survey in Western Transylvania: problems and potential. In: Raczkowski W, Bewley RH (eds) Aerial archaeology. Developing future practice. IOS Press, Amsterdam, pp 109–115, and 353–355

Lasaponara R, Masini N (2006) On the potential of QuickBird data for archaeological prospection. Int J Remote Sens 27(16/20):3607–3614

Lasaponara R, Masini N (2007) Detection of archaeological crop marks by using satellite QuickBird multispectral imagery. J Archaeol Sci 34:214–221

Mills J, Palmer R (eds) (2007) Populating clay landscapes. Tempus, Stroud

Oltean IA, Hanson WS (2007a) Cropmark formation on “difficult” soils in Romania. In: Palmer R, Mills J (eds) Populating clay landscapes. Tempus, Stroud, pp 73–87

Oltean IA, Hanson WS (2007b) Reconstructing the archaeological landscape of Southern Dobrogea: integrating imagery. In: Ehlers M and Michel U (eds) Remote sensing for environmental monitoring, GIS applications, and geology VII, 17–20 Sept 2007, Florence. Proceedings of SPIE, 6749. SPIE, Bellingham, p 674906

Palmer R (2005) If they used their own photographs they wouldn’t take them like that. In: Brophy K, Cowley D (eds) From the air. Understanding Aerial Archaeology, Stroud, pp 94–116

Parcak SH (2009) Satellite remote sensing for archaeology. Routledge, London/New York

Richards J, Jia X (2005) Remote sensing digital image analysis: an introduction, 4th edn. Springer, Berlin

Wilson DR (2000) Air photo interpretation for archaeologists. Tempus, Stroud

Chapter 14
Integrated Remote Sensing Approach in Cahuachi (Peru): Studies and Results of the ITACA Mission (2007–2010)

Nicola Masini, Rosa Lasaponara, Enzo Rizzo, and Giuseppe Orefici

Abstract ITACA (Italian heritage Conservation and Archaeo-geophysics) is an international mission of the Italian CNR which applies different scientific methodologies, strongly based on the use of Earth Observation, to contribute to the study of the precolombine archaeology and the cultural heritage conservation and management in Peru and Bolivia. From 2007 up to today the research activity of the ITACA, funded by the Italian Foreign Ministry Affairs, has been focused on the Nasca Ceremonial Centre of Cahuachi, the Nasca geoglyphs (both of them in Southern Peru), the Ceremonial center of Tiwanaku (Bolivia) and the archaeological sites of Arenal and Ventarron in the Lambayeque region (Northern of Peru). Most of the scientific investigations have been carried out in Cahuachi and in the drainage basin of the Rio Nasca, with the aim of supporting archaeological studies and excavations of the *Centro de Estudios Arqueológicos Precolombinos* directed by Giuseppe Orefici. The main activity has been the archaeo-geophysics based on the integration of ground, aerial and satellite remote sensing methods, thus allowing the archaeologists to find buried walls, tombs and ceremonial offerings in Cahuachi and to discover a large buried settlement in the Nasca riverbed.

N. Masini (✉)
Institute of Archaeological and Architectural Heritage, CNR-IBAM,
C. da S. Loya, 85050 Tito Scalo, PZ, Italy
e-mail: n.masini@ibam.cnr.it

R. Lasaponara • E. Rizzo
Institute of Methodologies for Environmental Analysis, CNR-IMAA,
C. da S. Loya, 85100 Tito Scalo, PZ, Italy
e-mail: lasaponara@imaa.cnr.it

G. Orefici
Centro de Estudios Arqueológicos Precolombinos,
Avenida de la Cultura, 600 (Bisambra), Nasca, Peru
e-mail: cahuachi@terra.com.pe

R. Lasaponara and N. Masini (eds.), *Satellite Remote Sensing: A New Tool for Archaeology*, Remote Sensing and Digital Image Processing 16,
DOI 10.1007/978-90-481-8801-7_14,

Keywords Satellite remote sensing • Archaeo-geophysics • Georadar • Magnetometry • Nasca • Cahuachi • Peru

14.1 Introduction: Study Area, State of Art and Motivation of Itaca Mission Investigations in Cahuachi

Cahuachi has many reasons to be considered one of the most important cultural resources of South America, such as: (i) to be the largest adobe ceremonial center built in the world, (ii) the monumentality of its architecture and the complexity of its urban fabric, (iii) to have been the cultural, religious and administrative core of the Pampa region of Peru during the Nasca age, in particular from the first to the fourth century AD.

The Nasca civilization is well-known for its refined and colourful pottery, characterized by a rich iconographic repertory, and, above all, for the huge geoglyphs drawn on the arid plateaus of the Rio Grande de Nasca Basin, characterized by different patterns and shapes from geometrical to biomorphic ones.

The mystery which has been surrounding the Nasca geoglyphs, since its discovery in the 1920s, has favoured the postulation of several different hypotheses on their cultural meaning and function and the way to have been drawn.

Among them (by omitting the imaginative theories related to landing sites for aliens !), for sake of brevity we cite only the studies of Paul Kosok and Maria Reiche (Kosok and Reiche 1949) who proposed an astronomical meaning to the lines, that is a sort of astronomy book which possibly enabled the Nasca to predict when to plant and harvest their crops, the hypotheses by Aveni which found a similarity of some lines with the Incas *ceque* system (Aveni 1986), the theories of Johan Reinhard (1996), according to which the lines were "sacred pathways to a place from which the people worshiped the mountains as the source of water, and invoked the mountain gods", and finally other studies which point out the function of the lines as signal or maps of subterranean aqueducts and wells.

Whichever hypothesis increases the interest for the Nasca civilization and so, for Cahuachi which was the theocratic capital of the Nasca region, up to its abandonment due to natural disasters (flooding and earthquakes) in the fifth century AD. The pyramids of Cahuachi and the geoglyphs are linked to each other by a common vision of the religion, the nature and the relationship between the human beings and Gods, by means of ritual activity, thus generating an extensive zone of ideological influence, not only on the coastal areas but also in the upper valleys and in the Andean vicinity (Orefici 2009b, c).

14.1.1 Geographical and Historical Setting of the Drainage Basin of Nasca: Brief Notes

The archaeological evidence of Cahuachi mainly spread out on a Pre-Montane desert formation on the left of the Nasca River, at an elevation of 365 m a.s.l.

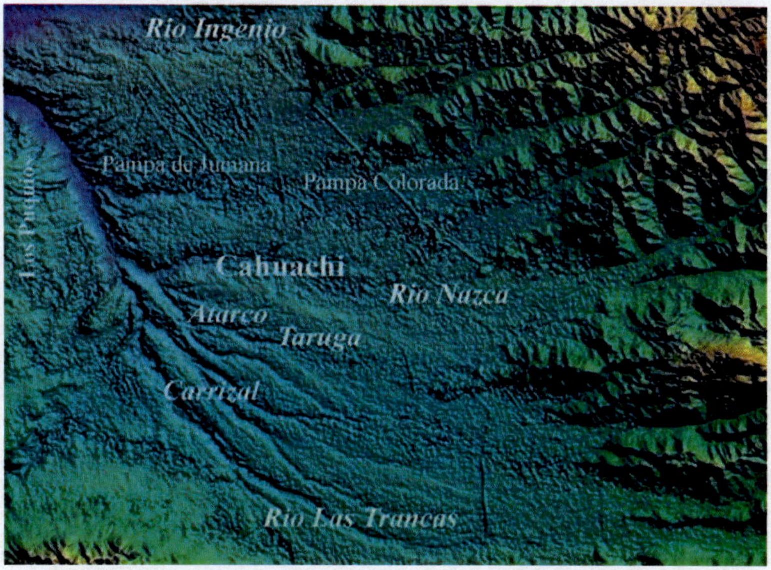

Fig. 14.1 DEM derived from ASTER data: detail of the hydrographical basin of *Rio Grande* with *Rio Ingenio*, *Rio Nasca* and its tributaries. The digital map puts in evidence some linear geoglyphs of *Pampa de Jumana* and *Pampa Colorada* at North of Cahuachi

In particular, they are around 40 tells (earthen mounds), some of which have been completely excavated, located on Quaternary sedimentary rock formations (that are riverine and riverine alluvial) of a tectonic depression (Ica-Nasca Depression). A few kilometres away are the Andean foothills, composed of Jurassic and Cretaceo formations (Montoya et al. 1994).

With regard to the climate, it is hot and arid as well as it is characterized by very little measurable rainfall recorded each year, due to the confluence of a cold ocean current (the Humboldt Current) along with other climatic factors (Schreiber and Lancho Rojas 2009).

Nevertheless, the ecosystem of the valley of the Nasca River and its main tributaries (see Fig. 14.1) has been essential in forming the first complex societies, since about 5000 BC.

This territory has been characterized by a long and intense human activity since the Formative Period (2000–800 BC), to the Early Horizon period (800–200 BC, characterized by Paracas Culture), to the Early Intermediate Period (200 BC–500 AD) when the region flourished under the Nasca Culture and Cahuachi was founded and developed. In particular, at the end of the Early Horizon Period, several regional centers with a religious function developed. Cahuachi stood out among the other centers due to the unity of the religious creed, the control of water sources and the formation of an incipient autonomous social organization.

As time passed, the area of Cahuachi and its adobe pyramids and platforms were continually growing until reaching 24 km^2, probably also including the riverbed and the hills on the right bank of the river.

14.1.2 The Ceremonial Center of Cahuachi from the Archaeological Record

The first studies in Cahuachi date back to the 1950s of the twentieth century. Between 1952 and 1953 W.D. Strong investigated an early village occupation. From the archaeological record, Strong credited Cahuachi a housing function—although limited to some periods of its development (from a report of W.D. Strong cited by Orefici 2009b). John H. Rowe supposed a transition process of Cahuachi: from a sacred place to a city (Rowe 1963).

More data have been provided by the studies and excavations of Silvermann and Orefici since the 1980s, which allowed us to identify a prevailing ceremonial functional of Cahuachi (Silverman 1993; Orefici 1993). In particular, systematic excavations, carried out on more than 150 sectors by Orefici, in two zones named A (0.16 km2) and B (0.10 km2), unearthed five historical building phases from the fourth century BC to the fifth century AD (see Fig. 14.2a, b). Such historical phases (400–100 BC; 100 BC–100 AD; 100–350 AD.; 350–400 AD; 400–450 AD), respond to climatic and environmental anomalies, as well as they reflect the functional and cultural evolution of the site (Orefici and Drusini 2003).

At the beginning (400–100 BC), it was a sanctuary or *huaca*. The architecture was composed of two step platforms and walls of *quincha*, cane or *guarango* tree branches interwoven by ropes of vegetal fiber, covered in mud and plaster, which supported a straw or mat roof.

In the second phase, (8100 BC–100 AD) Cahuachi becomes a ceremonial centre. The urban landscape is the expression of the functional change of the site. The architecture built with conic shaped adobes becomes more monumental and wider than in the past. The new platforms were composed of more than two steps and pre-existing structures were transformed into large storehouses. Moreover, the distribution of spaces is more complex, thus reflecting the hierarchic structure of the society, dominated by a religious elite. The new adobe architecture is also the result of works of remodelling and filling with material brought from other constructions that included remains of votive ceramics used in ritual sacrifices.

The third phase (100–400 AD) coincides with the apogee of the ceremonial center. The zone A is the monumental core of Cahuachi. It is composed of three big pyramids known as, *Gran Piramide*, *Grande Templo* and *Piramide Naranja* (see A1, A2, and A3 in Fig. 14.2; see also Fig. 14.3) surrounded by several minor temples that are connected through plazas, enclosures, stairs, passages and intermediate spaces enclosed by big walls. These buildings are examples of both the opulence that the Nasca society enjoyed during this period and the maximum specialization reached by artisans.

The building system, that used bread-shaped adobes, follows the same evolution: the interior structures were transformed into platforms filled with materials used in the rituals; while large perimeter walls hid the temples and the interior buildings friezes (Orefici 2009b).

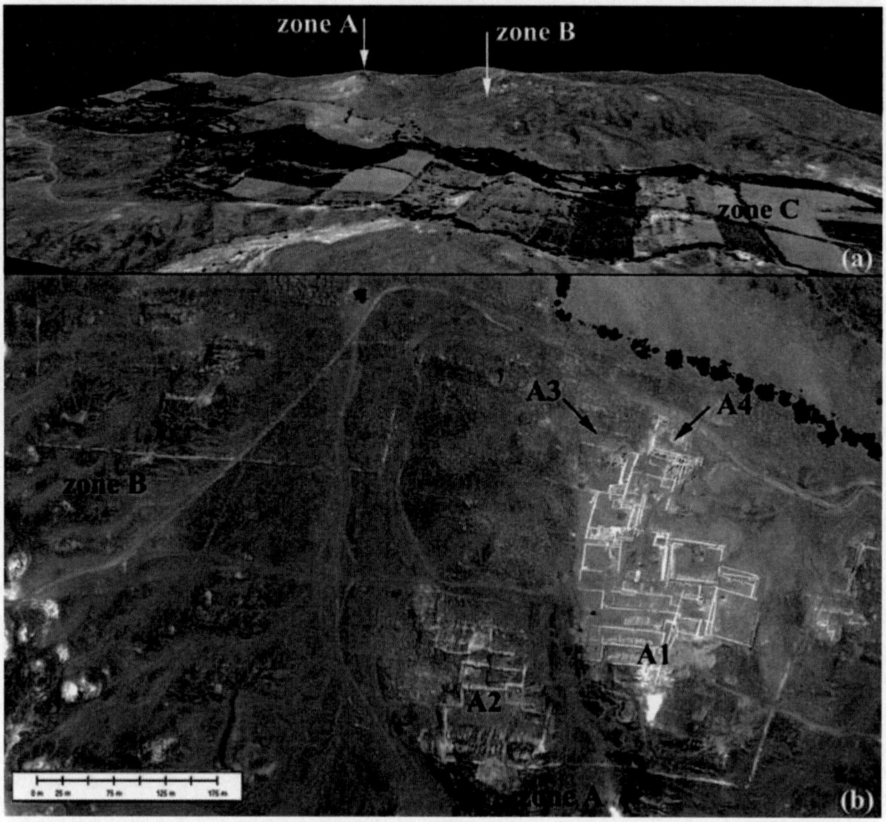

Fig. 14.2 GeoEye panchromatic image acquired in March 2011. (**a**) 3d visualization of the Nasca River and the monumental complex of Cahuachi composed of the two zones A and B; (**b**) 2D image: *A1*, *A2*, *A3* and *A4*, indicate *Gran Piramide*, *Grande Templo*, *Piramide Naranja* and *Grande Templo*, respectively

Fig. 14.3 Cahuachi landscape view from the Nasca River

Possibly during this phase, the excessive use of wood to fire pottery, to build, and to perform other activities, the intensive exploitation of agriculture and environmental resources cause an ecological change, partially faced by the construction of irrigation canals for the distribution of water. Some temples were used as storehouses thus proving the overproduction of food to feed the multitude of artisans and the specialists working in the transformation of Cahuachi.

Probably in this phase Cahuachi reached the maximum urban expansion, with two monumental complexes with U-shaped courtyards (the so-called zone A and B, see Fig. 14.2b), oriented along the southeast-northwest direction, characterized by temples facing North, towards the Rio and the geoglyphs of *Pampa de Jumana* and *Pampa Colorada* (see Fig. 14.1).

The Fourth Phase (350–400 AD) was rich in cultural expressions but also was a time of profound and quick changes that underline a crisis determined by a series of mudslides and a very destructive earthquake. The access entries to the temples were changed and several of the most important buildings lost their monumental functions. The large enclosures were filled so that new embankments could be built and a big part of the terraces with columns and decorated ceilings were demolished in order to make room for the large platforms. During this period, the *Grande Templo* was completely uncovered due to the elimination of several rows of columns, after a ritual in which thousands of antara pipe flutes were destroyed and placed under a new clay floor. The archaeological records evidence a series of mudslides that left profound marks in the structures as well as in the layers of natural clay, due to a strong water flowing.

The fifth phase (400–450 AD) was the moment when the site was effectively abandoned. As in the previous phase, the ceremonial enclosures were filled after an intense series of big collective sacrifices of animals, ceremonial objects and human beings. The offering holes, intensely used during the Fourth Phase, were emptied; the tops of the temples were covered by layers of earthy materials and sealed with clay. The structures of Cahuachi were altered into big platforms and a large necropolis. The sacred character of the ceremonial center was maintained but its primary functions were modified as it was transformed into a large cemetery that was used until the Late Intermediate Period. The collapse of Ceremonial Centre was also due to the loss of power by the sacerdotal class after the mudslides and the earthquake that struck in the previous phase. The theocratic regime of the Nasca culture was lost in the last 50 years of its time, which gave place to the fragmentation of power among local lords (*curacas*), whose political system ruled the urban centers then spread in the valleys, right where the ideology of a central government developed in a homogeneous manner before (Orefici 2009b).

After the abandonment of the temples, the ritual activities moved to Estaquería, at 3.5 km NW from *Gran Piramide*. Estaqueria was occupied during the hegemony of Cahuachi and perhaps it functioned as a small temple. Its importance and its cultural influence on the surrounding territory had been strongly increasing after the collapse of Cahuachi, thus contributing to transmit the Nasca ideology, towards west, in the lower valley.

Fig. 14.4 Northern side of *Gran Piramide*

14.1.3 Archaeological and Research Issues Addressed into ITACA Mission

After more than the 25 years of investigations, excavation trials and systematic excavations focused prevailingly on the sector A, just one pyramid has been unearthed and restored (the so called *Gran Piramide*, see Fig. 14.4), other two temples are almost completely excavated and the restoration are in progress (*Piramide Naranja* and *Templo del Escalonado*). The archaeological findings allowed us to reconstruct the historical outline of Cahuachi from the Paracas period to its abandonment.

Still many issues need to be resolved regarding: (i) the extension of Cahuachi (Did it include also the riverbed and the hills on the right bank of the *Rio Nasca*?); (ii) the continuity of the ceremonial activity after the abandonment of Cahuachi (Did it move to Estaqueria? Why in that place? Is it possible that other settlements have been built to the west of Cahuachi as Estaqueria?); (iii) the function and the meaning of *Piramide Naranja* (Did the last ceremonies and sacrifices take place there?).

Since 2007 ITACA mission has been requested to support the archaeological investigations with the use of Earth Observation (EO) methods (satellite remote sensing and geophysics) in order to provide additional information on the history of Cahuachi and its surrounding territory.

In particular, the issues to be addressed have been the following:

(1) the detection of buried structures and ritual offerings in the *Piramide Naranja* and near *Templo del Escalonado*, which mainly conserve witnesses dating back to the decadence of Cahuachi (end of the III and IV phase);

(2) the survey of settlements to the west side of Cahuachi, where likely ceremonial activity moved as the archaeological records in Estaqueria proved (Sánchez Borjas 2009).

For both tasks 1 and 2 integrated remote sensing techniques have been applied. A complementary activity has been the evaluation of spectral capability and of the most reliable process methods of satellite images both for the vegetated and non-vegetated areas (see Sect. 14.3.1).

Moreover, before performing the geophysical investigations, the different technologies to be employed have been applied to test sites where the archaeological excavations were in progress, and so, the results would have been available after short time. This allowed us to provide information useful for the interpretation of geophysical outputs in presence of buried walls, tombs and ritual offerings.

14.2 Archaeogeophysics Rational Basis

14.2.1 Satellite Dataset

A multitemporal and multiscale satellite dataset has been used both to archaeological purposes and for the study of the environmental setting. In particular, we used: (i) 90 and 30 m Digital Elevation Models (see Fig. 14.1) maps derived from SRTM (http://srtm.csi.cgiar.org) and ASTER (Advanced Spaceborne Thermal Emission and Reflection Radiometer), respectively; (ii) multispectral ASTER images to study land cover and the hydrographic pattern of the Rio Grande drainage basin. For the first results of these studies, the reader is referred to Chap. 12, in this book.

For the detection of archaeological features, we used Very high resolution (VHR) satellite commercially available imagery acquired from: (i) QuickBird2 (QB2), (ii) WorldView-1 (WV1), (iii) WorldView-2 (WV2), (iv) GeoEye-1 (GE1) sensors.

QB2 has panchromatic and multispectral sensors with resolutions of 61–72 cm and 2.44–2.88 m, respectively, depending upon the off-nadir viewing angle (0–25°). The panchromatic sensor provides images in a bandwidth ranging from 450 to 900 nm. The multispectral sensor acquires data in four spectral bands from blue to near infrared (NIR) (http://www.satimaging corp.com/satellite-sensors/quickbird.html).

WV1 provides panchromatic imagery with a ground sample distance (GSD) varying from 50 to 59 cm, depending on the off-nadir viewing angle. Additional details can be found at http://www.satimagingcorp.com/satellite-sensors/worldview-1.html.

WV2 acquires panchromatic imagery with 50 cm GSD, and eight-band multispectral imagery with 1.8 m resolution (http://www.ulalaunch.com/launch/WorldView-2/WV-2_MOB.pdf)

GE1 collects images with a GSD of 0.41-m or 16 inches in the panchromatic and multispectral imagery at 1.65-m resolution. The available GeoEye panchromatic images are re-sampled at 0.5 m (see http://launch.geoeye.com/LaunchSite/about/).

The VHR data used for this study have been:

(i) QB2 scene acquired on the 16th September 2002 at 15:17 with an off-nadir of 7.90, 61.90 cm and 2.48 GSD at panchromatic and multispectral, respectively;
(ii) QB2 scene acquired on the 25th March 2005 at 15:29 with an off-nadir view angle of 11.9°, 63.40 cm and 2.54 m GSD at panchromatic and multispectral, respectively;
(iii) WV1 image acquired on 31th July 2008 at 15.26 with an off-nadir view angle of 23.9° and 58.10 cm GSD;
(iv) WV2 image provided on 11th September 2010, at 15.30, with an off-nadir view angle of 2.7°, 50 cm and 2.0 m GSD at panchromatic and multispectral, respectively;
(v) GE1 image provided on 28th February 2011 at 15.09, 50 cm and 2.0 m GSD at panchromatic and multispectral, respectively

14.2.2 Satellite Image Processing Approach

The image processing approach used for exploiting VHR satellite data has been aimed at improving the potential of imagery both in the spatial and in the spectral domain, by means of algorithms of image fusion (pan-sharpening), enhancement and edge detection.

14.2.2.1 Feature Enhancement Based on Pan-Sharpening Techniques

The first step of data processing has been the pan-sharpening, which provides a spatial enhancement of the lower resolution multi-spectral (MS) data (i.e. the four bands of QB2, WV2 and GE1). Equivalently, we can observe that pan-sharpening increases the spectral resolution of the panchromatic (Pan) image having a higher spatial resolution, but a lower spectral resolution bearing no spectral information.

The general protocol employed, can be summarised in two steps: (1) extraction of high-resolution geometrical information from the panchromatic image; (2) injection of such spatial details to the interpolated low-resolution MS bands through proper models.

According to this protocol, the pan-sharpening techniques can be divided into two main classes: (i) component substitution (CS) techniques, which are based on a spectral transformation of the MS data followed by the substitution of the first transformed component with the Pan image and reverse transformation to yield back the sharpened MS bands (the most widely used CS-based methods are Intensity-Hue-Saturation (IHS), Principal Components Analysis (PCA) and the

Gram-Schmidt (GS) orthogonalisation procedure (Laben and Brower 2000)); (ii) techniques that employ multi-resolution analysis (MRA) such as wavelets and Laplacian pyramids. These methods extract from the Pan image the geometrical information that will be added to the MS bands.

Due to the variety of surface covers in Cahuachi (bare-ground surface in the archaeological area, vegetated in the Nasca riverbed), and archaeological features (crop-marks, microrelief, surface archaeological structures), more algorithms, such as GS, enhanced GS (Aiazzi et al. 2007) with context adaptivity (GSA) and generalised Laplacian pyramid (GLP) with context adaptivity (Aiazzi et al. 2006), were used.

The comparative analysis performed in a qualitative way put in evidence that the methods based on CS technique (GS and GSA) provide fused images with higher geometrical quality with some spectral impairments, whereas the GLP is spectrally accurate, but is unsatisfactory in terms of spatial enhancement, as already experienced in other archaeological test sites (Aiazzi et al. 2008).

14.2.2.2 Feature Enhancement Based on Linear Combinations of Spectral Bands

The spectral indices are generally computed by a linear combination of different spectral bands in order to obtain quantitative measures of surface properties. Spectral indices, attempt to quantify surface properties such as brightness, moisture, biomass cover, or vegetative vigour. The widely used index is the Normalized Difference Vegetation index (NDVI) obtained by using formula 14.1:

$$NDVI = \frac{(NIR - R)}{(NIR + R)} \tag{14.1}$$

The NDVI operates by contrasting intense chlorophyll pigment absorption in the red against the high reflectance of leaf mesophyll in the NIR. On the basis of the vegetation spectral properties, the NDVI provides a quantitative measure, suitable to assess biomass, vegetation type and vigour. NDVI is indicative of plant photosynthetic activity and it has been found to be related to the green leaf area index and to the fraction of photosynthetically active radiation absorbed by vegetation. High values of the vegetation index identify pixels covered by substantial proportions of healthy vegetation, whilst disease or stressed vegetation exhibits lower NDVI values.

PCA is a linear transformation which decorrelates multivariate data by translating and/or rotating the axes of the original feature space (Richards and Xiuping 2006). In this way, the data can be represented without correlation in a new component space. In order to do this, the process firstly computes the covariance (Unstandardized PCA) or the correlation matrix (standardized PCA) among all input spectral channels. Then eigenvalues and eigenvectors are calculated in order to obtain the new feature components. So, the PCA transforms the input multispectral bands in new components whose number is equal (or less) to the number of the input channels. In detail, the first component contains the major portion of the variance and provides a sort of average of all the input channels. Each successive component contains less of the total dataset variance and may represent information for a small area or essentially noise; in this case, it must be disregarded. The PCA should be able to make the identification of distinct features and surface type easier. This is a direct result of different facts: (i) only the meaningful low correlated data can be considered, (ii) the effect of noise can be easily identified and strongly reduced because it is in the later components. Moreover, the PCA can be used to obtain a new color enhancement technique, "Decorrelation Stretching", based on the following steps: (i) PCA transformation, (ii) each PCA component is contrast stretched, (iii) stretched components are rotated back using the inverse PCA transformation (iv) finally, the resulting components are generally shown using RGB compositions.

Finally, in order to better identify spectral signature anomalies on both bare ground and vegetated areas, Minimum Noise Fraction (MNF) transformation has been applied. By examining the images and eigenvalues, MNF allowed to determine which bands contain the coherent images, thus segregating and removing noise from the pansharpened multispectral channels (see Boardman and Kruse 1994). By using a routine of ENVI, based on MNF transformation modified by Green et al. (1988), the procedure is two cascaded Principal Component transformations.

The first transformation, based on an estimated noise covariance matrix, decorrelates and rescales noise in the data. This first step results in transformed data in which noise has unit variance and no band-to-band correlations. The second step is a standard Principal Component transformation of the noise-whitened data. For the purposes of further spectral processing, the inherent dimensionality of the data is determined by examination of the final eigenvalues and the associated images. The data space can be divided into two parts: one part associated with large eigenvalues and coherent eigen images, and a complementary part with near-unity eigenvalues and noise-dominated images. By using only the coherent portions, noise is separated from the data, thus improving spectral processing results (ENVI user's guide 1999).

14.2.2.3 Feature Enhancement Based on Spatial Filtering

Once pan-sharpening and spectral combinations computed, in order to further improve the edges of objects (surface archaeological structures) and marks (possibly related to archaeological deposits), some convolution filters, including high pass, low pass, Laplacian, directional, Gaussian High Pass, Gaussian Low Pass, median, Sobel and Robert filters, have been used (for additional information on spatial filtering see Chap. 2, Sect. 2.4.5).

The best results in terms of edge enhancement of the edges have been obtained by using "High pass filtering" which removes the low frequency components of the

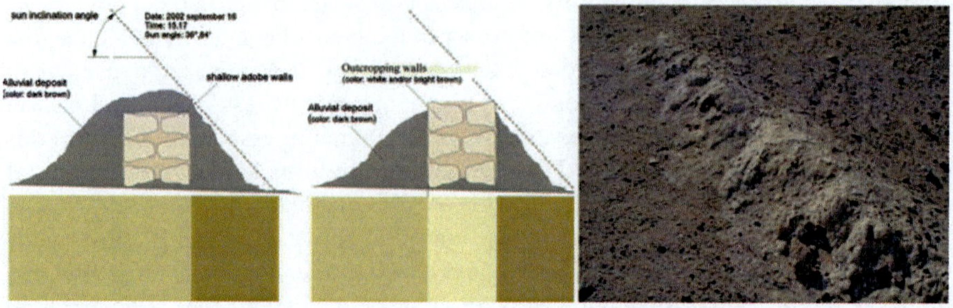

Fig. 14.5 (*Left* and *medium*) Shallow and surface walls visible from remote by means of the shadow and the reflectance of the clay of adobe walls

images, thus retaining the high frequencies (local variations in the presence of surface archaeological structures). The filter has been applied by adopting a 3×3 kernel with a value of 8 for the center pixel and values of -1 for the exterior pixels.

In case of rectilinear features (i.e. we refer to the edges of terraced platforms of some mounds), good results have been observed by employing a directional filter, which is a first derivative edge enhancement filter that selectively enhances image features having specific direction components (gradients).

The above-said image processing routines have been applied to different test areas in order to select those to be systematically adopted in the areas to be investigated (see Sects. 14.3.1 and 14.3.3).

Unfortunately, the identification of buried and/or shallow remains (i.e. walls) in the desolate area of Cahuachi by using optical imagery is a very complex challenge, due to the building material, which is adobe, whose composition (sun-dried earth) is quite similar to the soil which covers the archaeological remains. This has been confirmed by a multitemporal observation of aerial and satellite images which allowed us to compare some areas before and after the excavations (Masini et al. 2008, 2009a, b).

The comparison put in evidence that: (i) buried adobe structures are generally not visible from an aerial and/or satellite view; (ii) surface wall remains were more easily identified due to the high contrast in brightness between the clay of the surface adobe walls and the surrounding alluvial deposits which cover a large part of the mound; (iii) microrelief related to shallow walls are enough detectable thanks to the small shadow produced on the ground (see Fig. 14.5).

To face the challenge of the detection of archaeological features, once selected and used the most suited pan-sharpening and filtering algorithms, further processing methods have been applied to some test sites shown in Sect. 14.3.1 which put in evidence a good performance of the PCA for the bare-ground sites and vegetation indices for the vegetated and damp areas of the Nasca riverbed.

14.2.3 Geophysical Approach

The variety of archaeological features in the subsoil (buried structures, tombs, and ritual offerings), the low geophysical contrast between some archaeological features (adobe walls) and the soil, the complex stratigraphies unearthed by archaeologists made necessary the employ of different geophysical methods.

In some cases more methods have been used in order to integrate the information content, in other cases only one method has been employed due to logistic problems related to the application of all geophysical techniques or for the great extension and the complex morphology which suggest to perform only the least time consuming one, such as the geomagnetic one. In the following subsections, each geophysical method will be briefly described.

14.2.3.1 Georadar Method

The GPR is an electromagnetic (EM) method used for several kinds of applications and with different investigation depths (Fig. 14.6). The archaeological targets are generally investigated by using medium frequency antennas. The EM-wave frequencies along with the electrical characteristics of the subsoil (permittivity and electrical conductivity) determine the investigation depth. The GPR radiates short EM pulses into the ground and detects the signals reflected from subsurface structures. The reflected signal is generated in the presence of a dielectric contrast between potential targets and surrounding soil.

This has been the crucial topic in Cahuachi because the adobe (made from sand, clay, fibrous material and water) have characteristics similar to the alluvial soil, in terms of resistivity and dielectric constant.

GPR measurements were performed by using the Subsurface Interface Radar (SIR) 3000 manufactured by GSSI: SIR 3000. It consists of a digital control unit with a keypad, VGA video screen, connector panel and is powered by a 12-V DC battery. The system is connected by fibre-optic cables at a monostatic type antenna (400 MHz) manufactured by Geophysical Survey Systems.

Due to the incoherent sand of soil surfaces the survey was acquired without a ‘wheel accessory’, so that a speed variation could occur. To reduce uncertainties on the antenna position, a reference metre rule was located along each profile and marked at each metre. In order to estimate the average electromagnetic wave velocity, a wall of known thickness fitted with a metallic plate was used to gain a maximum reflection of EM waves. A velocity of about 0.1 m/ns has been estimated observing the half two-travel time of the corresponding metallic plate reflection. The radar scans have been acquired in continuous mode with a 2-way time range of 40 ns. An interval band-pass filter of 100–800 MHz was used to reduce electronic, antenna-to-ground coupling noise as well as other low- and high-frequency noise.

The Reflex W software was used to process the data. The high quality of the traces only required standard analysis techniques for data processing and for

Fig. 14.6 Georadar system

reducing background noise, linked to trace editing, normalization, acquisition gain removal, zero time correction and a background removal filter. High-frequency noise was attenuated by means of a 2D average filter. An interactive velocity adoption, based on EM reflection waves, was used to estimate the average EM-wave velocity of the geological material that covers the archaeological deposits.

14.2.3.2 Geomagnetic Prospections

In Cahuachi geomagnetic method with a gradiometric configuration has been applied. The measurements were performed using an optical pumping magnetometer G-858 (by Geometrics) in gradiometric configuration, with two magnetic probes set in a vertical direction at a distance of around 1 m each other (Fig. 14.7). Such a configuration allowed the automatic removal of the diurnal variations of the natural magnetic field. Before defining the acquisition modalities, it was necessary to set up the proper orientation of the two magnetic sensors of the Caesium Magnetometer. Such an orientation depends on the survey direction and site location in the world. To do this CSAZ software (by Geometrics) has been used. It provides information about the Earth's magnetic field parameters including total field, inclination and declination anywhere in the world, using the IGRF (International Geo-Magnetic Reference Field). After entering latitude and longitude of the archaeological site and indicating the

Fig. 14.7 Geomagnetic prospections

survey direction, the software provides the orientated caesium sensor to have the maximum signal and best performance. Therefore, the instrument was set with a tilt angle of 45° and the survey was defined along parallel profile in N-S direction.

Data were acquired along parallel profiles 0.5 m apart with a sampling rate of 10 Hz, obtaining a mean spatial resolution of 0.5 m × 0.125 m. The surveyed areas were geocoded through GPS differential measurements. All the acquired data were processed and interpolated to create regular grids. Afterwards, the final matrixes were mosaicked. In some cases, where the area to be investigated was very large (i.e. in the Nasca riverbed, see Sect. 14.3.3) or characterized by steep slopes, magnetic prospections have been performed by means of GPS.

The rough magnetic data have been visualized as 8 bit raster images, after having created regular grids using a Kriging interpolator with a linear variogram. Then, the rough magnetic data have been filtered to increase the signal/noise ratio by using MagMap software which provides pass-band filter, spikes elimination and destripe.

Finally, the filtered have been visualized as a shaded relief image, in order to highlight the main magnetic anomalies.

14.2.3.3 Georesistivity

The third method employed has been the georesistivity, by using the RM-15 Resistance Meter (by Geoscan). It is a technique that measures the resistance encountered by an electrical current in passing through the subsoil between four probes. One pair of probes is fixed far from the investigated area, while the second one is mounted on a frame and inserted into the ground at regular distance. In the investigated area the interval distance was 1 m along parallel lines with a regular distance of 1 m. Each reading is a measure of the amount of the resistance encountered by the current as it passes through the soil.

In presence of adobe walls in the subsoil, low resistivity values are expected. On the contrary, electrical current should encounter greater resistance in tombs and ritual offering, due to the presence of voids and high content of pores.

14.3 Experimental Section: Test Sites, Archaeo-Geophysical Investigations, Results

14.3.1 VHR Satellite Image Processing Approach: Preliminary Tests

Before the first systematic archaeo-geophysical campaign in 2008, some preliminary investigations have been carried out in 2007 on some test sites (TS) into the archaeological area and in its surrounding, which allowed to develop the most suited image processing approach for different archaeological features and land cover.

In this chapter we show the results obtained on four tests sites (see Fig. 14.8) which are: (i) the zone B located at West of zone A (named TS1); (ii) some lineal geoglyphs sited at South of *Gran Piramide*, at North of *Atarco* (TS2); (iii) a vegetated area north the Nasca River (TS3). The satellite images used for the preliminary investigations have been the 2002 QB data (see additional information of the image in Sect. 14.2.1).

The test site TS1 includes mounds which likely conserve pyramids, terraced platforms, a square and a U-shaped enclosure. The U-shaped courtyard is an urban scheme which typically characterizes the ceremonial town since Early Horizon period (Gavazzi 2009, 2010). In Cahuachi the U-shaped courtyard of sector B, faces North, in direction of the geoglyphs of the Pampa. The mounds and the platforms compose a chessboard shaped layout with some offset which allowed to look to east in direction of the sacred mountain of *Cerro Blanco*.

All these archaeological features in zone B are enough visible from the panchromatic image (Fig. 14.9a), however the shadow produced by the micro-relief (terraces, platforms, walls), the circular looted area and the presence of surface chaotic material, residues of adobe structures make difficult their identification and survey. So, TS1 has been visualized and interpreted by using not only the data input (panchromatic and pan-sharpened bands, but also the results of the post processing approach described in Sect. 14.2.2).

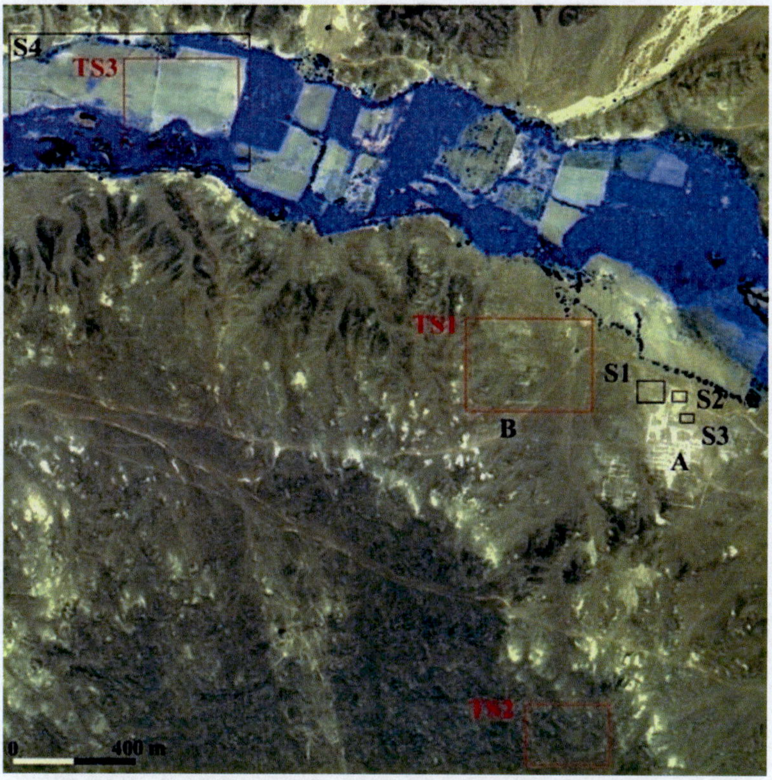

Fig. 14.8 Location of test sites (TS1, TS2, TS3) and the investigates sites (S1, S2, S3 and S4)

The comparative observation of known archaeological features from the data input put in evidence a better performance of the panchromatic images respect to multispectral pan-sharpened data, including the RGB visualization. This has been found also in the other test site characterized by bare and arid soil surface (TS2). So, the post processing has been focused on the panchromatic data, showed in Fig. 14.9. The best results have been obtained by using PCA and MNF. For sake of brevity we show only the results of PCA (the percentage of the computed eigenvalues have been 79% for PC1, 16% for PC2, 5% for PC3 and PC4). In Fig. 14.9, related to the panchromatic image of the sector B, the result of PC1 has been overlaid in the lower left corner has, thus revealing a slight enhancement of some details in particular of the looted areas, characterized by circular holes.

A better discrimination of archaeological features has been obtained by PC2 (Fig. 14.9b). The latter makes the identification of edges related to mounds and platforms easier (see 3 and 7, in Fig. 14.9a, b), and some parts of the U-shaped enclosure (see 1, in Fig. 14.9a, b). This is in perfect accordance with the fact that the first PCA substantially provides an average of all the input spectral channels, e.g. a map very close to the brightness of the scene quite similar to the panchromatic image.

Fig. 14.9 Test site TS1. (**a**) Panchromatic 2002 QB, in the corner lower right the PC1 result has been overlaid on the panchromatic scene; (**b**) PC2 result

Whereas the second component cannot be associated with a general meaning since it is obtained from the statistic computation and depends on the current dataset.

Still better results of PC2 could be observed in TS2 (see Fig. 14.10).TS2 is an area located between Cahuachi and *Atarco* River, about 1 km South of *Gran Piramide*, where several geoglyphs could be observed (Fig. 14.10a, c). They are geoglyphs less known and rich from the iconographical point of view respect to those of *Pampa Jumana* and *Pampa Colorado*. Here we do not find any biomorphic geoglyph. However they are important because they clearly seem to be part of a ritual complex closely related to Cahuachi. They are likely pathways used by pilgrims on their way to the Ceremonial Centre.

As in the Pampa this area could be another place of gatherings and ritual activities. Such geoglyphs have been produced removing stones from their original

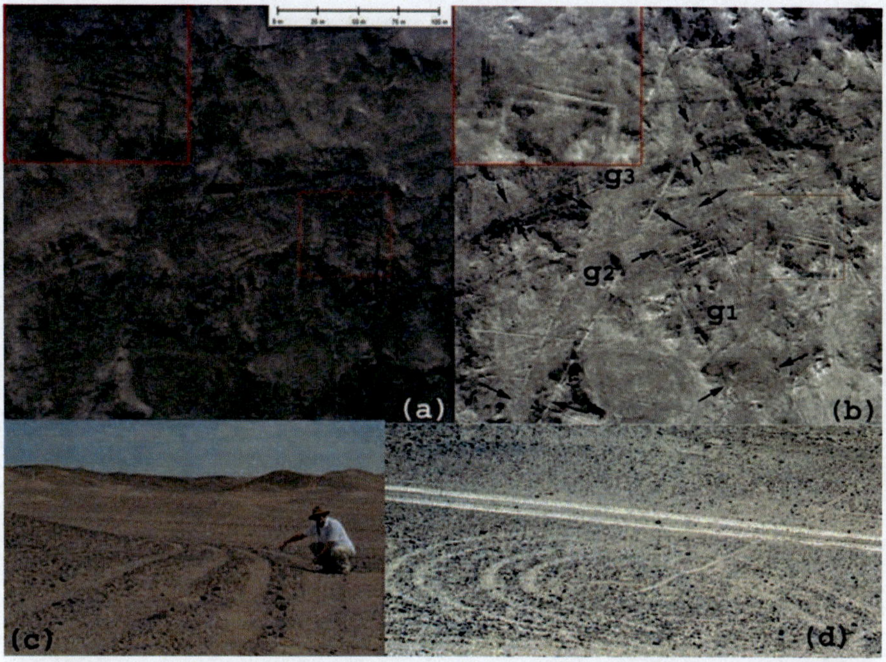

Fig. 14.10 Test site 2: geoglyphs at south of Cahuachi (for their location see Fig. 8). (**a**) 2002 panchromatic image with a zoom in the corner upper *left*; (**b**) the visualization of PC2; (**c–d**) ground truth

place in order to create different motifs such as lineal (straight lines, U-shaped, meandering, zig-zag, spiral, areal trapezoidal and triangular), as showed in Fig. 14.10a–c. In some cases the contrast clear-dark had been obtained by removing darker stone material (pebble and gravel) and unearthing the clearer subsoil. The image processing of QB images has been aimed to improve the contrast of the lines. The test site has been selected because it is characterized by the superimposition of more geoglyphs. The convolution filters and PC2 allowed us to identify the temporal sequence of the drawn lines. The visualization of PC2 in Fig. 14.10b puts in evidence at least three different tracking phases (g1, g2 and g3). Such superimposition did not necessarily mean the destruction of a line (or pathway) and the creation of a new one. In some cases the overlapped geoglyphs created an intersection of active pathways to be walked.

The final test site herein described is a vegetated area at 1.8 NW from *Gran Piramide*, located on a cultivated strip on the right of the Nasca River. The river is close to the test site (around 200 m) and supplies enough water for the cultivations.

Some aerial images captured during in-flights on April 2007 had put in evidence some interesting crop marks. For this reason it is selected to test different image processing methods in order to exploit the spectral features of QuickBird images.

Fig. 14.11 Test site TS3 selected in a vegetated area. (**a**) 2002 QB2 panchromatic image; (**b**) *red* pan-sharpened channel; (**c**) NIR pan-sharpened channel; (**d**) NDVI map. The latter put in evidence a quadrangular shape signal (indicate by *black arrows*)

Unfortunately the month of acquisition satellite data (September) was not ideal for crop marks. Panchromatic image and NIR channels do not show significant information (Fig. 14.11a–c).

On the contrary, red channel (Fig. 14.11b) put in evidence a signal apparently irregular at west of the field division. The NDVI (Fig. 14.11d) map has strongly enhanced this signal, thus revealing a quadrangular shape and some linear marks parallel to the sides of the areal anomaly. The geometric pattern suggests a possible anthropogenic cause to the signal observed from NDVI map. On site some ceramic fragments and bread-shaped adobe refer to a possible archaeological interest of the place. This test site was very useful in the perspective to perform an archaeo-geophysical investigation, based on the integration of geophysics and satellite remote sensing, carried out the following year (2008).

Section 14.3.3 shows the results of this research which lead to the discovery of a large buried settlement.

14.3.2 *Archaeo-Geophysical Investigations and Results in the Sector A*

14.3.2.1 Site S1. From Satellite to Geophysics: The Discoveries in *Piramide Naranja*

The study of *Piramide Naranja* has been the last as well as crucial investigation phase within the multi-year archaeological programme of the mission directed by Giuseppe Orefici. The first archaeological findings before ITACA investigations had clearly put in evidence the richness of tombs and ritual offerings and the presence of walls dating back to third and fourth phase.

For archaeologists, the expected results of excavations in the *Piramide Naranja* would have contributed in casting new light on the last historical phase of the Ceremonial Centre: its abandonment.

So, archaeo-geophysical investigations have been starting since 2008 with twofold aims: (1) the identification of surface and shallow archaeological features referable to walls in order to spatially characterize the trunk-pyramidal architecture; (2) the detection of buried archaeological deposits related to tombs and ritual offerings (the first results of geophysical investigations have been published in Lasaponara et al. 2011).

For the first aim three different approaches have been adopted. One is multitemporal, that is the observation of a time series of aerial and VHR satellite images (1955, 2003, 2005, 2007, and 2008). The second has been based on the analysis and processing of VHR satellite imagery. The third approach has been the integration of geomagnetic and georadar methods. The geomagnetic method has been employed to confirm the presence of shallow walls already visible by the optical data set as well as to add further information. The georadar has been used to obtain information on the depth of buried walls. This latter task has not been fully obtained, since the attenuation of radar signal capable to survey only shallow reflections referable to the upper part of buried walls.

Figure 14.12a–f show the 3d visualization of *Piramide Naranja* from the available aerial and satellite images, performed by using a DEM (a) derived from a topographical map (1:1.000 scale). The map has been done with photogrammetrical techniques on the base of the 1955 aerial images (one of which is showed in Fig. 14.12b) and a topographical survey carried out in 1988–1989.

Piramide Naranja has an asymmetric shape, as the other pyramids in Cahuachi. It faces to North as well as it is characterized by four terraced platforms. The multitemporal data set analysis put in evidence some changes of the site, such as the increasing of the looting activity from 1955 to 2003 (Fig. 14.12b, c) and the walls unearthed on the eastern slope which are visible since 2007 (Fig. 14.12e, f).

For the mapping of the edges of terraced platforms all the data set of Fig. 14.12 have been assessed in terms of discrimination of the mentioned archaeological features. From the comparative qualitative assessment the best results have been obtained by the 1955 aerial photo and 2002 QB panchromatic scene (see Figs. 14.12b, 14.13a, and 14.14a). So, the enhancement processing has been focused on the above said data.

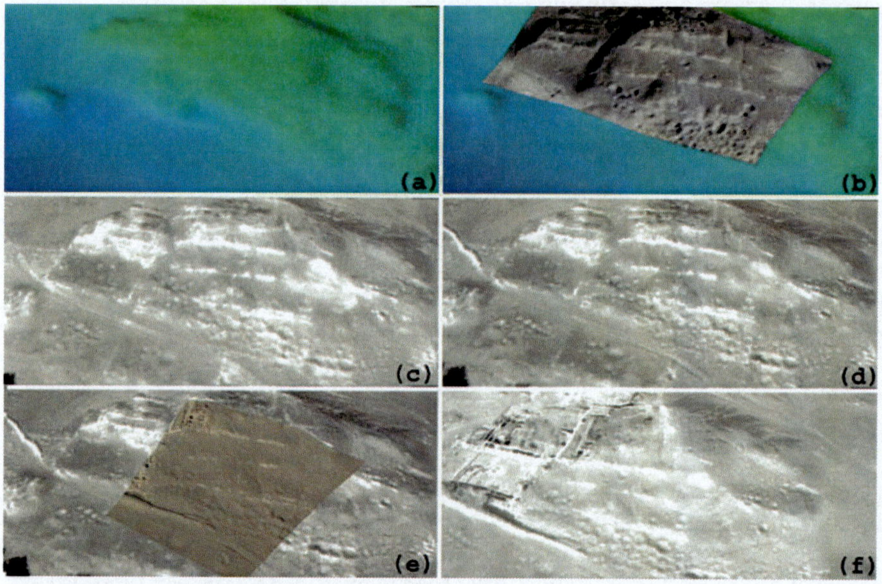

Fig. 14.12 *Piramide Naranja*: 3d visualizations of aerial and satellite data set by using a DEM (**a**) derived from a topographical map performed for scale 1:1.000. The 3d visualizations are related to an aerial photo taken in 1955 (**b**), two QuickBird images acquired in 2002 and 2005 (**c–d**), an aerial photo captured in 2007 (**e**) and WorldView-1 image acquired in 2008 (**f**)

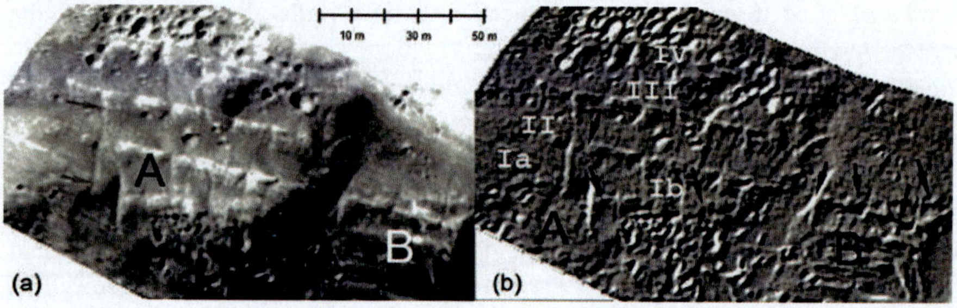

Fig. 14.13 (**a**) Aerial image taken in 1995; (**b**) result of convolution filter. *A* and *B* denote respectively the *Piramide Naranja* and the *Templo del Escalonado*

The Fig. 14.13a shows the 1955 aerial photo orthorectified on the base of the available DEM (Fig. 14.12a). The scene includes *Piramide Naranja* (labelled with letter A), divided by a moat from an adjacent terraced structure on the east, known as *Templo del Escalonado* (labelled with letter B).

The historical photo has been filtered by directional convolution, computed by applying a direction filter angle of 90° and kernel size = 3 × 3 (Fig. 14.13b), thus improving the contrast of the terraced platform edges *II*, *III* and *IV*. Moreover, it allows to identify the edge of the top platform I, characterized by a broken line

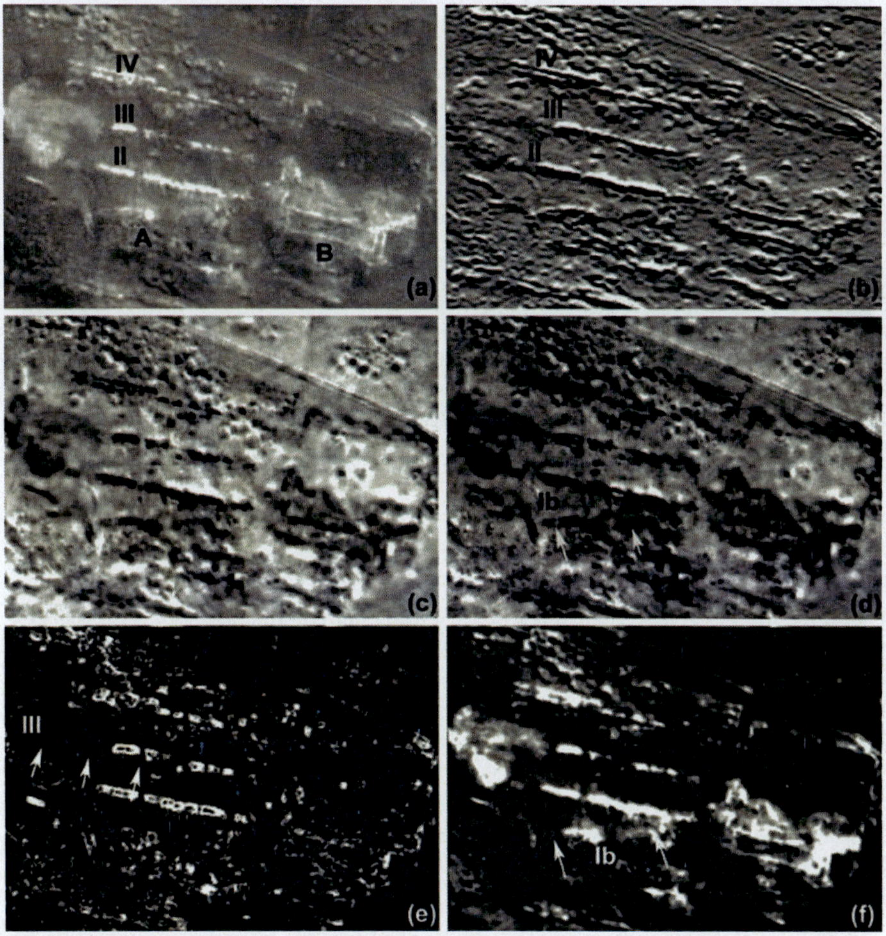

Fig. 14.14 Detail of northern slope of *Piramide Naranja* from 2002 QB2 panchromatic image **(a)** and the results of some post processing applications such as: **(b)** directional convolution; **(c)** PC3; **(d)** standard deviation; **(e)** Geary's C index; **(f)** Moran's index

composed of a segment parallel (*Ia*) to *II-IV* and an oblique segment (*Ib*). The filtered image shows also the alignment of II and *Ib* with some edges of the adjacent *Templo del Escalonado*. Such alignments are less visible in the non processed aerial image (Fig. 14.13a) as well as non visible in the aerial and satellite dataset, since 2002, due to clandestine excavation activity.

Figure 14.14a shows the 2002 QB2 panchromatic scene. On the QB2 imagery (panchromatic and pan-sharpened bands) directional and high pass filter, PCA, Standard deviation and spatial autocorrelation statistics has been computed. The best discrimination of linear archaeological features has been obtained by the directional convolution (Fig. 14.14b), PC2 and the standard deviation. Respect to the

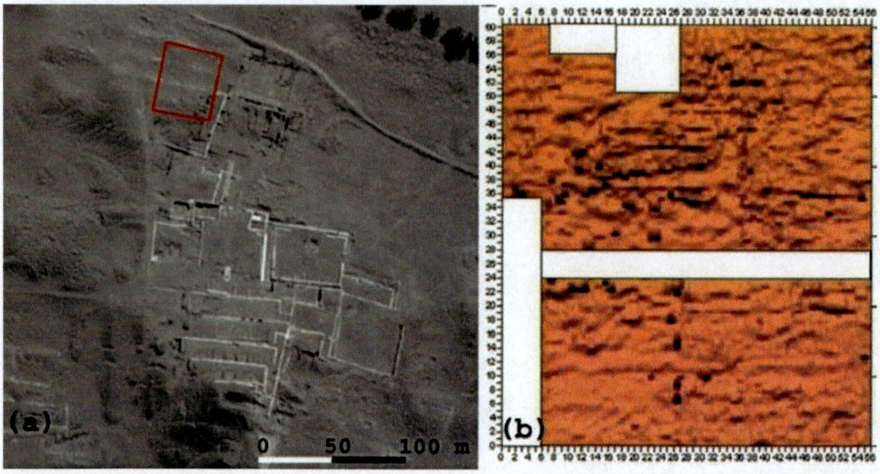

Fig. 14.15 (**a**) Location of the area of *Piramide Naranja* investigated by geomagnetic method; (**b**) geomagnetic map

panchromatic image (Fig. 14.14a) the directional convolution (Fig. 14.14b) reduces the noise due to the high reflectance of the clay which the surface and shallow walls are composed of, thus allowing to thin the archaeological features. PC2, Standard deviation (Std) and the result of Geary's C index show the above mentioned oblique feature IB, whereas Moran's index allow to identify a subtle linear feature aligned (at west) with the feature III.

After mapping the linear features related to the edges of terraced platforms geomagnetic investigations have been performed. The aim was to detect materials characterized by magnetic properties such as metals, ceramic and coal associated to possible tombs and ritual offerings.

Lower differences of magnetic field between the shallow walls and its surrounding were expected. The total map surveyed was around 56 × 60 m, thus covering three terraced platforms of northern slope of the pyramid.

As measurement acquisition modalities, we selected the mapped survey mode, which allows us to define and visualize the survey area as well as to move around within it in a non-continuous fashion by means of regular grids (for detail on data acquisition and processing, see Sect. 14.2.3.2).

In order to overcome logistic problems, due to the presence of excavation trials and old pits dug by grave looters, the investigated area was subdivided into ten sectors. The survey direction was always South-North. Figure 14.15a shows the location of the investigated area. On the right of the same figure (Fig. 14.15b) the geomagnetic map obtained from gradient data is showed.

The overlay on the satellite image (see Fig. 14.16) helped in interpreting the magnetic anomalies. In particular, lower but linear oriented anomalies (in E-W direction) correspond to the archaeological features, already visible from the optical dataset (see I, II and III in Figs. 14.14a and 14.16a, b), which have been thought to

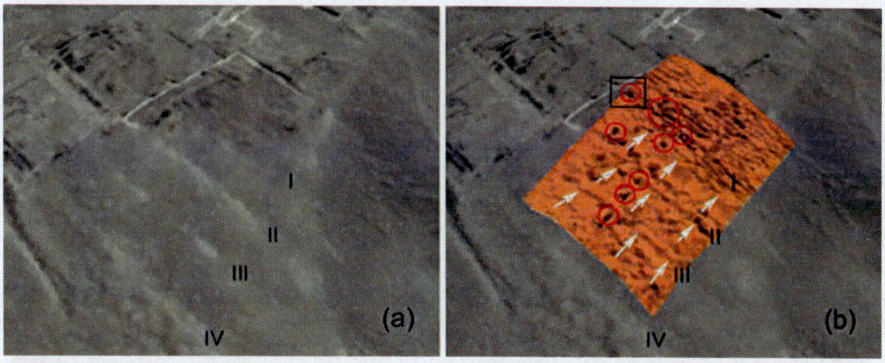

Fig. 14.16 (**a**) 3d satellite image (2008 WV1) visualization of northern slope of *Piramide Naranja*, and (**b**) geomagnetic map overlaid on the satellite image. The *white arrows* denote linear magnetic signals, the *red circles* indicate more intense magnetic signals, the *black block* evidence the anomaly selected in collaboration with archaeologists to be firstly investigated by GPR and then excavated

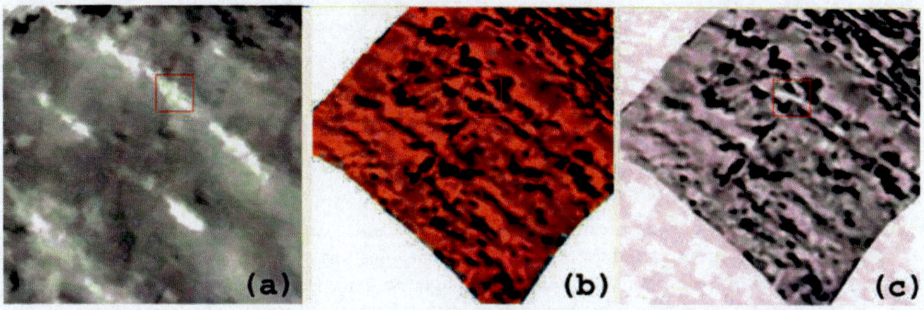

Fig. 14.17 *Piramide Naranja*: data fusion between WW-1 panchromatic image and geomagnetic map

be surface and shallow walls which limit the external edges of the platforms. Other linear anomalies are oriented in N-S direction, thus suggesting the presence of walls related to ramps and corridors which connect the different levels of the pyramid.

The fusion between panchromatic image and geomagnetic map has been performed by using Grahm-Schmidt spectral sharpening, with nearest neighbour resampling (see Fig. 14.17a–c). The image fusion product confirms such correspondence as well as improves the quality of information content. Moreover, small circular and intense magnetic anomalies could be observed (Figs. 14.16b and 14.17b). Most of them are over pits partially dug by grave looters, with a greater density in the top of the pyramid. Some other strong anomalies are aligned along a line between the platforms II and III (Fig. 14.16b).

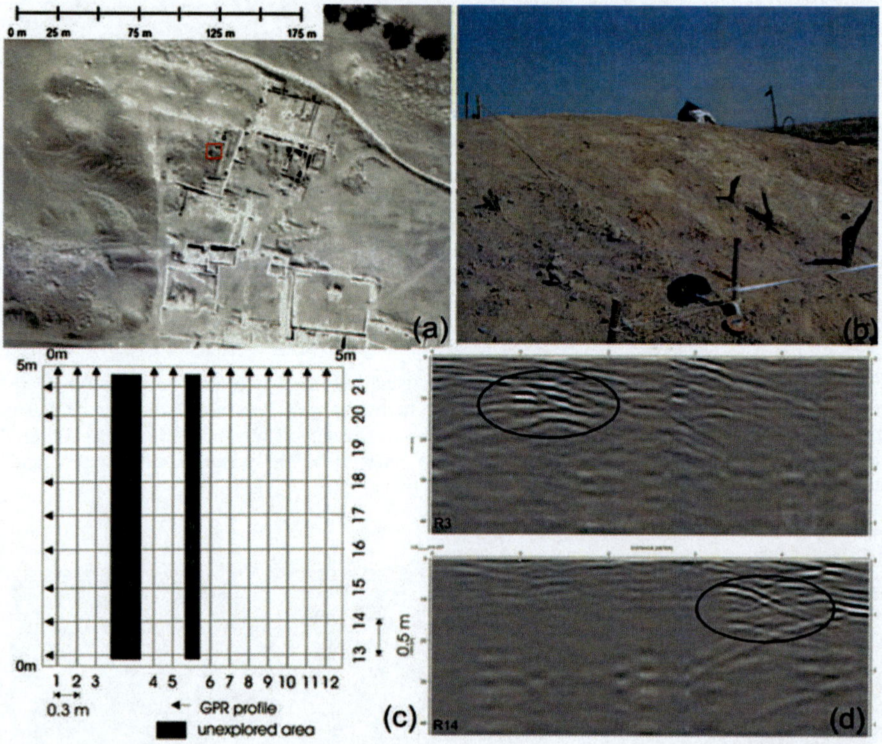

Fig. 14.18 (**a**) Location of the area investigated by georadar method, indicated by a *red square box*; (**b**) detail of the investigated area; (**c**) georadar profiles performed on a 5 x 5 m area where geomagnetic technique provided high values of magnetic field; (**d**) migrated radargrams R3 and R14, where it is possible identify some superficial reflection hyperbolas (indicated by *black ellipses*)

All these circular anomalies have been thought to be linked to tombs. To prove this hypothesis, before proceeding with excavations, georadar prospections on a test area have been performed. The selected area was on the top of the eastern slope of the pyramid (see Fig. 14.16b and 14.18a, b). The soil was mostly characterized by sand, thus explaining the higher electromagnetic wave average velocity respect to other sites (estimated by hyperbola analysis) and equal to 0.15 m/ns. Therefore, by exploiting the estimated *Vem* for the 400 MHz antenna, the time range of 30 ns corresponds to a maximum investigated depth around to 2 m.

Several GPR profiles were carried in the area divided into several grids of 5 m × 5 m, selected according to the dimensions of each square excavated by the archaeologists (Fig. 14.18c). The square area was investigated by 21 profiles carried out along two directions (nn. 12 and 9 in E-W and N-S direction, respectively).

For sake of brevity, herein we focus on some profiles, representatives of the final results: profiles R3 and R14 (shown in Fig. 14.18d). The profiles are perpendicular each other and have an investigation depth around 2 m (with $Vem = 0.15$ m/ns).

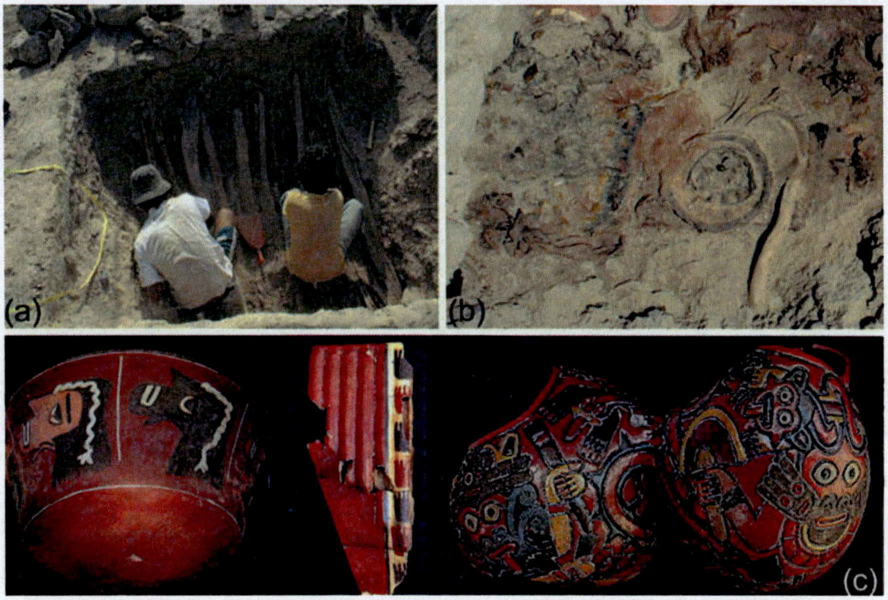

Fig. 14.19 (**a**) Archaeological layer at 50 cm depth: wood framework of a tomb is visible; (**b**) deeper archaeological layer which put in evidence a rich offering ritual material: (**c**) detail of some artefacts unearthed from the excavation: from *left* to *right*: a polychrome vessel, an antara and two painted pumpkins

The GPR profiles point out several reflections at depths ranging from 6 to 16 ns (0.45 and 1 m, respectively). In particular profiles R3 and R14 (Fig. 14.18d) show the presence of a strong reflector, ascribed to buried objects. The excavation of the first 50 cm unearthed remains of *huarango* trunks and branches, in agreement with the reflector of the radargrams (Fig. 14.19a). Such *huarango* elements belong to a wood framework, which typically covers a Nasca ceremonial offering and/or tomb.

The excavation of the subsequent layers revealed the existence of a rich ceremonial offering (Fig. 14.19b), which included 80 ceramics, painted textiles, precious metal objects and painted pumpkins of the Nasca Culture (see Fig. 14.19c). Finally, the archaeological deposit also revealed two human bodies belonging to a child and to an adult. Both of them were sacrificed and formed part of the ceremonial offering.

14.3.2.2 Site S2: *Templo del Escalonado*

Between 1984 and 1988 the Italian archaeological mission unearthed a monumental structure belonging to a transitional period between the architectural phases II and III of Cahuachi, and characterized by precious incised friezes with stepped motif on the walls, which the name of *Templo del Escalonado* derives from (Orefici 1988).

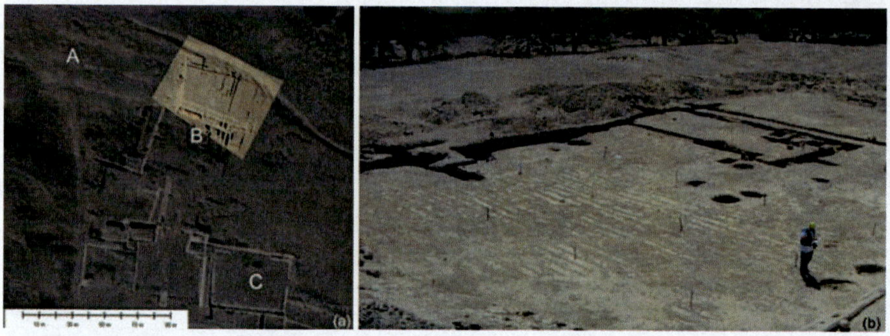

Fig. 14.20 (**a**) Location of the investigated area near *Templo del Escalonado*: the image is given by an orthorectified aerial photo on WW-1 satellite panchromatic image. *A* denotes *Piramide Naranja*, *B* indicates *Templo del Escalonado*, *C* is a *plaza undida* of *Gran Piramide*. (**b**) View of the prospected area

Since 2001 the investigations were enlarged to a flat area below the *Templo del Escalonado*. Trial excavations unearthed some walls. Unfortunately, the archaeological records till now available do not allow us to understand the spatial and functional relationships between this area and the *Templo del Escalonado*. This made necessary to conduct additional investigations by using geophysical methods (Rizzo et al. 2010).

Geophysical prospections have been carried out in two different field trips using: (i) a Ground Penetrating Radar (GPR) in a three dimensional survey (April 2008) and (ii) a geomagnetic survey with a gradiometer configuration (November 2008).

Figure 14.20a, b show the investigated area and its location. Figure 14.21b shows the areas investigated by Geomagnetic and georadar methods, contoured by black continuous and dot lines, respectively.

The GPR survey area of around 235 m^2 was investigated by 26 profiles 18 m long and with a line separation distance of 0.5 m. The two-way time acquisition range was of 40 ns, but after the processing the useful signals was only 20 ns, which corresponds to 1.5 m depth. In order to better interpret potential buried features, a 3D analysis has been planned using a developed GPR time slice, which required very closely spaced profiles (Goodman et al 1995; Conyers and Goodman 1997). Once the signal processes were applied to the radargrams, time slice analysis was defined. In this analysis, the recorded amplitudes of the reflections across the entire site are compared at different times to generate amplitude time slice maps.

Figure 14.21b, c show two GPR time slices where it is possible to detect two different kinds of potential archaeological features with circular (Fig. 14.21b) and linear (Fig. 14.21c) shapes, at around 15 and 30 cm depth, respectively.

The time slice of Fig. 14.21b shows two reflected zones, A1 and B1, with circular shape at high amplitude. The profile, which we do not show for sake of brevity, confirms the presence of several reflection hyperbolas due to some buried objects. In the time slice at around 0.3 m depth (Fig. 14.21c) several linear reflections are visible (see black arrows).

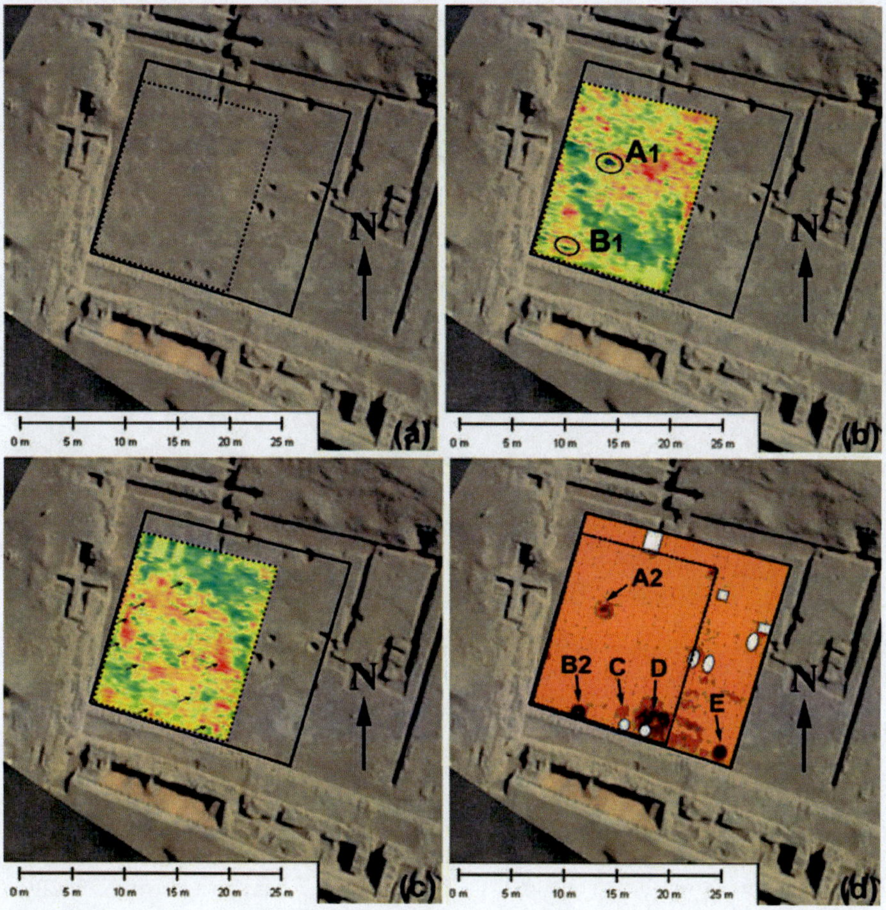

Fig. 14.21 **(a)** The investigated area: the two rectangles with *dot and continuous line* indicate the areas investigated by georadar and geomagnetic method, respectively; **(b–c)** GPR time slices at 15 and 30 cm, respectively; **(d)** geomagnetic map

In the same area a geomagnetic map has been carried out with a gradiometric configuration. The investigated area was enlarged compared to that of the GPR survey (see Fig. 14.21a, d). The data have been acquired on a regular grid 20 m × 20 m by an interspaced line of 0.5 m and a sampling rate of 10 Hz.

The data filtered in order to obtain the best signal/noise ratio, further processed using a kringing interpolation and visualized as a shaded relief image highlighted four main geomagnetic anomalies.

Figure 14.21d shows the final geomagnetic map, where several anomalies are visible. Black arrows indicate the main magnetic anomalies and white circles and squares put in evidence pits excavated by looters. In particular the gradiometric map shows two anomalies A2 and B2, which could be correlated

Fig. 14.22 After the localization on site of the anomaly A detected by both georadar and geomagnetic method (**a**), archaeologists excavated thus unearthing a ritual fire (**b**). Then another excavation has been performed in correspondence of the anomaly D. The result (**c**) has been a ceremonial altar

with ones localized on the GPR time slice at 0.15 m depth, A1 and B1 of Fig. 14.21b, respectively.

Moreover, other two circular geomagnetic anomalies, C and D, are detected at X = 1.5 m and Y = 18.5 m and at X = 7 m and Y = 2.5 m, respectively. Finally, a large anomaly, E, with a regular shape is between X = 4 and 10 m and Y = 4 and 10 m.

So, two anomalies of possible archaeological interest have been detected by both geophysical methods (A_1, A_2, B_1, and B_2). Geomagnetic method shows further anomalies not surveyed by georadar (see *C* and *D* in Fig. 14.21d). Moreover, the enlargement of geomagnetic prospection allowed us to identify another anomaly, indicated as *E* in Fig. 14.21d.

On august 2009, a trial excavation was carried out in correspondence of anomaly *A* detected from both GPR and geomagnetic methods (see Fig. 14.22a). A ceremonial offering was unearthed. It was characterized by the presence of coal and remains of a ritual fire made by fluvial stones with lens shape (Fig. 14.22b). Inside several coals are well defined and covered by leaves of *pacae*.

From the archaeological perspective, this ceremonial offering was very significant, because the hearth was located below a floor dated back to the end of the phase IV (known as phase IVc) and, therefore, archaeologically associated to a platform built after the earthquake and the mudslides described in Sect. 14.1.2. The phase IV

(350–400 AD) of Cahuachi was characterized by several offerings and sacrifices, as consequence of a crisis determined by the above said devastating natural disasters which determined profound and quick changes. The discovery of this ritual fire and its archaeological implications oriented the archaeologists to conduct further analysis also in correspondence of the other anomalies indicated in Fig. 14.21d.

On the base of archaeo-geophysical results, after a month archaeologists started excavating again in this area. They focused on the zone characterized by geomagnetic anomaly D. The excavation unearthed a ceremonial altar (Fig. 14.22d) dated back to the phase IV and composed of two large platforms symmetric respect to a rectilinear groove. The latter was axial to a well characterized by a mouth composed of two eccentric circles. In the well archaeologists found four gold bars (which are likely the cause of the magnetic field change) arranged according a square shape whose centre is characterized by the presence of a shell. Additional offers, such as necklaces and animals (bird and cuy) sacrificed to the divinities were also found.

14.3.2.3 Site S3

An 800 sqm plain area at North of the *Gran Piramide* was investigated by the georesistivity method (Fig. 14.23a). The survey was carried out with 1 m resolution using two different configurations of the probes which allow to provide georesistivity maps at the depths of 50 cm and 1 m. Figure 14.23b shows the geophysical system and the investigated site composed of four rectangular areas indicated with letters M1, M2, M3 and M4. Figure 14.23d shows the resistance map at 1 m depth, which visualizes low resistance values ($<300\ \Omega$) measured on a wide area, including most of M3 and M4 (at East of the map). The irregular shape does not help in identifying possible archaeological features. However the low resistivity is compatible with the presence of earthen buried walls and platforms. In the western part of the map (M1 and M2, in Fig. 14.23c, d), smaller areas with high resistivity values are evident. The latter could be related to tombs or ritual offerings.

14.3.3 Site S4. New Research Perspectives for the Study of Cahuachi: The Buried Settlement

The investigations on Site S4 followed the preliminary study of test site TS3 performed by using 2002 QB2 imagery which put in evidence some features characterized by geometric shapes of possible archaeological interest (see Sect. 14.3.1). In particular the NDVI map showed more info than panchromatic and multispectral pan-sharpened images (see such features depicted in Fig. 14.11d).

This area, as well as other zones of the Nasca riverbed, were focused to study and map *puquios* (see Chap. 12) and palaeoriverbeds. Then, the aim was enlarged to the

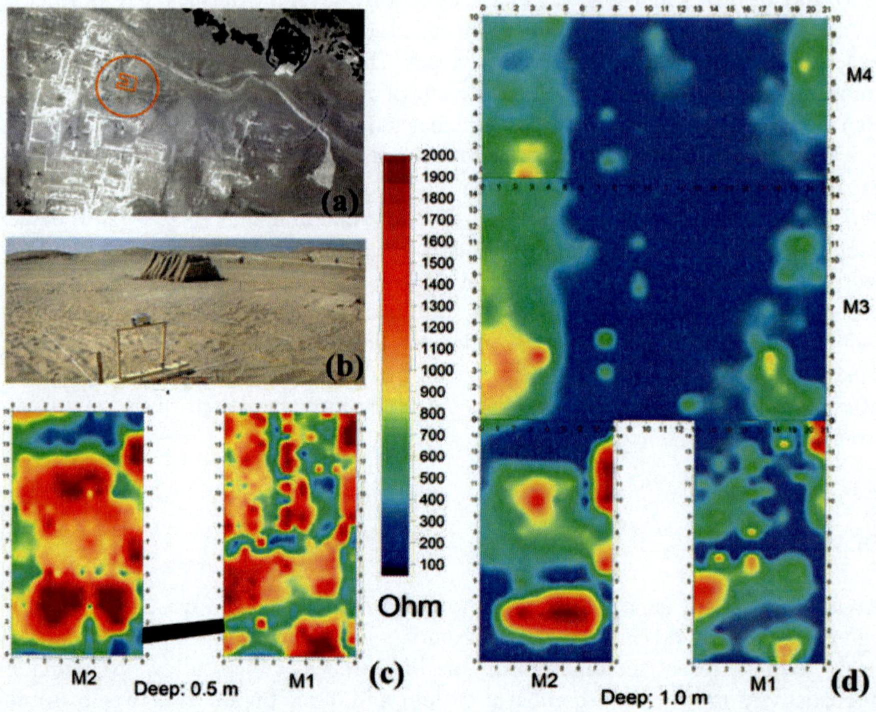

Fig. 14.23 (**a**) Location of the study area; (**b**) the study area; (**c**) georesistivity map at 50 cm depth; (**d**) georesistivity map at 1 m depth

identification of buried settlements at the west of the pyramids of Cahuachi, in the direction of Estaqueria, where ceremonial activities moved after the abandonment of the ceremonial centre.

Some aerial photos taken in April 2007 put in evidence the presence of several damp marks, some of them related to recent buried infrastructures, other referable to palaeoriverbeds and many other features not clearly referable to the above said features. In particular, our attention was focused on an area about 2 km NW of *Gran Piramide*. Herein ceramic fragments and adobes were found.

So, another QB2 image acquired on March 2005 was selected from the Digital Globe archive and purchased. The expected result was an improved visualization of the damp mark geometrical patterns, as it really was.

Panchromatic was very effective in highlighting several features, the typical crop-marks (see Lasaponara and Masini 2007), as the site was partially covered by grass and cotton plants at the time of satellite image acquisition. Moreover damp marks characterized the bare ground areas of the site.

So, particular attention was paid to enhance such features by applying different pan-sharpening algorithms which allowed us to exploit the higher spatial resolution of panchromatic image and the spectral content of the red and NIR bands.

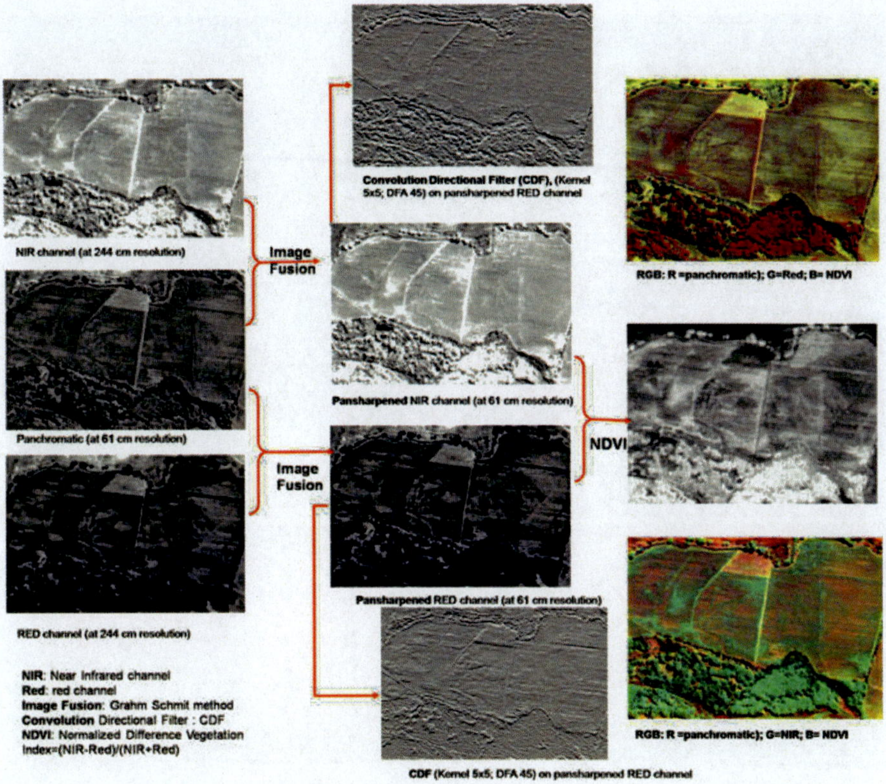

Fig. 14.24 Site S4: satellite image processing approach adopted for the detection and reconnaissance of archaeological features

Graham Schmidt algorithm has been considered the most effective image fusion algorithm in terms of discrimination of marks, for the red and NIR band. The RED pan-sharpened channel provides the greater content of information. A further improvement has been achieved by computing NDVI, as in the case of test site TS3. Finally the edge thinning was performed by using directional filters (see Fig. 14.24).

The identification of marks reported on the RGB composition NDVI, Pan and Red, respectively (see Fig. 14.25a), put in evidence four different patterns of features named with letters A, B, C and D.

Pattern A has a rectangular shape with dimensions 100 × 90 m and an orientation of about 45° respect to the North. The shape and parallel segments inside the rectangle suggest to us that it is probably a complex of terraced platforms.

Pattern B is composed of several linear features suggesting the edges of platforms, which represent quadrangular shapes intersected by oblique lines, for a total area equal to 1 ha.

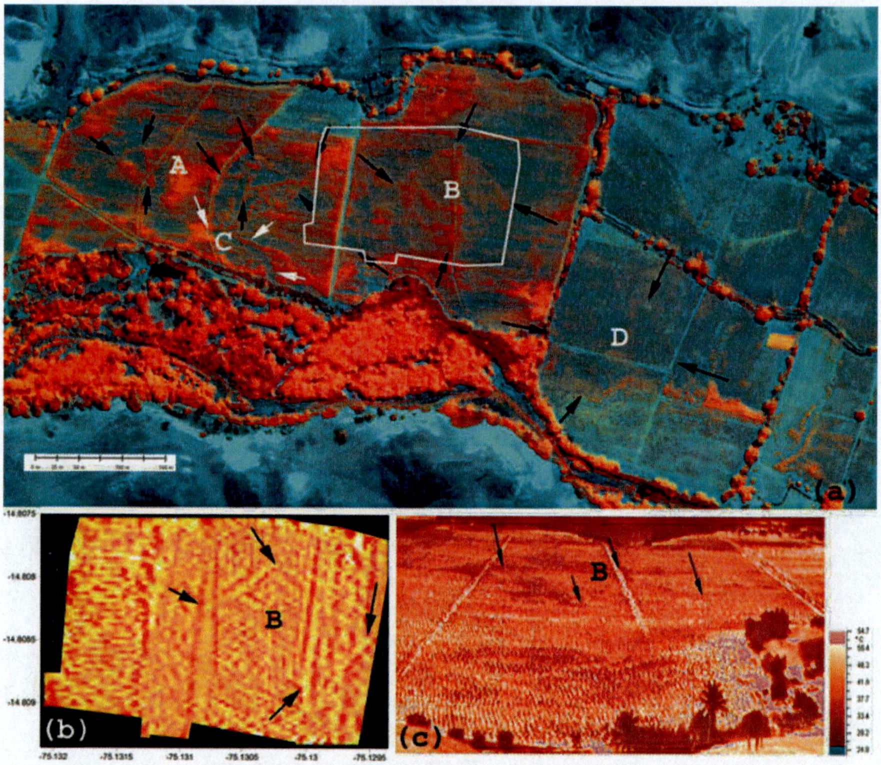

Fig. 14.25 Site S4. (**a**) RGB satellite image composition: R = NDVI; G = pan; B = red. (**b**) Letters *A*, *B*, *C* and *D* indicate the patterns of anomalies identified. The features of pattern *B* were confirmed by geomagnetic method and (**c**) infrared thermography

Feature C is a typical palaeoriverbed which along with other similar features indicate changes of the river over time. Finally, subtle marks of pattern D, visible from RED and NDVI images, seem to reveal a quadrangular shape referable to similar buried structures to the above mentioned A.

The pattern B was investigated by geomagnetic method and infrared thermography (Fig. 14.25b, c). The magnetic prospections were carried out using GPS over an area of 2.2 ha, on 2009 August. Linear magnetic signals oriented in correspondence to the four sides of the quadrangular pattern were identified (Fig. 14.25b).

Finally, Infrared thermography (2008 November) provided thermal anomalies in agreement with the features surveyed by the processing of satellite data and geomagnetic map.

From the archaeological point of view the discovery of this buried settlement opened new perspectives of research (Orefici 2009a) and new questions regarding its function and its spatial and temporal relation with the near Ceremonial Centre of Cahuachi.

14.4 Conclusions

This chapter provides an overview of the results from the scientific investigations carried out by ITACA Mission in Cahuachi and in *Rio Nasca* from 2007 up to today.

The main scientific activity of ITACA focused on archaeo-geophysics based on the integration of ground, aerial and satellite remote sensing methods, thus allowing the archaeologists to find buried walls ceremonial offerings in Cahuachi and to discover a large buried settlement in the Nasca riverbed.

The importance of these investigations was to address an open issue which is the detection of buried earthen structures. Such issue is a strategic challenge as crucial as complex. It is crucial because earthen archaeological remains are widely present through the world (in South America, Asia, Africa) and it is complex due to the subtle physical contrast between earthen remains and the surrounding subsoil.

To address this challenge we investigated some test sites selected from within: (i) *Piramide Naranja*, (ii) *Templo del Escalonado*, (iii) a 800 sqm plain non vegetated area north of the *Gran Piramide*, and (iv) a vegetated area located on the opposite river bank from the known archaeological area.

These sites were investigated using integrated non destructive analyses based on VHR satellite imagery (2002, 2005, 2008), geomagnetic (November 2008, August 2009), GPR (April 2008) and georesistivity methods (August 2010).

In particular for *Piramide Naranja*, we investigated: (i) North East side, not yet excavated, using satellite data and magnetic method and (ii) eastern side, where GPR has been used to support the ongoing excavation activities. In the North East side, results from the analysis of satellite images, based on spatial autocorrelation statistics and image filtering techniques, allowed the identification of unknown shallow and outcropping adobe walls, related to terraced platforms which compose the trunk-pyramidal structure. Additional information on the presence of buried and shallow walls were provided by geomagnetic surveys. The gradiometric map enabled also the identification of magnetic anomalies linked to tombs and ceremonial offerings. In the eastern side of the *Piramide Naranjada*, GPR allowed the discovery of an unknown rich ceremonial offering, including ceramics, painted textiles, precious metal objects and painted pumpkins, which belong to the Nasca civilization.

For the *Templo dell'Escalonado*, integrated geophysical investigations (GPR and geomagnetic) performed in two different field trips captured spatial anomalies which were confirmed by archaeological excavations. In particular, a ceremonial offering and an altar were unearthed. The ceremonial offering was characterized by the presence of coal and remains of a ritual fire made by fluvial stones with a lens shape. Inside several coals were well defined and covered by leaves of *pacae*. From the archaeological perspective, this ceremonial offering was very significant, because the hearth was located below a floor dated back to the end of the phase IV (known as phase IVc) and this, according to archaeologists, indicates that the fire was the "last ritual offering" before abandoning the Pyramid.

Therefore, it was ignited a symbolic funeral ritual for the Pyramid to close the ceremonial centre.

The ceremonial altar, dated back to the phase IV, is composed of two large platforms which are symmetric in respect to a linear groove. Archaeologists found four gold bars and additional offering, such a snacklaces and animals (bird and cuy) sacrificed to the divinities.

Investigations conducted using the georesistivity method have been performed over a 800 sqm plain, at north of the Gran Piramide. The maps obtained at the depths of 50 cm and 1 m showed anomalies probably linked to the presence of earthen buried walls, and platforms and tombs or ritual offerings.

Out of the archaeological area investigations focused on the vegetated areas located on the opposite river bank to study and map puquios and palaeoriverbeds. Then, the aim was enlarged to the identification of buried settlements at the west of the ceremonial centre in the direction of Estaqueria, where ceremonial activities moved after the abandonment of Cahuachi. Some aerial photos taken in April 2007 had put in evidence the presence of several damp marks, some of them related to recent buried infrastructures, other referable to palaeoriverbeds and many other features. The processing of different VHR satellite images and magnetic maps allowed us to discover several anomalies of archaeological interest in an area about 2 km NW *of Gran Piramide*. The rectangular shape of one of the detected anomalies has been interpreted as a complex of terraced platforms. From the archaeological point of view the discovery of this buried settlement opened new perspectives of research (Orefici 2009a, b) and new questions regarding its function and its spatial and temporal relation with the near Ceremonial Centre of Cahuachi.

As a whole, our results pointed out that the use of different remote sensing technologies can open new perspectives for the detection and documentation of adobe archaeological remains, not only for the ancient Andean civilizations but also for the earthen archaeology in Middle East, Northern Africa and Asia. Earth Observation can significantly contribute to the monitoring of the fragile earthen archaeological heritage, whose conservation policies must also address looting activity.

References

Aiazzi B, Alparone L, Baronti S, Garzelli A, Selva M (2006) MTF-tailored multiscale fusion of high-resolution MS and Pan imagery. Photogramm Eng Remote Sens 72(5):591–596

Aiazzi B, Baronti S, Selva M (2007) Improving component substitution pansharpening through multivariate regression of MS + PAN data. IEEE Trans Geosci Remote Sens 45 (10):3230–3239

Aiazzi B, Baronti S, Alparone L, Lasaponara R, Masini N (2008) Data fusion techniques for supporting and improving satellite-based archaeological research. In: Lasaponara R, Masini N (eds) Advances in remote sensing for archaeology and cultural heritage management. Aracne, Roma, pp 31–36

Aveni AF (1986) The Nazca Lines: patterns in the desert. Archaeology 39(4):32–39

Boardman JW, Kruse FA (1994) Automated spectral analysis: a geological example using AVIRIS data, north Grapevine Mountains, Nevada. In: Proceedings, ERIM tenth thematic conference on geologic remote sensing, Environmental Research Institute of Michigan, Ann Arbor, pp I-407–I-418

Conyers LB, Goodman D (1997) Ground penetrating radar: an introduction for archaeologists. AltaMira Press, Walnut Creek, London and New Delhi

ENVI User's Guide (1999) ENVI, Research Systems

Gavazzi A (2009) La arquitectura de Cahuachi, in Nasca. El desierto del los Dioses de Cahuachi. Graph, Lima, pp 114–131

Gavazzi A (2010) Arquitectura Andina: forma e historia de los espacios sagrato. Graph Ediciones, Lima

Goodman D, Nishimura Y, Rogers JD (1995) GPR time slices in archaeological prospection. Archaeol Prospect 2:85–89

Green AA, Berman M, Switzer P, Craig MD (1988) A transformation for ordering multispectral data in terms of image quality with implications for noise removal. IEEE Trans Geosci Remote Sens 26(1):65–74

Kosok P, Reiche M (1949) Ancient drawings on the desert of Peru. Archaeology 2(1):207–215

Laben CA, Brower BV (2000) Process for enhancing the spatial resolution of multispectral imagery using pan-sharpening. US Patent # 6,011,875, Eastman Kodak Company

Lasaponara R, Masini N (2007) Detection of archaeological crop marks by using satellite QuickBird multispectral imagery. J Archaeol Sci 34:214–221

Lasaponara R, Masini N, Rizzo E, Coluzzi R, Orefici G (2011) New discoveries in the Piramide Naranjada in Cahuachi (Peru) using satellite, ground probing radar and magnetic investigations. J Archaeol Sci 38:2031–2039

Masini N, Rizzo E, Lasaponara R, Orefici G (2008) Integrated remote sensing techniques for the detection of buried archaeological adobe structures: preliminary results in Cahuachi (Peru). Adv Geosci 19:75–82

Masini N, Lasaponara N, Orefici G (2009a) Addressing the challenge of detecting archaeological adobe structures in Southern Peru using QuickBird imagery. J Cult Herit 10S:e3–e9. doi:10.1016/j.culher.2009.10.005

Masini N, Rizzo E, Lasaponara R (2009b) Teledeteccion y Investigaciones geofísicas en Cahuachi: primeros resultados. In: Nasca. El Desierto de los Dioses de Cahuachi. Graph Ediciones, Lima, pp 250–277

Montoya M, Gracia W, Caidas J (1994) Geologıa de los Quadrangulos de Lomitas, Palpa, Nasca y Puquio. INGGEMET (Istituto Geologico Minero y Metallurgico), Lima

Orefici G (1988) Una expresión de arquitectura monumental Paracas-Nasca: El Templo del Escalonado, en Atti del Convegno Internazionale "Archeologia, Scienza e Società nell'America Precolombiana". Brescia 1988:191–201

Orefici G (1993) Nasca: arte e società del popolo dei geoglifi. Jaca Book, Milano

Orefici G (2009a) El Proyecto Nasca. In: Nasca. El desierto del los Dioses de Cahuachi. Graph, Lima, pp 18–35

Orefici G (2009b) Cahuachi, the largest adobe Ceremonial centre in the world. In: Nasca. El desierto del los Dioses de Cahuachi. Graph, Lima, pp 36–59

Orefici G (2009c) Los geoglifos: Espacios abiertos y ceremonias colectivas. In: Nasca. El desierto del los Dioses de Cahuachi. Graph, Lima, pp 94–113

Orefici G, Drusini A (2003) Nasca: hipótesis y evidencias de su desarrollo cultural, 2nd edn, Documentos e Investigaciones. CISRAP, Brescia

Reinhard J (1996) The Nazca lines: a new perspective on their origin and meaning. Los Pinos, Lima

Richards JA, Xiuping J (2006) Remote sensing digital image analysis: an introduction. Springer, Berlin/Heidelberg

Rizzo E, Masini N, Lasaponara R, Orefici G (2010) ArchaeoGeophysical methods in the Templo del Escalonado (Cahuachi, Nasca, Perù). Near Surf Geophys 8:433–439

Rowe JH (1963) Urban settlements in ancient Peru. Ñawpa Pacha 1:1–27, Berkeley, California
Sánchez Borjas A (2009) Estaquería: Sobreviviendo a la extinción. In: Nasca. El desierto del los Dioses de Cahuachi. Graph, Lima, pp 60–71
Schreiber H, Lancho Rojas J (2009) El control del agua y los puquios de Nasca. In: Nasca. El desierto del los Dioses de Cahuachi. Graph, Lima, pp 132–151
Silverman H (1993) Cahuachi in the ancient Nasca world. University of Iowa Press, Iowa City

Index of Names

R. Lasaponara and N. Masini (eds.), *Satellite Remote Sensing: A New Tool for Archaeology*, Remote Sensing and Digital Image Processing 16,
DOI 10.1007/978-90-481-8801-7, © Springer Science+Business Media B.V. 2012

Index of Topics

N

O

P

Index of Places